GRAPHING CALCULATOR MANUAL

ROBERT E. SEAVER
Lorain County Community College

BASIC TECHNICAL MATHEMATICS

EIGHTH EDITION

BASIC TECHNICAL MATHEMATICS WITH CALCULUS

EIGHTH EDITION

BASIC TECHNICAL MATHEMATICS WITH CALCULUS, METRIC VERSION

EIGHTH EDITION

Allyn J. Washington

Dutchess Community College

PEARSON

Addison
Wesley

Boston San Francisco New York
London Toronto Sydney Tokyo Singapore Madrid
Mexico City Munich Paris Cape Town Hong Kong Montreal

ISBN   0-321-19740-2

2 3 4 5 6 OPM 06 05 04

Contents

Contents

Preface

Since the first version of this manual was published in 1995, graphing calculators have continued to evolve. The main changes in the last few years have included the introduction of calculators that are faster, have more memory, are preloaded with application programs, and have a better interface with computers. Although procedures are included in this manual for working with several different models of calculators, the manual itself is generally calculator independent. The explanations, examples, and exercises apply to any graphing calculator. However, students using calculators other than those included, will need to refer to the user's manual for their calculator to obtain specific instructions.

Included in this edition are additional exercises, many new figures showing example screens, and a reordering of some of the material to better coordinate with Washington's texts. In addition, the newer TI-84 Plus model has been added and the older TI-81 model has been dropped. The TI-84 Plus is a faster model, has a built in USB port, has preloaded application programs, and has a larger memory.

This manual is unique in that it incorporates procedures for several different graphing calculators in one place. Integrated into the text and summarized in the appendix are procedures for Texas Instruments Graphing Calculator models TI-82, TI-83, TI-84 Plus, TI-85, and TI-86. A TI-83 Plus model is not included since it is an enhanced version of the TI-83 and the procedures are generally the same as those of the TI-83. It has not been listed separately except where appropriate. Appendix B provides corresponding procedures for the somewhat different and more powerful and advanced TI-89 calculator and Appendix C provides procedures for the Casio Power Graphic CFX-9850G calculator. The Casio CFX-9850G is included because of its several special features including the ability to graph in three different colors.

Whichever model is used, graphing calculators continue to provide the student with an excellent method of better understanding mathematical concepts, visualizing functions, doing matrix operations, and performing statistical calculations. This tool is particularly well suited to working problems in the engineering technologies. The graphing calculator takes the drudgery out of the graphing of functions (particularly of non-algebraic functions), out of operations with matrices, and out of statistical calculations. It allows the student to quickly discover how the changing the parameter in a function will change the graph. This discovery approach is taken in many of the exercises in this manual. Another advantage of these calculators is that they allow the investigation of realistic data in application problems.

The main purpose of this manual is to provide the information necessary to enhance the teaching of technical mathematics through the use of graphing calculators. A secondary purpose is to provide a background that will enable the student to efficiently use the graphing calculator in other classes and in future employment. Most of this material has been successfully used for several years in technical mathematics classes at Lorain County Community College.

In this manual, use of the graphing calculator is developed through the learning of procedures. The procedures are presented in an order that corresponds to their use in a typical technical mathematics course. Experience has shown that students acquire these procedures quite quickly and rarely need to refer back to procedures previously learned. However, if a reference is needed, an organized summary of calculator procedures is available in the appendices.

This manual is written at a level that should be understood by a technical mathematics student with a good basic knowledge of algebra. Each section contains a brief explanation of the material under consideration, example problems, and related exercises. The explanations are not intended to be a complete, in-depth coverage of topics, but only to serve as background material for the calculator procedures and exercises.

The material has been keyed to corresponding material in the Allyn J. Washington's texts: Technical Mathematics and Technical Mathematics with Calculus, Eighth Edition. Several sections in this manual correspond to material in the text and these sections are indicated by a notation this type: *(Washington, Section 3.4)*. In addition, certain examples and exercises correspond to and are very similar to examples and exercises in Washington's text. When such a correspondence exists, the example or exercise in the manual is indicated by the symbol ♦ in the left margin and identified as follows: ♦ Example 2.8: *(Washington, Section 3.5, Example 7)* or ♦ 22. *(Washington, Exercise 1.5, #45)*. In many cases changes were made in the exercises to better illustrate the use of the graphing calculator.

It has been enjoyable updating this material and it is always a challenge to keep up with technology. It is my desire that this manual be error free. However, being realistic, I understand that there may be some errors. I apologize in advance for any errors and it is my hope that they will cause only minor inconvenience.

I appreciate the help and encouragement given to me by Suzanne Alley and also for the help of Joe Vetere and others at Addison-Wesley. As always, it has been a pleasure to work with them. It is also a pleasure to work with Allyn J. Washington. He continues to update and enhance his great text and over the years I have come to appreciate him for the person that he is.

RES

This manual is dedicated to my wonderful wife Janet,
to my father who continues to enjoy life in his 90's
and to Jed, Cody, Ethan, Araenae, Kelsey, and Josiah .
May God Bless Them.

Chapter 1

Introduction to a Graphing Calculator

1.1 Introduction to Graphing Calculators.

Over the past several years, the graphing calculator has continued to evolve into an essential tool for the modern day student. Newer calculators have more memory, more built-in capabilities, and the ability to connect to computers for transfer of files. The graphing calculator, in addition to doing all the calculations normally done on a scientific calculator, readily graphs mathematical functions, does statistical calculations, and performs matrix operations. The graphing calculator is, in fact, a miniature computer and can also be programmed to solve mathematical problems in much the same way that a computer can be programmed.

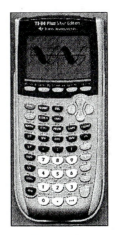

TI-84 Plus

Figure 1.1

      There are several differences between a graphing calculator and a standard scientific calculator. A typical graphing calculator is shown in Figure 1.1. An obvious difference is the larger screen that not only permits the viewing of a graph but also allows the user to see a complete problem and not just the answer. Changes can quickly be made in an entered expression if an error is detected and the calculation can be easily be performed a second time. Other differences are the additional keys on the keyboard and the ability to enter alphabet characters.

      The material in this manual applies specifically to the Texas Instruments Graphing Calculators models TI-82, TI-83, TI-83 Plus, TI-84 Plus, TI-85, and TI-86. Since the procedures for the TI-83 Plus are essentially the same as the TI-83, it is not listed separately in the manual. Students using the TI-83 Plus should follow the procedures for the TI-83 model. Most of the discussion may also be applied to other graphing calculators. Since other calculators may handle certain operations differently, the user is referred to the manual that came with a particular calculator for specific help. One calculator, the color Casio CFX-9850G, is unique since it has color capability on the screen and procedures for it are given separately in the Appendix C.

      The TI-83 Plus calculator has improved capabilities over the older TI-83 model including additional memory and an upgradeable operating system that allows it to download and store extra software applications. The newer TI-84 Plus is an enhanced version of the TI-83 Plus and offers connectivity to a computer via a USB port, has an improved display, and offers preloaded programs. For information regarding turning calculators on and off, clearing all memory, adjusting the contrast of the screen, and clearing the screen, the user is referred to Basic Procedures B1, B2, B3, and B4 in the appendices of this supplement.

1.2 On Working with the Calculator

The screen on the graphing calculator is made up of dots called **pixels**. Characters or graphs appear on the screen when certain of these pixels are turned on. A typical graphing calculator screen contains 96 pixels in the horizontal direction and 64 pixels in the vertical direction although some may contain as many as (or even more than) 160 pixels in the horizontal direction and 100 in the vertical direction. When the graphing calculator is turned on and the screen is clear, there should be a blinking rectangle or line on the screen. This is referred to as a **cursor** and indicates the location where a character (such as a letter or number) will appear when entered from the keyboard. On the graphics screen, the cursor may appear as a cross. On the upper right of the keyboard are four arrow keys called the **cursor keys.** These keys are used for moving the cursor around the screen and for selecting items on the screen. The screen that normally appears when the calculator is turned on and on which calculations are done is referred to as the **home screen**.

      Many of the keys on the keyboard, in addition to their primary use, are also used to display alpha characters or to perform secondary actions. A **character** is any number, letter, or symbol that may be displayed on the calculator screen. An **alpha character** is a letter or symbol that may be used as variables or in writing programs. The typical graphing calculator screen will display up to eight rows of characters. Each row may contain up to 16 (or more) characters. **Secondary operations** are the additional functions performed by many of the keys. The alpha characters appear above and to the right of a key and the secondary operations appear above and to the left of a key. The methods of accessing these alpha characters or secondary operations are given in Procedure C1.

__Procedure C1__.   **To perform secondary operations or obtain alpha characters.**

    1. To perform secondary operations:
      (a) Press the key: 2nd
      (b) Then, press the key for the desired action
          (For example: To find the square root of a number, first press the key 2nd followed by the key: x^2. (The square root symbol $\sqrt{}$ appears above and to the left of the x^2 key.)
          Then, enter the number and press ENTER.)

    2. To obtain alpha characters:
      (a) Press the key: ALPHA
      (b) Then, press the key for the desired character.
          (For example: To display the letter A on the screen, press the key: ALPHA then press the key that has the letter A above and to the right of the key.)
      On the TI-85 and 86, pressing the key: 2nd before pressing the ALPHA key will give lower case letters.
    Note: The alpha key may be locked down on the TI-82, 83, and 84 Plus by first pressing the key 2nd then pressing the key: ALPHA.
          Press the ALPHA key twice to lock down the alpha key on the TI-85 and 86.

Graphing calculators are menu oriented and when certain keys are pressed, menus appear on the screen. A **menu** is a list of possible options from which we may make a selection in much the same way as we would select an item from a dinner menu in a restaurant.

On occasion, a word, item, number, or area of the screen is highlighted. A **highlighted area** is that area with the background shaded.

Procedure C2. Selection of menu items

A menu on the TI-82, 83, and 84 Plus calculators appears as a list of numbers, each followed by a colon and a menu item. A menu item may be selected by:

1. Pressing the number of the item desired

or

2. Moving the highlight to the item desired by pressing the up and down cursor keys, then pressing the key ENTER.

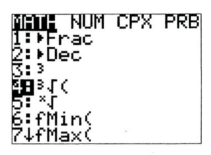

Figure 1.2

(Example: To find the cube root of 3.375, press the key MATH, leave MATH highlighted in the top row and select 4:$\sqrt[3]{}$. This screen is shown in Figure 1.2. Then enter the number 3.375 and press ENTER. The result is 1.5.)

In some cases pressing the same key may access more than one menu. This is indicated by more than one title appearing at the top of the screen with one of the words highlighted. The right and left arrow keys are used to move the highlight from one title to another. As this is done different menus appear on the screen. First select the correct menu, then the desired item.

A menu on the TI-85 and 86 calculators appears as a list on the bottom of the screen. Make a selection by pressing the function key just below the item desired. The function keys are the keys labeled F1, F2, F3, F4, and F5.

In some cases, selection of a menu item will produce a secondary menu. Items are selected from the secondary menu by pressing the appropriate function key. If the symbol $\rightarrow$ appears on the right of a menu, there are additional items in the menu. To see these additional items, press the key: MORE

To exit from a menu to the previous menu or to the home screen, press the key: EXIT. To exit to the home screen, press QUIT

(Example: To find the largest integer which is less than or equal to the number 2.15, first press the key MATH, move the highlight in the top row to NUM, select 5:int((or on the TI-82, select 4:Int). The screen appears as in Figure 1.2. Now enter the number 2.15, and press ENTER. The answer should be 2.)
To exit from a menu without selecting an option, press QUIT.

(Example: To find the largest integer less than or equal to 2.15, press the key: MATH, then press the key F1 for NUM, and from the secondary menu, select **int** by pressing the key F4. Then enter the number 2.15 and press ENTER.
The result is 2.)

1.3 Numerical Calculations and Error Correction

We are now ready to use the calculator to do numerical calculations in much the same way as on any scientific calculator. But first a few comments about the order of arithmetic operations.

If we enter the arithmetic expression $3^4 + 5*2 - 6 \div 3$ into a scientific calculator (or into a computer) there is a specific order in which the calculator will perform the operations of addition, subtraction, multiplication, division and of finding exponents and roots. This order should also be followed when doing a calculation by hand.

Order of operations:

First: Find all powers (exponents) and roots.

Second: Perform all multiplications and divisions in order from left to right.

Third: Perform all additions and subtractions in order from left to right.

If parentheses appear in the expression, the operations inside the parenthesis are performed first. If more than one set of parenthesis appears, the expression in the innermost parenthesis is performed first, then the second innermost parenthesis, etc. Within a set of parenthesis, the given order of operations is followed.

Example 1.1: Evaluate by hand $3^4 + 5 \times 2 - 6 \div 3$.

Solution: 1. First do the power to obtain
$$81 + 5 \times 2 - 6 \div 3$$

2. Second do the multiplications and divisions
$$81 + 10 - 2$$

3. Finally do additions and subtractions

89

On t he calculator, if more than one exponent operation appears in order, the operations are performed from left to right. (In a computer program, exponent operations are normally performed from right to left.) Thus, if you enter 2^2^3 on your calculator (the symbol ^ is used to find exponents on the calculator), and press ENTER, you will obtain 64 since 2^2^3 = 4^3 = 64. It is best when using more than one exponent operation in a row to use parenthesis to indicate which operation is to be done first.

To perform the calculation of Example 1.1 on a graphing calculator, the expression is first entered onto the home screen of the calculator using the arithmetic operation keys located on the right hand side. Procedure C3 indicates how to perform arithmetic operations.

Procedure C3. To perform arithmetic operations.

1. To add, subtract, multiply or divide press the appropriate key: $+, -, \times,$ or $\div$.
   On the graphing calculator screen, the multiplication symbol will appear as * and the division symbol will appear as /.
   If parentheses are used to denote multiplication it is not necessary to use the multiplication symbol. (The calculation $2(3 + 5)$ is the same as $2 \times (3 + 5)$.)
2. To find exponents use the key $\wedge$ (Example: 2^4 is entered as 2^4)
   For the special case of square we may use the key x^2.
3. After entering the expression press the ENTER key to perform the calculation.
   Notes: (a) On the graphing calculator, there is a difference between a   negative number and the operation of subtraction. The key (-) is used for negative numbers and the key $-$ is used for subtraction.
             (b) Enter numbers in scientific notation by using the key EE.
                  (Example: To enter 2.15×10^5, enter 2.15, press EE, enter 5.)

A graphing calculator has the advantage that the complete entered expression is visible on the screen. This allows us to check the expression to see if it has been entered correctly, to easily make corrections, and to quickly and easily change a value in the expression.

Procedure C4. To correct an error or change a character within an expression.

1. (a) To make changes in the current expression:
       Use the cursor keys to move the cursor to the position of the character to be changed.
   (b) To make changes in the last expression entered (after the ENTER key has been pressed) obtain a copy of the expression by pressing the key ENTRY. Then, use the cursor keys to move the cursor to the character to be changed.

2. Characters may now be replaced, inserted, or deleted at the position of the cursor.
   (a) To replace a character at the cursor position, just press the new character.
   (b) To insert a character or characters in the position the cursor occupies, Press the key: INS.
   Then, press keys for the desired character(s).
   (c) To delete a character in the position the cursor occupies, press the key: DEL

Example 1.2: Use the calculator to evaluate $3^4 + 5 \times 2 - 6 \div 3$ then, evaluate the expression $3^5 + 5 \times 2 - 6 \div 3$.

Solution: 1. Enter the first expression into the calculator exactly as given and perform the calculation (see Procedure C3). The answer of 89 should appear one line down and on the right of the screen.

2. The second expression differs from the first only in the exponent of the number 3. We need to change the exponent from a 4 to a 5. Use Procedure C4 to make changes in the last expression. Move the cursor to the 4 and replace with a 5. Press ENTER to perform the calculation. The value 251 should appear on the right of the screen.

In Example 1.2, we made use of Procedure C4 to replace a character in an expression. This same procedure may be used to correct an error within an expression that we have just entered.

We need to take special care when working with fractions. Consider the evaluation of the quantity $\dfrac{3+9}{2}$. If this is entered as $3 + 9 \div 2$, we obtain the value is 7.5 since the calculator is following the order of operations and first divides 9 by 2.. In order to have the 3 added to the 9 before division, it is necessary to insert parenthesis and enter the expression as $(3 + 9) \div 2$. This gives the correct answer of 6. In a fraction, expressions in the numerator or in the denominator must be evaluated first before doing division. For purposes of entering fractions into a calculator, a numerator or denominator containing more than one quantity should be thought of as being in parenthesis. Example 1.3 illustrates the use of parenthesis and of changing numerical values within an expression.

Example 1.3: Evaluate the expressions $\quad 6500\left(1 + \dfrac{0.06}{2}\right)^8$, $6500\left(1 + \dfrac{0.06}{4}\right)^{16}$, and

$6500\left(1 + \dfrac{0.06}{12}\right)^{48}$. Round answers to the nearest hundredth. These expressions

calculate the amount of money in a savings account after 4 years for an initial deposit of $6500, earning interest at the rate of 6% per year where the interest is compounded 2 times a year, 4 times a year, and 12 times a year.

Solution:

1. Use Procedure C3 to enter the first expression $6500\left(1+\dfrac{0.06}{2}\right)^{8}$ into the calculator and perform the calculation. The result should be: 8234.005529 which we round to 8234.01. The calculation screen is in Figure 1.3.

```
6500(1+0.06/2)^8
          8234.005529
```

Figure 1.3

2. To evaluate the second expression we need only change two of the numbers. Use Procedure C4 to change the 2 to a 4 and the 8 to 16 and perform the second calculation. The answer is 8248.40606 and we round this to 8248.41.

3. To evaluate the third expression, we now need to change the 4 to a 12 and the 16 to a 48. Again use Procedure C4 to make these changes and perform the calculation. The answer is 8258.179547 or 8258.18.

Special mathematical functions may require use of special menus. We will consider only those functions that are important to us at this time.

Procedure C5. Selection of special mathematical functions.

1. To raise a number to a power:
   Enter the number
   Press the key ^
   Enter the power
   Press ENTER.

2. To find the root of a number (other than square root):

   On the TI-82, 83, or 84 Plus: On the TI-85 or 86:
   Enter the root index   Enter the root index
   Press the key MATH   Press the key MATH
   With Math highlighted,   Select MISC
   Select $\sqrt[x]{}$ Press MORE,
   Enter the number   From the secondary menu,
   Press ENTER   Select $\sqrt[x]{}$
                                          Enter the number
                                          Press ENTER.

3. To find the factorial of a positive integer:

   On the TI-82, 83, or 84 Plus: On the TI-85 or 86:
   Enter the integer   Enter the integer
   Press the key MATH   Press the key MATH
   Highlight PRB   Select PROB
   Select !   From the secondary menu,
   Press ENTER   select !
                                          Press ENTER.

4. To find the largest integer less than or equal to a given value (This is called the greatest integer function and is indicated by square brackets [ ]):

On the TI-82, 83, or 84 Plus:
Press the key MATH
Highlight NUM
Select Int
Enter the value
Press ENTER

On the TI-85 or 86:
Press the key MATH
Select NUM
From the secondary menu,
select int
Enter the value
Press ENTER.

5. For the number π, press the key for π on the keyboard.

6. To find the absolute value of a given value:

On the TI-82:
Press the key ABS
on the keyboard.
Enter the value
Press ENTER

On the TI-85 or 86:
Press the key MATH
From the menu, select NUM
Select abs
Enter the value
Press ENTER.

On the TI-83 or 84 Plus:
Press the key MATH
Highlight NUM
Select abs(
Enter the value
Press ENTER

Example 1.4:   Find the square, square root, cube, and cube root of π. Also, find 8! and [-1.78]. (the greatest integer function.) Round answer to four decimal places.

Solution:
1. Find the square of π by:
   Enter π, press the key x^2, (π^2 appears on the screen), then press ENTER. The answer is 9.8696.
2. Find the square root of π by
   Pressing the key $\sqrt{}$, entering π, and pressing ENTER. The answer is 1.7725.
3. Find the cube of π by using Procedure C5 to raise a number to a power. The answer is 31.0063.
4. Find the cube root of π by using Procedure C5 to find the root of a number. The answer is 1.4646.
5. Find the factorial of the integer 8 by using Procedure C5. The answer is 40320.
6. Find the greatest integer function of -1.78 by using Procedure C5. The answer is -2.

1.4 About Significant Digits, Precision, and Accuracy.

Some numbers are exact and some are approximate. In approximate numbers, not all the digits may have meaning. If a technician were to measure the resistance of a resistor with an ohmmeter that was only accurate to a tenth of an ohm, it would be pointless to write down that the resistance is 60.125 ohms. Instead, the technician should write down that the resistance 60.1 ohms. The number is rounded to indicate the accuracy of the measurement. The digits 2 and 5 are meaningless in this case.

When working with approximate numbers it is assumed that only the digits that have meaning appear in the written number. Such digits are referred to as **significant digits**. We will assume that any not zero digit is a significant digit and the digit zero is significant if

(1) if it appears between two other significant digits

or

(2) if it is to the right of a significant digit *and* is to the right of the decimal point.

Example 1.5: Give the number of significant digits in each of the numbers (a) 1203.,
              (b) 2.3050, (c) 0.00200, and (d) 124000.

Solution: (a) The number 1203 has 4 significant digits. The zero is significant since it is between two other significant digits.
              (b) The number 2.3050 has 5 significant digits. The first zero appears between two significant digits and the second zero appears to the right of a significant digit and to the right of the decimal point.
              (c) The number 0.00200 has 3 significant digits. The zeros in front of the 2 are not significant and the two zeros behind the 2 are significant since they are to the right of a significant digit and to the right of a decimal point.
              (d) The number 124000 has 3 significant digits. In this case the three zeros are not significant even though they are to the right of a significant digit, since they are to the left of the decimal point.

The words accuracy and precision are closely tied to the idea of significant digits. **Accuracy** has to do with the number of significant digits in a value and **precision** has to do with the location of the rightmost significant digit. The larger the number of significant digits, the more accurate the number. The more to the right, in relation to the decimal point, the rightmost significant digit appears, the more precise the number.

Example 1.6: Consider the numbers 123., 0.00020, and 2034.2.
              (a) Which of the numbers is the most accurate and which is least accurate?
              (b) Which is most precise and which is least precise?

Solution: (a) The number 2034.2 is most accurate since it contains the most significant
                 digits (five). The number 0.00020 is least accurate since it contains the
                 fewest number of significant digits (two).
             (b) The number 0.00020 is most precise since the rightmost significant digit is
                 farthest to the right of the decimal point. The number 123. is least precise
                 since the rightmost significant digit is farthest to the left relative to the
                 decimal point.

When doing a calculation, the result often contains a large number of digits, many of
which are not really significant. This is particularly true when performing calculations on a
calculator where results may contain 8 or more digits. In order to obtain answers that have the
proper degree of accuracy or precision, we make use of certain guidelines.

Guidelines for Rounding:

1. When adding or subtracting approximate numbers, the result should
    not be more precise than the least precise of the numbers being added
    or subtracted.

2. When multiplying or dividing approximate numbers, the result should
    not be more accurate than the least accurate of the numbers being
    multiplied or divided.

3. When finding roots or powers, the result should not be more accurate
    than the numbers involved.

4. The following table gives comparative accuracy for working with
    trigonometric functions.

Angle (in degrees) to nearest	Angle (in radians) to	Value of trigonometric functions to
degree	2 s.d.	2 s.d .
0.1 degree	3 s.d.	3 s.d.
0.01 degree	4 s.d.	4 s.d.
0.001 degree	5 s.d.	5 s.d.

Some numbers are exact and not approximate. For example, when calculations involve
the numbers of quarts in a gallon or the number of meters in a kilometer, the numbers involved
(4 or 1000) are by definition **exact**. Other exact numbers may occur in formulas such as in
$F = \dfrac{9}{5}C - 32$. When performing calculations involving both exact and approximate numbers,
the rules above are used only in reference to the approximate numbers. Exact numbers are
assumed to be "infinitely accurate and precise".

Exercise 1.1

Use the graphing calculator to evaluate each of the following. Round answers to the proper number of significant digits. Consider integers as exact numbers.

1. $3.1526 + 233.4 + 122.156$

2. $23.125 - 7.24$

3. $-13.271 + 6.344 - 6.25$

4. $(0.034)(20.2) + 6.783$

5. $607 \div 28$

6. $\dfrac{23.2 + 16.7}{18.9 - 6.25}$

7. $\sqrt{23.4}$

8. $\sqrt{12.2^2 - 3.75^2}$

♦ 9. *(Washington, Exercise 1.2, # 55)* The daily average temperatures in degrees Celsius for each day of the first week of December were 12.5, 6.8, 3.2, 0, – 1.7, – 2.8, 3.2.

10. Find the average of 74.2, 63.5, and 91.2. After you make this entry, you discover that the number 63.5 should have been 68.5. Make the change and determine the new average without reentering all the numbers.

11. $6.2^2 + (3.5)(4.221)$

12. $6.514^3 + \sqrt{82.15}$

13. If $I = Prt$, where $P = 5225$ and $t = 0.750$, then find I for the following values of r: 0.050, 0.055, 0.060, and 0.065

14. If $F = \dfrac{9}{5}C + 32$, then find F for the following values of C: 10.2, 15.0, 20.0, 25.5, and 31.2.

15. Modify the formula in exercise 26 and find C for the following values of F: 32.0, 68.0, 98.6, 100.0, and 155.5.

16. If $R = \dfrac{R_1 R_2}{R_1 + R_2}$, then find R if

   (a) if $R_1 = 16.3$ and $R_2 = 152.5$.   (b) if $R_1 = 155.6$ and $R_2 = 226.3$.

17. $3.45^3 \pi$

18. $\sqrt[3]{625 - 24.235^2}$

19. $63.7\sqrt[4]{82.4}$

20. $(1.772)(3.1563^4)$

♦ 21. *(Washington, Exercise 1.3, #59)* The percent of alcohol in a certain car engine coolant is found by performing the calculation $\dfrac{100(42.6+51.9)}{103.2+51.9}$. Find this percent of alcohol. The number 100 is exact.

♦ 22. *(Washington, Exercise 1.8, #68)* In designing a building, it was determined that the forces acting on an I-beam would deflect the beam an amount (in cm.) given by $\dfrac{x(3723.9-31.0x^2+x^3)}{1850}$, where x is the distance (in m) from one end of the beam. Find the deflection for x = 4.5 m, 6.8 m, and 9.2 m.

23. 12!

24. $\dfrac{8!}{7!2!}$

25. The number of ways in which n digits can be arranged in an order (with no digits repeating) is n!. How many ways can 7 digits be arranged?

26. In how many ways can the 26 letters of the alphabet be arranged (see previous exercise)?

27. Evaluate $\dfrac{(4\times62)}{(3\times125)}$. Are both sets of parentheses needed?

28. Evaluate $\dfrac{(23.0\div3.12)}{(3.5\div12.7)}$. Are both sets of parentheses needed?

29. Find [2.999]

30. Find [−1.999]

31. Find $\left[\dfrac{3.5}{0.81}\right]$

32. Find $\left[114.3-\dfrac{64.2}{0.27}\right]$

Notes: 1. Functions entered by this procedure will remain in the calculator's memory until changed or erased or until the memory is cleared.

2. While a graph is on the screen, press HOME or QUIT to return to the Home Screen. The graph remains in the calculator's memory and can be seen again by pressing GRAPH.

3. When a function has been entered, a check mark ($\checkmark$) will appear to the left of that function indicating that it is turned on.
To turn a function on or off, highlight the function and press the key F4. Only functions that are turned on (checked) will be graphed.

Procedure G2. To change or erase a function.

1. Press $Y=$ to display the $Y=$ Editor on the screen.

2. (a) To change a function: Use the cursor keys to move the cursor to the desired function, press ENTER to place the function on the Entry Line, and make the changes by inserting, deleting, or changing the desired characters.

   (b) To erase a function: With the cursor on the same line as the function, press the key: CLEAR

3. Select GRAPH to graph the function or HOME or QUIT to exit to the home screen.

Procedure G3. To obtain the standard viewing rectangle.

To obtain the standard viewing rectangle:

1. With the graph on the screen, press the key F2 (ZOOM).

2. Select 6:ZoomStd

For the standard viewing rectangle: The x-value at the left of the screen is -10 and at the right of the screen is 10. The y-value at the bottom of the screen is -10 and at the top of the screen is 10. Each mark on the axes represents one unit.

Procedure G4. To use the trace function and find points on a graph.

1. With the graph on the screen, press the key F3 (Trace) for the trace function.

2. As the right and left cursor keys are pressed, the cursor moves along the graph and, at the same time, the coordinates of the cursor are given on the bottom of the screen.

3. If more than one graph is on the screen, pressing the up or down cursor keys will cause the cursor to jump from one graph to another. The number of the function will appear in the upper right of the screen.

4. If the graph goes off the top or bottom of the screen, the cursor will continue to give coordinates of points on the graph.

5. If the cursor is moved off the right or left of the screen, the graph scrolls (moves) right or left to keep the cursor on the screen.

Note: If the cursor is moved by using the cursor keys without first pressing the trace key, then the cursor can be moved to any point on the screen. The coordinates of this screen point will be given at the bottom of the screen.

Procedure G5. **To see or change viewing rectangle.**

1. To see values for the viewing rectangle, press the key WINDOW.
2. To change viewing rectangle values:
   Make sure the cursor is on the quantity to be changed and enter a new value for that quantity. To keep a value and not change it, either (a) just press the key ENTER or (b) use the cursor keys to move the cursor to a value that is to be changed.
3. To see the graph, after the new values have been entered, press GRAPH
   To return to the home screen, press the key HOME or QUIT.

In changing the scales for the axes, Xmin must be smaller than Xmax and Ymin must be smaller than Ymax. If either Xscl or Yscl is 0, no marks will appear on that axis.

Note: The standard viewing rectangle has the following values:
$$\text{Xmin} = -10, \text{Xmax} = 10, \text{Xscl} = 1,$$
$$\text{Ymin} = -10, \text{Ymax} = 10, \text{Yscl} = 1$$

Procedure G6: **To obtain preset viewing rectangles.**

1. Press F2 (Zoom) to bring the zoom menu to the screen while either a graph is on the screen or the values for the viewing rectangle is on the screen.
2. Select the desired preset viewing rectangle.
   (a) For the **standard viewing rectangle** select 6:ZoomStd.
   $$\text{Xmin} = -10, \text{Xmax} = 10, \text{Xscl} = 1,$$
   $$\text{Ymin} = -10, \text{Ymax} = 10, \text{Yscl} = 1$$
   (b) For the **square viewing rectangle** select 5:ZoomSqr.
   The square viewing rectangle keeps the y-scale the same and adjusts the x-scale so that one unit on x-axis equals one unit on y-axis. For a square viewing rectangle the ratio of y-axis to x-axis is about 1:2.
   (c) For the **decimal viewing rectangle** select 4:ZoomDec.
   The decimal viewing rectangle makes each movement of the cursor (one pixel) equivalent to one-tenth of a unit.
   (d) For the **integer viewing rectangle** select 8:ZoomInt.
   The integer viewing rectangle makes each movement of the cursor (one pixel) equivalent to one unit. After selecting the integer viewing rectangle, move the cursor to the point that is to be located at the center of the screen. Then press ENTER.
   Note: To have one movement of the cursor differ by k units set:
   $$\text{Xmin} = -158k/2 \qquad \text{Xmax} = 158k/2$$
   (e) For the **trigonometric viewing rectangle** select 7:ZoomTrig.
   The trigonometric viewing rectangle sets the x-axis up in terms of π or degrees (depending on mode) and sets each pixel movement in the x-direction to $\pi/24$ (0.13...) and xscl to $\pi/2$ (or 90°).

Procedure G7. To graph functions on an interval.

To graph functions on an interval use the *when* instruction.

The form is: **when**(*condition, fntrue, fnfalse*)

The instruction *when* is obtained from the catalog (see note at beginning of this appendix).

If *condition* is true, the function *fntrue* is graphed and if *condition* is not true, the function *fnfalse* is graphed.

The *condition* may involve one of the mathematical symbols $<, \le, >, \ge, =,$ or $\ne$ to compare two variables and/or constants. These symbols as well as *and* and *or* are obtained by pressing the key MATH, selecting 8:Test, then selecting from the menu.

Notes: 1. For more complex functions, the when function may be nested as in the following example:

(Example: Y2= when($x<2$,when($x<-3,-2x,-2$),x^2))

    2. The condition may also contain the words *and* and *or* as in the following example.

(Example: Y3= when($x>-3$ and $x \le 2, x^2, 3$))

1. To graph a function on the interval $x < a$ or on the interval $x \le a$ for some constant a:

In the Y= editor, enter the function $f(x)$ as: when($x<a, f(x)$))

(Example: To graph $y = x^2$ on $x<2$, enter the function: when($x<2, x^2,0$))

2. To graph a function on the interval $a < x < b$ or on the interval $a \le x \le b$ for some constants a and b:

In the Y= editor, enter the function as: when($x>a$ and $x<b, f(x)$), 0)

(Example: To graph $y = x^2$ on $-3 \le x \le 2$, enter the function:
when($x \ge -3$ and $x \le 2, x^2,0$)

3. To graph a function on the interval $x > a$ or on the interval $x \ge a$ for some constant a:

In the Y= Editor, enter the function $f(x)$ as when($x>a, f(x)$), 0).

(Example: To graph $y = x - 5$ on $x>2$, enter for the function when($x>2,x-5,0$)

Note: The above forms may be combined by writing their sum.

Procedure G8. Changing graphing modes.

To draw graphs connected or with dots:

With the Y= Editor on the screen,

1. Move the highlight to one of the functions.
2. Press the key F6.   (Hold down the key 2nd, and press F1.)
3. Select form the menu:

1:Line	Draws the graph with all plotted points connected.
2:Dot	Draws on the points corresponding to pixels on x-axis.
3:Square	Draws the graph as little squares.
4:Thick	Draws the graph using thick lines.
5:Animate	A cursor moves along the path of the graph.
6:Path	A cursor moves along the path and traces out the graph.
7:Above	Shades the area above the graph.
8:Below	Shades the area below the graph.

4. Graph the function or press ESC to exit.

Procedure G9. **To zoom in or out on a section of a graph.**

1. With a graph on the screen, Press F2 (Zoom)
   To zoom in, select 2:Zoom In

   or

   To zoom out, select 3:Zoom Out
2. Use the cursor keys to move the cursor to near the point of interest. This point will relocate to the center of the screen.
3. Press the key: ENTER

Note: It is a good idea to use the TRACE function together with ZOOM.

Procedure G10. **To change zoom factors**

1. With the Y= Editor or the graph on the screen, press F2 (Zoom)
2. Select C:SetFactors   (To highlight, scroll down.)
3. Change the factors as desired.
   There are factors for magnification in the x-, y- and z-directions.
   The preset factors are 4. Make changes to XFact for the magnification in the x-direction and to YFact for magnification in the y-direction. ZFact is used for 3 dimensional graphing and changes the magnification in the z-direction.
4. To exit this screen press ENTER to save changes or ESC to exit without changes.

Procedure G11. **Solving an equation in one variable.**

1. Write the equation so that it is in the form $f(x) = 0$, let $y = f(x)$, graph the function and adjust the viewing rectangle. Use a viewing rectangle so that the x-intercept of interest appears on the screen.
2. Determine a solution by:
   (a) Using Procedure G9 to zoom in on the x-intercept.

   or

   (b) Using the built in procedure:

      (1) With a graph on the screen, Press F5 (Math)
      (2) Select 2:Zero from the menu. The words Lower Bound? appear.
      (3) Move the cursor near, but to the left of the x-intercept and press ENTER. The words UpperBound? appear.
      (4) Move the cursor near, but to the right of the intercept and press ENTER. The zero (x-intercept) will appear at the bottom of the screen.

Procedure G12. **Finding maximum and minimum points.**

1. Graph the function and adjust the viewing rectangle so that either the desired local maximum or local minimum point is on the screen.
2. Determine the local maximum or local minimum point by:
   (a) Using Procedure G9 to zoom in on the point

or

    (b) Use the built-in procedure:

        (1) With the graph on the screen, press F5 (Math)

        (2) From the menu, select 3:minimum or 4:maximum. The words LowerBound? appear.

        (3) Move the cursor near, but to the left of the desired point and press ENTER. The words UpperBound? appear.

        (4) Move the cursor near, but to the right of the desired point and press ENTER. The coordinates of the minimum or maximum point will appear at the bottom of the screen.

Procedure G13.   Find the value of a function at a given value of *x*.

1. Graph the function and adjust the viewing rectangle so that the given *x*-value is on the screen.

2. (a) Using Procedure G9 to zoom in on the point

      or

    (b) Use the built-in procedure:

        (1) With the graph on the screen, press F5 (Math)

        (2) From the menu, select 1:value. The x-coordinate (*xc*) will be highlighted at the bottom of the screen. With this highlighted, enter the *x*-value.

        (3) Press ENTER. The cursor will move to the point and the coordinates will appear at the bottom of the screen.

Procedure G14.   To zoom in using a box.

1. With the Y= Editor or the graph on the screen, press F2 (Zoom)

2. Then select 1:ZoomBox

3. Use the cursor keys to move the cursor to a location where one corner (first corner) of the box is to be placed and press ENTER.

4. Use the cursor keys to move the cursor to the location for the opposite corner (second corner) of the box and press ENTER. As the cursor keys are moved a box will be drawn and when ENTER is pressed, the area in the box is enlarged to fill the screen.

Procedure G15.   Finding an intersection point of two graphs.

Enter the functions into the calculator and adjust the viewing rectangle so the region of the graph containing the intersection point or points is on the screen. Then, either

1. Graph the function and adjust the viewing rectangle so that the region containing the intersection point is on the screen.

2. Determine the point of intersection by:

   (a) Use Procedure G9 to zoom in on the point of intersection. To obtain the desired degree of accuracy, use the trace function to move the cursor just to the left and then just to the right of the intersection point after each zoom. A solution to the desired accuracy is obtained when the *x*-values, on either side of the intersection point, rounded off to the desired number of significant digits are equal.

   or

   (b) Use the built-in procedure:

      (1) With the graph on the screen, press F5 (Math)
      (2) From the menu, select 5:Intersection
          The words 1st Curve? appear.
      (3) If the cursor is on one of the intersecting curves, press ENTER, if not use the top and bottom cursor keys to move to the correct curve, then, press ENTER. The words 2nd Curve? appear
      (4) If the cursor is on the second of the intersecting curves, press ENTER, if not use the top and bottom cursor keys to move to the correct curve, then, press ENTER. The words Lower Bound? appear.
      (5) Move the cursor near, but to the left of the desired point and press ENTER. The words UpperBound? appear.
      (6) Move the cursor near, but to the right of the desired point and press ENTER. The coordinates of the intersection point will appear at the bottom of the screen.

Procedure G16. To graph each of two functions and then their sum.

1. Set the correct values for the viewing rectangle.
2. Enter the first function into the $Y=$ Editor (e.g. $y1$).
3. Enter the second function into $Y=$ Editor (e.g. $y2$).
4. Move the cursor after the equal sign for another function (e.g. $y3$) and enter the sum of the first and second function in function notation.($y3 = y1(x) + y2(x)$).
5. Press GRAPH

Procedure G17. To change graphing modes for regular functions, parametric, or polar equations.

1. Set the graphing mode by pressing the key MODE.
2. With the cursor blinking after the word Graph:
   Press the right arrow key and select:

   1:FUNCTION   for graphing in regular rectangular coordinates.
   2:PARAMETRIC   for graphing parametric equations.
   3:POLAR   for graphing polar equations.

   Then, press ENTER.

Notes: 1. For graphing parametric and polar equations involving trigonometric functions, Angle should be set to radians.

2 For more advanced work there is also the options of graphing sequences (4:SEQUENCE), three dimensional graphs (5:3D), and differential equation graphs (6:DIFF EQUATIONS).

Procedure G18.   To graph parametric equations.

1. Make sure the calculator is in correct mode for parametric equations. (Procedure G17)
2. Press the $Y=$ key to enter the equations:

    The screen will appear as: $xt1=$
    $yt1=$
    $xt2=$
    $yt2=$
    (etc.)

    (Up to 99 pairs of equations may be entered.)

    Functions are not turned on (the equal signs are not highlighted) until functions have been entered for both x and y variables. Functions may be turned on and off as with functions in rectangular coordinates (see Note 3, Procedure G1). Only functions turned on will be graphed.
3. Enter the first equation for x in the first row and the corresponding equation of y in the second row. If there are other sets of functions you wish to graph, enter them for the other x's and y's.
    To obtain the variable t, use the key T.
4. Press GRAPH to see the graph or QUIT or HOME to exit.

Procedure G19.   To change values for the parameter t.

With the calculator in the mode for parametric equations:
1. Press the key WINDOW.
2. Enter new values for tmin, tmax, and tstep and press ENTER or move to a next item by using the cursor keys.
    The standard viewing rectangle values for t are:

    tmin = 0, tmax = 6.28...(2π) , and tstep = 0.13...$(\pi/24)$
    (or tmax = 360 and tstep = 7.5 if in degree mode).

    tstep determines how often values for x and y are calculated.
    The size of tstep will affect the appearance of the graph.
    tmax should be sufficiently large to give a complete graph.
3. Press GRAPH, QUIT or HOME to exit.

Procedure G20.   To graph equations involving y^2.

1. Solve the equation for y. There will be two functions: $y = f(x)$ and $y = g(x)$. In many cases it will be true that $g(x) = -f(x)$.
2. Store $f(x)$ under one function name and $g(x)$ for the second function name (see Procedure C10).
    (Example: $y1 = f(x)$ and $y2 = g(x)$)
    or

If $g(x) = -f(x)$, we may do the following:

(a) Store $f(x)$ after a function name (Example: $y1$).

(b) Move the cursor to a second function name (Example: $y2$).

(c) Enter the negative sign: (–)

(d) Enter the name of the first function. (Example: $y2 = -y1$)

3. Graph the functions.

Procedure G21. **To obtain special *Y*-variables.**

Generally it is acceptable to enter variable names such as xmin, xmax, ymin, or ymax from the keyboard. Function names may also be entered from the keyboard, but must include a variable name in parentheses following the function name (Example: $y1(x)$).

Procedure G22. **To shade a region of a graph.**

1. While in the Entry Line on the Home screen, select the Shade command from the Catalog. (See note at beginning of the appendix.) Make sure all functions not wanted are deleted or turned off.

2. The form of the Shade command is:

Shade *L, U, Lt, Rt, Pat, D*

Where (relative to shaded area):

L is lower boundary (value, function, or name of function).

U is upper boundary (value, function, or name of function).

Pat is a digit from 1 to 4 indicating the type of shading pattern.

D is the density value (an integer from 1 to 10) that determines the spacing of the lines the calculator draws to shade the region. The higher the value for the density, the more widely spaced the shading lines. If the density value is omitted, the shading is solid.

Lt is the left most *x*-value.   *Rt* is the right most *x*-value.

[*Lt, Rt, Pat,* and *D* are optional.]

Enter the appropriate values after the word Shade, then, press ENTER.

3. To clear the shading:

With the graph on the screen, press F6 (2nd key plus F1) and Select 1:ClrDraw

(Example: Shade –8, $y1(x)$,xmin, 5, 2

shades in the area above $y = -8$, below the function $y1$, to the right of the minimum *x*-value, to the left of $x = 5$ and with pattern number 2.)

Note: To shade above the function represented by $y1$, use Shade $y1(x)$, *ymax*.

To shade below the function represented by $y1$, use Shade *ymin*, $y_1(x)$

Procedure G23. **To graph in polar coordinates**

1. Make sure the calculator is in correct mode for polar equations. (Procedure G17)

2. Press the *Y=* key to enter the equations:

The screen will appear as: $r1=$

$r2=$

$r3=$

(etc.)

(Up to 99 equations may be entered.)

Functions may be turned on and off as with functions in rectangular coordinates (see Note 3, Procedure G1). Only functions turned on will be graphed.

3. Enter an equation for *r*.
   To obtain the variable θ, use the key θ.
4. Press GRAPH to see the graph or QUIT or HOME to exit.

Procedure G24. To change viewing rectangles values for θ.

With the calculator in the mode for polar equations:

1. Press the key WINDOW.
2. Enter new values for θmin, θmax, and θstep and press ENTER or move to a next item by using the cursor keys.
   The standard viewing rectangle values for θ are:

   θmin = 0, θmax = 6.28...(2π) , and θstep = 0.13...($\pi/24$)
   (or θmax = 360 and θstep = 7.5 if in degree mode).

   θstep determines how often points are calculated and may affect the appearance of the graph. Leaving θstep at approximately 0.1 is generally sufficient but θstep may have to be changed if there is a major change in the viewing rectangle. θmax should be sufficiently large to give a complete graph.
3. Press GRAPH, QUIT or HOME to exit.

Procedure G25. Drawing a tangent line at a point.

1. Store the function for a function name and graph the function.
2. With a graph on the screen,
   (a) Press F5 (Math)
   (b) Select A:Tangent   The words Tangent at? appear on the screen.
   (c) Move the cursor to the point where the tangent is to be drawn and press ENTER.

To clear tangent lines from the screen use the same process as in Procedure G22, Step 3.

Procedure G26. Determining the value of the derivative at points on a graph.

1. Store the function for a function name and graph the function.
2. With a graph on the screen,
   (a) Press F5 (Math)
   (b) Select 6:Derivatives and select a:d*y*/d*x*.
       The words d*y*/d*x* at ? appear on the screen.
   (c) Move the cursor to the desired point and press ENTER.

Procedure G27. To find the constant of integration for an indefinite integral.

This procedure assumes that the general antiderivative is known.
1. Store a value of 0 for C. ($0 \rightarrow C$)
2. Store the antiderivative function with C attached for a function name and graph.
3. Given the information that $y = y_0$ when $x = x_0$ and with the graph of the antiderivative (for $C = 0$) on the screen, evaluate the antiderivative function at $x = x_0$ (Procedure G13).   With the cursor on the graph of the antiderivative, read the corresponding y-value, y_a. The constant of integration is the difference between the given y-value y_0 and the value from the graph y_a. That is $C = y_0 - y_a$.
4. To check your answer, store the value obtained for C, graph the antiderivative function, and evaluate at $x = x_0$.

Procedure G28. Function to determine the area under a graph.

1. Store the function for a function name and graph the function.
2. With a graph on the screen,
    (a) Press F5 (Math)
    (b) Select 7:$\int f(x)dx$
         The words Lower Limit? appear on the screen.
    (c) Move the cursor to first x-value (Lower Limit) and press ENTER.
         The words Upper Limit? appear on the screen.
    (d) Move the cursor to the second x-value (Upper Limit) and press ENTER.
         The area between the curve, the x-axis, and the two x-values will be shaded and will be given at the bottom of the screen.

Procedure G29.   For drawing lines.

On the Entry Line of the Home screen,
To draw a line:
1. Select Line from Catalog   (See note at beginning of this appendix.)
2. The form is: **Line** x_1, y_1, x_2, y_2
    where x_1, y_1 and x_2, y_2 are the beginning and ending points of the line.
or
On the graphics screen,
1. Clear any lines or drawing from the graph screen:
    With the graph on the screen, press F6 (2nd key plus F1) and select 1:ClrDraw
2. With the graph on the screen, press F6 (2nd key plus F1) and select 3:Line
    The words 1st Point? appear on the screen.
3. Move the cursor to a point on the screen and press ENTER.
    The words 2nd point appear on the screen. Move the cursor to the second point and press ENTER. The line will be drawn.
4. Press ESC to quit drawing lines.

B.4 Programming Procedures

Procedure P1. **To enter a new program.**

1. Press the key: APPS and select 7:Program Editor
2. Select 3:New...
3. Move the cursor down to the box following the word Variable and enter a name. This will be the name of the program. The name may be up to eight characters long. Then, press ENTER, then press ENTER again.
4. The name of the PROGRAM appears at the top of the screen. Also appearing are the commands :Prgm and :EndPrgm. The entire program must be between these two commands. Move the cursor to the line below the :Prgm command. A colon precedes the cursor (:). The calculator is now in programming mode and is ready for the first statement in the program.
5. Enter the steps of the program one at a time, pressing ENTER after each step. More than one instruction may be placed on a line if they are separated by a colon (:).
6. When finished entering the program, press HOME or QUIT to return to the home screen.

Procedure P2. **To execute, edit, or erase a program.**

1. To execute a program:
   A program may be executed from the home screen by entering the name of the program on the Entry Line followed by parentheses and pressing ENTER.

   Note: In more advanced work, values may be placed in the parentheses. However, we will not consider this at this time.

2. To edit a program:
   (a) If the program is not the current program:
       (1) Press the key APPS
       (2) Select 7:Program Editor
       (3) Select 2:Open
       (4) Then, move the cursor down to the word following the word variable and press the right cursor key. Select a program name from the list.

       or

       If the program is the current program (last program worked with):
       Select 1:Current
   (b) Make changes in the program as desired. The cursor keys may be used to move the cursor to any place in the program. The calculator is now in programming mode.

3. To erase a program:
   Press the key VAR-LINK.
   Listed are all variable names including matrices, expressions, functions, program names, etc.
   (a) To see just the list of programs:
       (1) Press F2 (View...)
       (2) Press the bottom cursor to move the highlight to the word following the words Var Type:

(3) Press the right cursor key.
(4) Select the type of variable, in this case 6:Program
(5) Press ENTER.

(b) Select the programs to be erased by moving the highlight to the program and pressing F4. This will place a check mark in front of the program name.

(c) Press F1 (Manage)

(d) Select 1:Delete
A list of the checked programs will appear.
To erase press ENTER otherwise press ESC.

(e) Press HOME or QUIT to return to the home screen.

Procedure P3. To input a value for a variable.

Entering a value for a variable uses the keywords: Input or Prompt.

The form of the statement is: **Input v** or **Prompt v**
(**v** represents the variable name)

With the calculator in programming mode:

1. To access the word Input or Prompt: Press F3 (I/O)

2. Enter the variable name and press ENTER.
(Example: Input C or Prompt C)

When the program is executed the word Input followed by a variable will cause a question mark to appear on the screen while the word Prompt will cause the name of the variable followed by the question mark to appear. Any number entered is stored for that variable and when the variable is later used in the program that value is used in the calculation.

Procedure P4. To evaluate an expression and store that value to a variable.

With the calculator in programming mode:
Enter the expression, press the key STO▷, and then, enter a variable name.
(Example: $(9/5)C + 32 \to F$)

Procedure P5. To display quantities from a program.

Either the value for a variable or an alpha expression may be displayed by using the keyword Disp.

The form of the statement is: **Disp v**   (v represents the variable name)

The quantity v may be a variable name, a number, or a set of characters enclosed inside of quotes. If v is a quantity enclosed in quotes, the quantity will be printed exactly as it appears.

With the calculator in programming mode:

1. To access the word Disp: Press F3 (I/O) and select 2:Disp

2. Enter the variable name and press ENTER.   (Example: Disp F)

When the program is executed, the value stored for the variable or the expression will be displayed on the screen.

Procedure P6. To add or delete line from a program.

With the calculator in programming mode:
1. To add a line:
    (a) To insert a line before a given line move the cursor to the first character of the line and to insert a line after a given line move the character to the last character in the given line.
    (b) If the calculator is in insert mode (indicated by a narrow cursor), just press ENTER. If not (indicated by a wide cursor), pressing the key INS will place it in insert mode. (To take it out of insert mode, press INS again.) This will create a blank line either before or after the given line
    (c) Place the cursor in the desired line. An instruction may now be entered.
2. To delete a line
    (a) Move the cursor to the desired line
    (b) Press the key CLEAR, then press DEL.

Procedure P7. Making a decision (If... command).

The form of the If command is **If** *condition*

To access If...
While in programming edit mode (at a step in program):
Press F2 (Control) and select 1:If

If *condition* is true, the statement following the If statement is executed next. If *condition* is not true, the second statement following the If statement is the next statement executed.
The *condition* may involve one of the mathematical symbols $<, \leq, >, \geq, =$, or $\neq$ to compare two variables and/or constants. These symbols are obtained by pressing the key MATH, selecting 8:Test and selecting from the menu.

Procedure P8. Making a decision with options (If...Then...Else...EndIf commands).

The form of this combination is

(program statements before)
If *condition* **Then**
(program statements)
Else
(program statements)
EndIf
(program statements after)

To access If... Then... Else...EndIf
While in programming edit mode (at a step in program):
Press F2 (Control) and select 2:If...Then
From the menu, select either 1:If...Then...EndIf **or** 2:If...Then...Else...EndIf

If the *condition* is true, the statements following the Then statement are executed. If the *condition* is not true, the statements following the Else statement (if included) are executed. After either, the statements after EndIf are executed.

The *condition* involves one of the mathematical symbols $<, \leq, >, \geq, =$, or $\neq$ to compare two variables and/or constants. These symbols are obtained by pressing the key MATH, selecting 8:Test and selecting from the menu.

Procedure P9.   Finding the area under a graph.

1. The desired function should be stored under the function name in the function table. (Procedure C6)
2. Enter the following program into your calculator to find the area between two values B and C:
   (It is assumed here that the function is stored under the function name $y1$)

```
area()
Prgm
Disp "Enter first Z"
Input B
Disp "Enter second Z"
Input C
∫(y1(x),x,B,C)→I
Disp "AREA IS:"
Disp I
EndPrgm
```

To obtain the definite integral function ∫(press the key ∫.
The form of the definite integral function is
    ∫(*function or function name, variable, smaller value, larger value*)
(This function may also be used without being in a program.)

Procedure P10.   Creating a While... loop (While, EndWhile commands).

The construction of the loop is of the form:

    ... (program statements before loop)
    While *condition*
    ... (program statements within loop if condition true)
    EndWhile
    ... (programming statements after loop)

To access While... and EndWhile... statements,
While in programming edit mode (at a step in program):
Press F2 (Control) and select 5:While...EndWhile

In this case, if *condition* is a true statement the statements following the word While are performed and if the condition is not true, the program will then transfer to the statements following the word EndWhile.

Procedure P11.   Creating an infinite Loop (Loop, EndLoop commands).

The construction of the loop is of the form:

    ... (program statements before loop)
    Loop
    ... (statements within loop repeated until condition is true)
    EndLoop
    ... (programming statements after loop)

To access Loop... and Endloop statements,
While in programming edit mode (at a step in program):
Press F2 (Control) and select 6:Loop...EndLoop.

This loop will continue forever unless a statement is placed within the loop to exit from it.   Two ways of exiting from the loop at to use an IF... statement (see Procedure P7) followed by an Exit statement or by a Goto... and a Label command. For Goto and Label commands see Procedure P13.
For the Exit command, press F2 (Control), select 8:Transfers, and select 5:Exit.

Procedure P12. Creating a For... loop (For, ForEnd commands).

The construction of the loop is of the form:

> ... (program statements before loop)
> **For** (*var, beg, end, inc*)
> ... (statements within loop repeated if *end* not exceeded)
> **EndFor**
> ... (programming statements after loop)

Where *var* is a variable name
beg is a beginning value
end is an ending value
inc is an increment value and is optional
(Increment is 1 if this is omitted.)

To access For.. and End statements,
While in programming edit mode (at a step in program):
Press F2 (Control) and select 4:For...EndFor

In this case, the first time through the loop, *var* has the beginning value, and the next time through the loop *var* has the value of *beg* + *inc*. Each time through the loop the value for *var* is increased by *inc*. The loop stops when the value for *var* exceeds *end* and the program will then transfer to the statements following the word End.
Note: One space is needed after the word For. (Example: For *k*,1,6,2)

Procedure P13. Creating an If...Goto... loop (Lbl, If, Goto commands).

The construction of the loop is of the form:

> ... (program statements before loop)
> **Lbl** *label*
> ... (program statements within loop)
> **If** *condition*
> **Goto** *label*
> ... (programming statements after loop)

Lbl and Goto commands are accessed from the CATALOG (see note at beginning of this appendix). For the If ... command see Procedure P7.
While in programming edit mode (at a step in program):
Note: A *label* may be a name of up to 8 characters, beginning with a letter.

The statements within the loop are executed as long as condition is true. When condition ceases to be true, the program will transfer to statements following the Goto statement.

Procedure P14. To display a graph from within a program (DispG command).

While in programming edit mode (at a step in program):
The DispG command is accessed from the CATALOG and displays the current graph screen.   (See note at beginning of this appendix.)

Procedure P15. Drawing a line on the screen from within a program (Line command).

The form is: **Line** x_1, y_1, x_2, y_2

While in programming edit mode (at a step in program):
The Line command is accessed from the CATALOG and draws a line from x_1, y_1 and x_2, y_2.   (See note at beginning of this appendix.)
Example: Line $(0,2,x,y1(x))$ draws a line from the point $(0,2)$ to the point $(x, f(x))$.

Procedure P16. To temporarily halt execution of a program (PAUSE command).

While in programming edit mode (at a step in program):
Press F2 (Control), select 8:Transfers, and select 1:Pause.
The Pause statement when used in a program halts execution of the program until the ENTER key is pressed.

B.5 Matrix Procedures

Procedure M1. To enter or modify a matrix.

1. Press the key APPS.
2. Select Data/Matrix Editor.
3. Select 3:New... to enter a new matrix.
   (a) With the word after Type: highlighted, press the right cursor key and select 2:Matrix.
   (b) Move the cursor down to the box after the word Variable: and enter a name for the matrix. The name may be up to eight characters long and the first character must be a letter. Press ENTER.
   (c) Use the bottom cursor key to move the cursor to the box following the words Row dimension: and enter the number of rows in the matrix.
   (d) Use the bottom cursor key to move the cursor to the box following the words Col dimension: and enter the number of columns in the matrix.
   (e) Press ENTER, then press ENTER again. A table (matrix) will appear with the requested number of rows and columns. Enter the elements of the matrix. Elements are entered row at a time. After each entry, press ENTER.

   or

Select 2:Open... to open an old matrix,

(a) With the word after Type highlighted, press the right cursor key.

(b) Select 2:Matrix

(c) Move the cursor down so the word after Variable: is highlighted and press the right arrow key. Select a list from the names and press ENTER.

or

Select 1:Current for the last matrix accessed.

4. To modify a matrix, move the cursor to the elements to be changed and make the changes.

5. Press HOME or QUIT to exit to the home screen.

Procedure M2. To view a matrix.

Enter the name of the matrix on the Entry Line of the Home screen and press ENTER.

Procedure M3. Addition, subtraction, scalar multiplication or multiplication of matrices.

On the Entry Line of the Home screen,

1. For addition of matrices:
    Enter the name of the first matrix, press the addition sign, enter t name of the second matrix, then press ENTER.   (Example: $M1 + M2$)

2. For subtraction of two matrices:
    Enter the name of the first matrix, press the subtraction sign, enter the name of the second matrix, then press ENTER.   (Example: $M1 - M2$)

3. Multiplication of matrix by a scalar:
    Enter the scalar, enter the name of the matrix and press ENTER.
    (Example: $2.5\,M1$ or $2.5 \times M1$)

4. Multiplication of two matrices:
    Enter the name of the first matrix, press the multiplication symbol, and enter the name of the second matrix, and press ENTER. (Example: $M1 \times M2$)

Many of these operations may be combined into one statement. The result of a matrix calculation is stored under ANS. An error will occur if the operation is not defined for the matrices selected.

Procedure M4. To find the determinant, inverse, and transpose of a matrix.

On the Entry Line of the Home screen,

1. To find the determinant:
    (a) Press the key MATH and select 4:Matrix.
    (b) Select 2:det(
    (c) Enter the name of the matrix and the closing parenthesis and press ENTER.
         (Example: det($M1$))

2. To find the inverse:
    (a) Select the name of the matrix.
    (b) Press the key ^
    (c) Enter (–)1 and press ENTER.   (Example: $M1\,{}^{\wedge}-1$)

3. To find the transpose:
    (a) Enter the name of the matrix.
    (b) Press the key MATH and select 4:Matrix.
    (c) Select 1:^{T}
    (f) Press ENTER.   (Example: $M1^T$)

Procedure M5. To store a matrix.

1. Select the name of the first matrix or enter a matrix expression.
2. Press the key STO▷
3. Select the name of the second matrix and press ENTER.
    (Example: $M1 + M2 \rightarrow M3$)

Note: Storing a matrix for a second matrix, erases the previous contents of the second matrix.

Procedure M6: Row operations on a matrix.

To do matrix row operations, first select the row operation to be preformed.
Do this with the cursor on the Entry Line of the Home screen. Either:

Press the key MATH and then, select 4:Matrix, then Select J:Row ops. From the menu select the desired operation.

or

Select the desired row operation from CATALOG.

1. To interchange two rows on a matrix, select rowSwap(
        Form: **rowSwap**(name, row1, row2)
        (Example: rowSwap($M1$, 1, 3)   Interchanges rows 1 and 3 of matrix $M1$.)
2. To multiply a row by a constant, select mRow:
        Form: **mRow**(constant, name, row)
        (Example: mRow(1/2, $M1$, 3)   Multiplies row 3 of matrix $M1$ by ½.)
        To divide a row by a constant, multiply by the reciprocal.
3. To multiply a row by a constant and add to another row, select mRowAdd(
        Form: **mRowAdd**(constant, name, row1, row2)
        Example: mRowAdd(−5, $M1$, 1, 3) (Multiplies row 1 of matrix $M1$ by
            −5 and adds the result to row 3 and replaces row 3 with the result.)
4. Enter the closing parenthesis and press ENTER.

Note: When 4:Matrix is selected, the matrix operation *ref* gives the row echelon form directly for a matrix and the operation *rref* gives the reduced row echelon form directly for a matrix. The forms are: ref ($M1$) and rref($M1$) where $M1$ is the name of the matrix to be reduced.

A.6 Statistical Procedures

Procedure S1. **To enter or change single variable statistical data.**

1. Press the key APPS.
2. Select Data/Matrix Editor.
3. Select 3:New... to enter a new set of data.
   (a) With the word after Type highlighted, press the right cursor key.
   (b) Select 1:Data
   (c) Move the highlight down to the box following the word Variable.
   (d) Enter a variable name in the box. The name can be up to eight characters and the first character must be a letter. It cannot be a name used for anything else. Press ENTER, then press ENTER again.
   (e) Place the data elements in one column and the frequencies (if any) in a second column

   or

   Select 2:Open... to open an old set of data,
   (a) With the word after Type highlighted, press the right cursor key.
   (b) Select 1:Data
   (c) Move the cursor down so the word after Variable: is highlighted and press the right arrow key. Select a list from the names and press ENTER.

   or

   Select 1:Current to open last set of data.
4. The data elements will be in one column and the frequencies (if any) in a second column. Changes may be made to any element by moving the cursor to the element and making the changes. Each frequency should be in the same row as the corresponding data element.
5. To exit, after all data have been entered or changed, press HOME or QUIT.
   Note: When there are intermediate data values with a frequency of zero, it is best, for graphing purposes, to enter these data values and the zero frequencies.

Procedure S2. **To graph single variable statistical data.**

1. With the statistical data on the screen, press F2 (Plot Setup)
2. Highlight an unused plot number and press F1 (Define).
3. With the word after Plot Type highlighted, press the right cursor key and select the type of graph.
4. Move the cursor so that the word after Mark is highlighted, press the right cursor key, and select the type of mark to be used.
5. Move the cursor down to the box following x... and enter the column name (c1, c2, etc.) for the column containing the data.
6. For scatter graphs or histograms, move the cursor down to the box following y... and enter the name of the column containing the frequencies.
   For other types of graphs, the frequencies are entered in the box following the word Freq... after Step 7 is performed.

7. Move the cursor so that the word following Freq and Categories is highlighted and press the right cursor key and select 1:NO if all the frequencies are 1 or the plot is a scatter graph or xyline graph and 2:YES for other types of graphs.

8. Press ENTER.

9. Make sure the viewing rectangle is appropriate for the data to be graphed, that the graph display has been cleared, and that all functions in the function table have been turned off.

10. Press the key GRAPH.

Note: Plots may be turned on or turned off by pressing the key F4 while the screen listing the plots (under plot setup) is showing.

Procedure S3. To clear the graphics screen.

1. All functions and plots should be deleted or turned off.
2. With the graph on the screen, press F6 (Draw).
3. Select 1:ClrDraw

Procedure S4. To obtain mean, standard deviation and other statistical information.

Be sure one-variable data has been stored as in Procedure S1.
With the data showing on the screen,

1. Press F5 (Calc) to set up the calculations.
2. With the word following the words Calculation Type highlighted, press the right cursor key.
3. Select 1:OneVar
4. Move the cursor down to the box following x... and enter the column name (c1, c2, etc.) for the column containing the data.
5. Move the cursor so that the word following Freq and Categories is highlighted and press the right cursor key and select 1:NO if all the frequencies are 1or 2:YES if frequencies are not all 1.
6. Move the cursor down to the box following Freq... and enter the column name (c1, c2, etc.) for the column containing the frequencies and press ENTER.
7. Press ENTER.

The following statistical quantities appear:

$\bar{x}$ is the mean of all the x-values

$\sum x$ is the sum of all the x-values

$\sum x^2$ is the sum of all the squares of the x-values

Sx is the s-standard deviation of the x-values
nStat is the total number of x-values
minX is smallest data value
$q1$ determines the first quartile
medStat is the median

8. Press ENTER to return to data screen.

Procedure S5. To clear statistical data from memory.

On the Home screen, press the key VAR-LINK.
Listed are all variable names including matrices, expressions, functions, program names, etc.

1. To see just the lists of statistical data:
    (a) Press F2 (View...)
    (b) Press the bottom cursor to move the highlight to the word following Var Type:
    (c) Press the right cursor key.
    (d) Select the type of variable, in this case B:Data
    (e) Press ENTER.
2. Select the data names to be erased by moving the highlight to a name and pressing F4. This will place a check mark in front of the data name. Press F4 again to remove the check mark.
3. Press F1 (Manage)
4. Select 1:Delete
    A list of the checked programs will appear.
    To delete press ENTER otherwise press ESC.
5. Press HOME or QUIT to return to the home screen.

Procedure S6. To enter or change two variable statistical data.

1. Press the key APPS.
2. Select Data/Matrix Editor.
3. Select 3:New... to enter a new set of data.
    (a) With the word after Type highlighted, press the right cursor key.
    (b) Select 1:Data
    (c) Move the highlight down to the box following the word Variable.
    (d) Enter a variable name in the box. The name can be up to eight characters and the first character must be a letter. It cannot be a name used for anything else. Press ENTER, then press ENTER again.
    (e) Place the x-values in one column, the y-values in a second column, and the frequencies (if any) in a third column

    or

    Select 2:Open... to open an old set of data,
    (a) With the word after Type highlighted, press the right cursor key.
    (b) Select 1:Data
    (c) Move the cursor down so the word after Variable: is highlighted and press the right arrow key. Select a list from the names and press ENTER.

    or

    Select 1:Current to open the last set of data.
4. The x-values will be in one column, the y-values in a second column, and the frequencies (if any) in a third column. Changes may be made to any element by moving the cursor to the element and making the changes. The x-values, y-values, and frequencies that correspond should be in the same row.
5. To exit, after all data have been entered or changed, press HOME or QUIT.

Procedure S7. To determine a regression equation.

Be sure one-variable data has been stored as in Procedure S6.
With the data showing on the screen,

1. Press F5 (Calc) to set up the calculations.
2. With the word following Calculation Type highlighted, press the right cursor key.
3. Select the type of regression equation desired.
   (a) For linear regression ($y = ax + b$), select 5:LinReg
   (b) For logarithmic regression ($y = a + b \ln x$), select 6:LnReg
   (c) For Exponential Regression ($y = ab^x$), select 4:ExpReg
   (d) For Power Regression ($y = ax^b$), select 8:PowerReg
   (e) For Quadratic Regression ($y = ax^2 + bx + c$), select 9:QuadReg
        (Note: Need at least 3 data points.)
   (f) For Cubic Regression ($y = ax^3 + bx^2 + cx + d$), select 3:CubicReg
        (Note: Need at least 4 data points.)
   (g) For Quartic Regression ($y = ax^4 + bx^3 + dx^2 + ex + f$), select A:QuartReg
        (Note: Need at least 5 data points.)
   (h) For sinusoidal regression ($y = a\sin(bx+c)+d$), select B:SinReg
   (i) For logistic regression ($y = \dfrac{a}{1 + be^{cx}} + d$), select C:Logistic.
4. Move the cursor down to the box following $x...$ and enter the column name (c1, c2, etc.) for the column containing the x-values.
5. Move the cursor down to the box following $y...$ and enter the column name (c1, c2, etc.) for the column containing the y-values.
6. Move the cursor so the word following Store RegEQ to.. is highlighted and press the right cursor key. If desired select a function to which to store the regression equation. Caution: This will replace any other function stored for that function name.
7. Move the cursor so that the word following Freq and Categories is highlighted and press the right cursor key and select 1:NO if all the frequencies are 1 or 2:YES if frequencies are not all 1.
6. If not all frequencies are 1, move the cursor down to the box following Freq... and enter the column name (c1, c2, etc.) for the column containing the frequencies and press ENTER.
7. Press ENTER.

A screen will appear showing the regression equation with values for the constants and the correlation coefficient r.

Procedure S8. **To graph data and the related regression equation.**

1. Make sure the viewing rectangle is set properly to graph the set of data, that the graph display has been cleared, and that all functions in the function table and other plots have been removed or turned off.

2. If the data has not already been entered into the calculator, it will need to be entered (Procedure S6) and the regression equation calculated (Procedure S7).

3. The regression must be stored to a function name in the Y= Editor. Do this when determining the regression equation (Procedure S7, Step 6.

4. Define the plot:

    (a) With the statistical data on the screen, press F2 (Plot Setup)
    (b) Highlight an unused plot number and press F1 (Define).
    (c) With the word after Plot Type highlighted, press the right cursor key and select the type of graph -- in this case a scatter graph.
    (d) Move the cursor so that the word after Mark is highlighted, press the right cursor key, and select the type of mark to be used.
    (e) Move the cursor down to the box following x... and enter the column name (c1, c2, etc.) for the column containing the x-values.
    (f) Move the cursor down to the box following y... and enter the column name (c1, c2, etc.) for the column containing the y-values.
    (g) Press ENTER, then press ENTER again.

5. Press the key GRAPH.

Note: Plots may be turned on or turned off by pressing the key F4 while the screen listing the plots (under plot setup) is showing.

Appendix C

Procedures for Casio CFX-9850G Calculator

The Casio CFX-9850G Color Power Graphic Calculator is menu driven. When first turned on a main menu appears with several menu choices. These choices are indicated by various icons. A choice may be made by using the cursor keys to move the highlight to the desired item and then pressing the key EXE or by entering the symbol in the lower right-hand corner of each icon. Selecting a choice enters a particular calculator mode.

The EXIT key is normally used to return to a previous screen and the MENU key is used to return to the Main Menu.

C.1 Basic Calculator Procedures

Procedure B1. To turn calculator on and off.

1. To turn the calculator on, press the key: AC$^{/ON}$
2. To turn the calculator off, you will need to first press the key: SHIFT then the key: AC$^{/ON}$

Procedure B2. To clear memory in the calculator.

1. Turn the calculator on.
2. Select MEM from the Main Menu by using the cursor keys to highlight the MEM box and pressing the key EXE **or** by pressing the key Alpha, then the key for E (the key labeled **cos**).
3. Use either the top or bottom cursor key to select Reset and press the key EXE. (The other choice is Memory Usage, which will indicate the amount of memory used in different areas and the amount of memory free.)
4. Press the F1 key to reset all memory. This will clear all memory.
Note: The Casio 9850G has approximately 32000 bytes of memory available.

Procedure B3. Adjusting the contrast and colors of the screen.

The Casio 9850G Calculator will display items, including graphs, in three different colors: orange, blue, and green. The intensity of these colors as well as the overall contrast of the screen may be changed by the following procedure.

1. Select CONT from the Main Menu by using the cursor keys to highlight the CONT box and pressing the key EXE **or** by pressing the key Alpha, then the key for D (the key labeled **sin**).
2. Move either the top and bottom cursor key till the symbol ▶ is to the left of the item to be adjusted. Then use the right and left cursor keys to adjust the

color or contrast to the desired level. Pressing the F1 key at this point will return these settings to the original (initial) settings.
To change the contrast at any time, press the SHIFT key then the right cursor key to make screen darker or the left cursor key to make the screen lighter.
3. When finished, press the MENU key.

Procedure B4. To Clear the Calculator Screen

To clear the calculator screen when in RUN mode, press the key: AC
Pressing the MENU key will always return you to the Main Menu.

C.2 General Calculation Procedures

The following procedures generally apply when the calculator is in Run Mode. Run Mode is the mode in which most normal calculations are performed. To enter Run Mode, from the Main Menu select RUN by using the cursor keys to highlight the RUN icon and pressing the key EXE **or** by pressing the key for the number 1.

Procedure C1. To perform secondary actions or obtain alpha characters.

1. To perform secondary actions of keys:
(The actions above and to the left of a key.)
(a) Press the key: SHIFT
(b) then, press the key for the desired action
(Example: To find the square a number we would first press SHIFT then the key x^2 ($\sqrt{}$), enter the number, then press EXE)
2. To obtain alpha characters:
(The characters above and to the right of a key.)
(a) Press the key: ALPHA
(b) then, press the key for the desired character.
(Example: To display the letter B on the screen, press the key: ALPHA then press the key log (for B))
Note: The alpha key may be locked down by first pressing the key: SHIFT then pressing the key: ALPHA.

Procedure C2. Selection of menu items

When certain keys are pressed the Casio 9850G lists menu items at the bottom of the screen. To select a menu item, press the function key (F1, F2, F3, F4, F5, F6) just below the item.
(Example: Pressing the ALPHA key also make available the single quotation marks (F1 key) and the double quotation marks (F2 key) and the symbol ~ (F3 key).)
To exit a screen menu, press the key EXIT.

Procedure C3. To perform arithmetic operations.

1. To add, subtract, multiply or divide press the appropriate key: +, −, ×, or ÷.
If parentheses are used to denote multiplication, it is not necessary to use the multiplication symbol.
(The calculation 2(3 + 5) is the same as 2×(3 + 5).)

2. To find exponents use the key ^.   (Example: 2^4 is entered as 2^4)
   For the special case of squaring a number we may use the key x^2.
   Note: On the Casio calculator, there is a key labeled (-) which is used to enter
   a negative number.
3. After entering the expression press the key EXE to perform the calculation.

Procedure C4. To correct an error or change a character within an expression.

1. (a) To make changes in the current expression: Use the cursor keys to move
   the cursor to the position of the character to be changed.
   (b) If it is desired to make changes in the last expression entered, after pressing
   the EXE key: Press the left or right cursor keys to obtain a copy of the last
   expression and use the cursor keys to move the cursor to the character to
   be changed.
2. Characters may now be replaced, inserted, or deleted at the cursor position.
   (a) To replace a character at the cursor position, just press the new character.
   (b) To insert a character or characters in the position the cursor occupies, press
   the key INS. Then, press keys for the desired character(s).
   (c) To delete a character in the position the cursor occupies, press DEL

Procedure C5. Selection of special mathematical functions.

1. To raise a number to a power:
   (a) Enter the number and press the key ^
   (b) Enter the power and press EXE
2. To find the root of a number:
   (a) For cube root, press the key $\sqrt[3]{\ }$ **or**

   For any root, enter the root index and press the key $\sqrt[x]{\ }$
   (b) Enter the number and press EXE
3. To find the factorial of a positive integer:
   (a) Enter the number
   (b) Press the OPTN key for Options Menu
   (c) Press F6 for the more options
   (d) Press F3 for the Probability (PROB) Menu
   (e) Press F1 to select x! from the menu and press EXE.
4. To find the largest integer less than or equal to a given value (This is the
   greatest integer function and is indicated by square brackets []):
   (a) Press the OPTN key for Options Menu
   (b) Press F6 for the more options
   (c) Press F4 for the Numerical (NUM) Menu
   (d) Press F2 to select Int from the menu
   (e) Enter the value and press EXE.
5. To find use the number π, press the key for π on the keyboard.
6. To find the absolute value of a number.
   Press the OPTN key for Options Menu
   Press F6 for the more options
   Press F4 for the Numerical (NUM) Menu
   Press F1 to select Abs from the menu
   Enter the value and press EXE.

Procedure C6. Changing the numerical display mode of your calculator.

Press the key: SET UP
Press the bottom cursor key to move the highlight down to Display

1. To fix the number of decimal places displayed:
   Press the key F1 to select Fix from the menu.
   Enter the number of decimal places desired and press EXE

2. To display a fixed number of significant digits in scientific notation:
   Press the key F2 to select Sci from the menu.
   Enter the number of significant digits desired and press EXE

3. To display a number in normal notation:
   Press the key F3 to select Nrm from the menu and press EXE

4. To display numbers in engineering notation:
   Press the key F4 to select Eng from the menu and press EXE

   Note: Engineering notation uses M for millions, k for thousands, m for 1/1000
   and μ for 1/1000000.

Procedures C7. Storing a set of numbers as a list.

From the Main Menu, select the LIST icon to enter LIST mode. Up to six lists
may be stored. Use the right or left cursor key to select a list, then enter the
numbers in the list by entering the number and pressing EXE after each entry.

To change a value, move the highlight to that value and enter a new value.

To delete a value, move the highlight to that value and pressing the F3 key (DEL).

An entire list may be deleted by highlighting a value in a list and press the F4 key
(DEL-A). Then press F1 (YES).

Procedures C8. Viewing a set of numbers.

From the Main Menu, select the LIST icon to enter LIST mode and move the
highlight to the list to be seen.

Procedures C9. To perform operations on lists.

1. From the Main Menu, select RUN icon to enter RUN mode.

2. To select a list:
   (a) Press the key OPTN
   (b) Press F1 (List)
   (c) Enter the number of the list

3. To add, subtract, multiply or divide lists, select the list (Step 2), press the
   appropriate key (+, −, ×, or ÷), select the second list and press EXE. Lists must
   be of the same length in order for these operations to be performed.
   (Example: List1 + List 2)

4. To perform an operation on all the elements of a list, enter a value, press the
   appropriate key (+, −, ×, or ÷), select the list (Step 2) and press EXE.
   Functions such as finding the squares or square roots, may be performed on
   each element of a list by performing the function of the selected list.
   (Example: 5×List 1)

Procedure C10. **To store or recall a function.**

The Casio 9850G will store functions in either regular Function Memory or in Graph Function Memory.

1. In Graph Function Memory
   (a) From the Main Menu, select the Graph icon. The graphing function menu will appear in the form *Y*1=, *Y*2=, etc. Up to 20 functions or expressions may be stored and viewed this way.
   (b) Move the highlight to one of the function names, enter the function, and then press EXE. Use the key *X, θ, T* for the variable X.

   Note: For rectangular coordinates *Y*= should appear at the top of the screen. If not, then press F3 (TYPE) and then press F1 (*Y*=).

or

2. In Regular Function Memory
   (a) To store a function.
       (1) From the Main Menu, select the Run icon.
       (2) Press the key OPTN (for Options).
       (3) Press F6 twice, then press F3 (FMEM).
       (4) Enter the function.
       (5) Press the key F1 for STO
       (6) Press one of the function keys to determine which function name this function is to be stored for. (The calculator will store up to 6 functions.)

       The function will now be stored "permanently" in the calculator. To exit, press QUIT or MENU

   (b) To recall a stored function.
       (1) From the Main Menu, select the Run icon.
       (2) Press the key OPTN (for Options).
       (3) Press F6 twice, then press F3 (FMEM).
       (4) Press the key F2 (RCL) or press the key F3 (fn).
       (5) Press one of the function keys to see a specific function.

       The function or the function name will appear on the screen at the location of the cursor. To exit, press QUIT or MENU

Procedure C11. **To evaluate a function.**

After a function has been stored,
1. Store the x-value at which we wish to evaluate the function.
   (a) Enter the value.
   (b) Press the key →
   (c) Press the key *X, θ, T*
2. Evaluate the function.
   (a) If the function is stored in regular Function Memory.
       (1) Recall the a stored function (see Procedure C10)
       (2) Press EXE to evaluate the function.
       The value of the function will appear at the right of the screen.
   (b) If the function is stored in Graph Function Memory.
       (1) Press the key VARS
       (2) Press F4 (GRPH)
       (3) Press F1 (*Y*)

(4) Enter the function number for the desired function.
(5) Press EXE to evaluate the function.
The value of the function will appear on the right of the screen.

Note: If an error message is obtained, press the key AC$^{/ON}$. This likely indicates that the value of x is not in the domain of the real function.

To evaluate the same function for different x-values, repeat Steps 1 and 2.

Procedure C12. To store a constant for an alpha character.

1. Enter the value to be stored.
2. Press the key →
3. Enter the alpha character.
4. Press EXE.

(Example: 6.35→H)

Procedure C13. To obtain relation symbols (=, ≠, >, ≥, <, and ≤).

1. Press the key: PRGM
2. Press the key F6, the press F3 to select REL from the menu.
3. Select the desired symbol from the menu by pressing a function key.

Procedure C14. To change from radian mode to degree mode and vice versa.

1. Press the key SET UP
2. Move the highlight to Angle.
    Press F1 for degrees.
    Press F2 for radians.
    Press F3 for gradients.

Procedure C15. Conversion of coordinates.

Make sure calculator is in correct mode -- radians or degrees.
1. Press the key OPTN
2. Press F6, then F5 (ANGL)
3. Press F6 again.
    A. To change from **polar to rectangular**.
       (1) Press the key Rec(
       (2) Enter the value of r, a comma, then the value for θ.
       (3) Press EXE

A pair of number is displayed. The top number (value 1) is the x-value and the bottom number (value 2) is the y-value.

(Example: Rec(2.0, 30) gives 1.732 for x, 1 for y. [in degree mode])

    B. To change from **rectangular to polar**.
       (1) Press the key Pol(
       (2) Enter the value of x, a comma, and then the value for y.
       (3) Press EXE

A pair of number is displayed. The top number (value 1) is the r-value and the bottom number (value 2) is the θ-value.

(Example: Pol(1.0, 1.0) gives 1.414 for r, 45 for θ. [in degree mode])

Procedure C16. **Finding the sum of vectors in polar form.**

1. Find the x- and y-components of each vector (see Procedure C15).
2. Find the sum of the x-components and the sum of the y-components.
3. Find the magnitude and direction of the resultant vector (see Procedure C15).

Procedure C17. **Conversion of complex numbers.**

1. To change from **polar to rectangular** form:
   Think of the complex number as r and θ in polar form and use Procedure C15 to convert to rectangular form.
2. To change from **rectangular to polar** form:
   Think of the complex number as x and y in rectangular form and use Procedure C15 to convert to polar form.

Procedure C18. **Operations on complex numbers.**

Operations are performed by entering the complex numbers directly in rectangular complex form.

1. Press the key OPTN
2. Presa the key F3 (CPLX)
3. Enter the complex numbers in the form $a + bi$ inside of parentheses. ($i = \sqrt{-1}$)
4. Perform an operation of addition, multiplication, subtraction, division, squaring or square roots on complex numbers.

(Example: $(3 + 5i) \div (2 + i)$ gives $2.2 + 1.4i$)

Procedure C19. **To determine the numerical derivative from a function.**

1. Press the key OPTN
2. Press the key F4 (CALC)
3. Press the key F2 (d/dx)

The form is **d/dx** (*fn, value, delx*)
   Where *fn* is the function that may be entered or recalled from memory, *value* is the x-value at which the derivative is to be determined, and *delx* is a small value of Δx used in the calculation. (*delx* is optional.)

(Example: $d/dx(x^2, 4)$ gives the value 8)

Procedure C20. **Function to calculate definite integral.**

1. Press the key OPTN
2. Press the key F4 (CALC)
3. Press the key F4 ($\int dx$)

The form is $\int$ (*fn, a, b, n*)
   Where *fn* is the function which may be entered or recalled from memory, a and b are the lower and upper limits of integration, and n is an optional number from 1 to 9 and relates to the accuracy of the calculation. (The higher the value, the more accuracy.)

(Example: $\int (x^3, 1, 3)$ is used to evaluate $\int_1^3 x^3\, dx$. The value is 20.)

C.3 Graphing Procedures

Procedure G1. To graph functions.

    1. From the Main Menu, select the Graph icon to enter Graph mode.

    2. Enter the function into Graph Function Memory (see Procedure C10, Step 2) Use the key X, θ, T for the variable x.

    3. Press F6 (DRAW) to see the graph.

    Notes: 1. Functions stored in memory will remain in the calculator's memory until changed or erased, or until the batteries run down or are removed.

              2. While the graph is on the screen, pressing the key: G↔T will clear the graph off the screen. However, the graph is still in the calculator's memory and can be seen again by pressing the key: G↔T

              3. When a function has been entered, the equal sign for that function becomes highlighted indicating that the function is turned ON - that is, it is an active function and will be graphed if the key for DRAW is pressed. Only functions that are turned on will be graphed. To turn a function on or off, move the highlight to the function and press F1 (SEL) to select or unselect the function.

              4. To use a particular color for a function, highlight the function, select F4 (COLR) and press F1 (blue), F2 (orange), or F3 (green). Press EXIT to return to function list.

              5. Functions may also be graphed from the Run mode. In Run mode, press the SHIFT key and select F4 (SKTCH). Press F5 (GRPH), then F1 ($Y=$). The words Graph $Y=$ appear on the screen. After the equal sign, enter the function to be graphed and press EXE.

Procedure G2. To change or erase a function from Graph Function Memory.

    1. From the Main Menu, select the Graph icon.

    2. In the Graph Function Memory (see Procedure C10, Step 2) move cursor to a function and make desired changes or press F2 (DEL), then F1 (YES) to delete a function.

Procedure G3. To obtain the standard viewing rectangle.

    1. Press the key V-Window.

    2. Press the key F3 (STD) to obtain the standard viewing rectangle.

Procedure G4. To use the trace function and find points on a graph.

    1. With a graph on the screen press the key TRACE key (F1). This will cause the cursor to appear on the graph.

    2. As the right and left cursor keys are pressed, the cursor moves along the graph and the coordinates of the cursor are given on the bottom of the screen.

    3. If more than one graph is on the screen, pressing the up and down cursor keys will cause the cursor to jump between the graphs.

    4. If the cursor is moved off the right or left of the screen the graph scrolls (moves) right or left to keep the cursor on the screen. The graph will not scroll

up or down but will give coordinates of points off the top or bottom of the screen.

5. Pressing the cursor keys without first pressing TRACE, will cause the viewing rectangle to move in that direction.

6. Pressing the F4 key while a graph is on the screen, then pressing F1 (Cls) will clear the screen and redraw function stored in Graph Function Memory.

Procedure G5. To see or change viewing rectangle.

With the graph on the screen.
1. Press the key V-Window.

2. For each of the quantities you wish to change, enter the value for that quantity. To keep a value and not change it, use the top and bottom cursor keys to move the cursor to another value. When finished, press EXE.
The standard viewing rectangle has the following values:

$$Xmin = -10 \qquad Ymin = -10 \qquad T,\theta min = 0$$
$$Xmax = 10 \qquad Ymax = 10 \qquad T,\theta max = 2\pi \ (\text{or } 360°)$$
$$Xscale = 1 \qquad Yscale = 1 \qquad pitch = 2\pi/100 \ (\text{or } 3.6°)$$

T,θmin, T,θmax and ptch are needed for parametric and polar graphing. For normal graphing they may be left unchanged.

Procedure G6. To obtain special viewing rectangles.

1. For a preset **"initial"** viewing rectangle with the following values

$$Xmin = -6.3 \qquad Ymin = -3.1$$
$$Xmax = 6.3 \qquad Ymax = 3.1$$
$$Xscale = 1 \qquad Yscale = 1$$

Press the key V-Window and press F1 (INIT).

2. For a **square viewing rectangle**, press the key ZOOM. Then press F6, then F2 (SQR). The x-scale adjusts so that one unit in the x-direction is the same distance as one unit in the y-direction.

3. To return to the **previous viewing rectangle**, press the key ZOOM. Then press F6, then F5 (PRE) for the last viewing rectangle or F1 (ORIG) for the original viewing rectangle.

4. For a viewing rectangle in which one movement of the cursor in the x-direction differs by k units, set the values to:

$$Xmin = -126k/2 \qquad Xmax = 126k/2$$

5. For the **trigonometric viewing rectangle**: Press the key V-Window, then press the F2 (TRIG)
Values for the trigonometric viewing rectangle are:

$$Xmin = -3\pi \ (\text{or } -540°) \qquad Ymin = -1.6$$
$$Xmax = 3\pi \ (\text{or } 540°) \qquad Ymax = 1.6$$
$$Xscale = \pi/2 \ (\text{or } 90°) \qquad Yscale = 0.5$$

Procedure G7. To graph functions on an interval.

    1. To graph a function on the interval $x < a$ or on the interval $x \leq a$ for some constant a:
Enter the function $f(x)$ followed by ,$[k, a]$ where k is a value less than Xmin.
(Example: To graph $y = x^2$ on the interval $x<2$, enter: $x^2, [-100, 2]$)

    2. To graph a function on the interval $a < x < $ b or on the interval $a \leq x \leq b$ for some constants a and b:
Enter the function $f(x)$ followed by ,$[a, b]$.
(Example: To graph $y = x^2$ on the interval $3 \leq x \leq 2$, enter: $x^2,[-3, 2]$)

    3. To graph a function on the interval $x > a$ or on the interval $x \geq a$ for some constant a:
Enter the function $f(x)$ followed by ,$[a, k]$ where k is greater than Xmax.
(Example: To graph $y = x - 5$ on the interval $x>2$, enter: $(x-5),[2, 50]$)

    Note: Piecewise functions need to be graphed as separate functions on separate intervals.

Procedure G8. Changing graphing modes.

    Press the key SET UP

    1. Highlight the word Draw Type, then
Press F1 (Con) for plotted points to be connected.

    or

    Press F2 (Plot) to plot individual points.

    2. Highlight the word Simul Graph, then
Press F1 (on) to graph all selected function at the same time.

    or

    Press F2 (off) to graph each function in sequence.

Procedure G9. To zoom in or out on a section of a graph.

    1. Graph a function(s).
    2. Press TRACE and move the cursor near the desired point. After zooming this point will be near the center of the screen.
    3. With a graph on the screen, press the key ZOOM
    4. To zoom in: Press F3 (IN).
To zoom out: Press F4 (OUT).
    5. To continue zooming, repeat step 4.

Procedure G10. To change zoom factors

    1. Press the key: ZOOM
    2. Press F2 (FACT).
    3. Change the factors as desired.
The initial factors are 2 in both directions. For the magnification in the x-direction, change XFact and for magnification in the y-direction, change YFact.
    4. To exit this screen press EXE

Procedure G11. Solving an equation in one variable.

1. Write the equation so that it is in the form $f(x) = 0$, let $y = f(x)$, and graph the function. Use a viewing rectangle so that the x-intercept of interest appears on the screen.

2. (a) Use Procedure G9 to zoom in on the x-intercept until the desired degree of precision is obtained.

    or

    (b) Use the built-in procedure:
    (1) With the graph of the function on the screen, press the key G-Solv
    (2) Press F1 (ROOT) and wait. The calculator will locate a root and show its x- and y-values at the bottom of the screen.
    (3) To obtain another root, press the right cursor key.

    Note: If more than one function is on the screen, after the F1 key is pressed the cursor will appear on one of the graphs. Use the top and bottom cursor keys to move to the correct graph and press EXE to obtain the root.

Procedure G12. Finding maximum and minimum points.

1. Graph the function and adjust the viewing rectangle so that the desired local maximum or local minimum point is on the screen.

2. (a) Use Procedure G9 to zoom in on the point until the coordinates are determined to the desired precision. It helps to use the trace function to move the cursor along the curve to determine the maximum or minimum point.

    or

    (b) Use the built-in procedure:
    (1) With the graph of the function on the screen, press the key G-Solv
    (2) Press F2 (MAX) or F3 (MIN) for maximum or minimum point and wait. The calculator will locate a point and show its x- and y-values at the bottom of the screen.
    (3) To obtain a second maximum or minimum point, press the right or left cursor key.

    Note: If more than one function is on the screen, after the F2 or F3 key is pressed the cursor will appear on one of the graphs. Use the top and bottom cursor keys to move to the correct graph and press EXE to obtain the point.

Procedure G13. Find the value y at a given value of x or a value of x at a given y.

1. Graph the function and adjust the viewing rectangle so that the desired point is on the screen.

2. (a) Use Procedure G9 to zoom in on the point to the desired degree of precision. Use the trace function to move the cursor to the value of x.

    or

    (b) Use the built-in procedure:
    (1) With the graph of the function on the screen, press the key G-Solv
    (2) Press F6 then F1 (Y-CAL) to find y or F2 (X-CAL) to find x.

(3) Enter the value for *x* after *X=* or the *y*-value after *Y=* and press EXE. To find another *y*-value press the right cursor key.

Note: If more than one function is on the screen, after the F1 or F2 key is pressed the cursor will appear on one of the graphs. Use the top and bottom cursor keys to move to the correct graph and press EXE to obtain the point.

Procedure G14. To zoom in using a box.

1. Press the key: ZOOM
2. Press F1 to select Box from the menu.
3. Use the cursor keys to move the cursor to a location where one corner of the box is to be and press EXE
4. Use the cursor keys to move the cursor to the location for the opposite corner of box and press EXE. As the cursor is moved, a box will be drawn and when EXE is pressed, the area in the box is enlarged to fill the screen.

Procedure G15. Finding an intersection point of two graphs.

1. Enter and graph both functions at the same time. Adjust the viewing rectangle so that the intersection points appear on the screen.
2. (a) Use Procedure G9 to zoom in on the intersection point to the desired degree of precision. Use the trace function to aid in determining the point.

   or

   (b) Use the built-in procedure:
      (1) With the graph of the function on the screen, press the key G-Solv
      (2) Press F5 (ISCT) and an intersection point will be indicated by the cursor and the coordinates displayed at the bottom of the screen.
      (3) If there are other intersection points press the right cursor key to obtain the next one.

   Note: If more than two functions are on the screen, after the F5 key is pressed the cursor will appear on one of the graphs. Use the top and bottom cursor keys to move to the correct graph and press EXE. The cursor will then move to a second graph. Again use the top and bottom cursor keys to indicate the second graph and press EXE.

Procedure G16. To graph two functions and their sum.

1. Store the first function. (Procedure C10)
2. Store the second function.
3. For a third function store the sum of the function names for the first two functions. (To enter function names, see Procedure 11, Step 2(b).) (Hint: Use a different color for this graph.)
4. Draw the graphs of these functions.

Procedure G17. To change graphing modes for regular functions, parametric equations, or polar equations.

1. In Graph Mode with the Graph Function Memory on the screen:
   (a) Highlight a function and press F3 (TYPE).

(b) Select the graphing mode desired:
Press F1 ($Y=$) for rectangular coordinates.
Press F2 ($r=$) for polar coordinates.
Press F3 (Parm) for parametric equations.
Press F4 ($x=c$) for graphing the equation x = constant.
Pressing F6 will give graphs of inequalities (see Procedure G22).

or

2. In Run Mode:
(a) Press the key SET UP and highlight Func Type
(b) Select the graphing mode desired (see this Procedure, Step 1(c)).

Procedure G18. To graph parametric equations.

1. From the Main Menu, select the Graph icon to enter Graphing mode.
2. Delete or turn off other functions.
3. Press F3 (TYPE) and select F3 (Parm).
4. Enter the pair of functions into Graph Function Memory. Use the key X, θ, T for the variable T.
5. Press F6 (DRAW) to see the graph.

Notes: 1. Both functions must have the = sign highlighted to be graphed.

2. See notes in Procedure G1.

3. Parametric functions may also be graphed from the Run mode. In Run mode, press the SHIFT key and select F4 (SKTCH). Press F5 (GRPH), then F3 (Parm). The words Graph $(X,Y)=($ appear on the screen. After the equal sign, enter the parametric functions inside of parentheses, separated by a comma like GRAPH$(X,Y)=(f_1, f_2)$.

4. It may be necessary to adjust view window values, particularly the pitch size for the parameter T.

Procedure G19. To change values for the parameter T.

With the graph on the screen:
1. Press the key V-WINDOW
The first screen gives the familiar values for x and y.
2. Move the bottom cursor key to scroll down to the next screen, which will give value for T, θ. Now, we can change values for Tmin, Tmax, and ptch. Ptch is the pitch and determines how far apart the points are plotted.
Standard viewing rectangle values are:

Tmin = 0, Tmax = 6.28(2π) or 360°, and ptch = 0.0628..($2\pi/100$) or 3.6°.

3. To exit, press EXE .

Procedure G20. To graph equations involving y^2.

1. Solve the equation for y. There will be two functions: $y = f(x)$ and $y = g(x)$.
2. Store $f(x)$ for one function and $g(x)$ for a second function.
or

If $g(x) = -f(x)$, we may do the following:

(a) Store $f(x)$ for a function name $(Y1)$.

(b) After a second function name store the negative of the first function name $(-Y1)$

3. Graph the functions.,

Procedure G21. **To obtain special Y-variables.**

1. To obtain minimum and maximum values of X and Y:
   In Run Mode or with the Graph Function Memory on the screen:
   (a) Press the key VARS and press F1 (V-WIN).
   (b) Press F1 (X) for x or F2 (Y) for y.
   (c) Press F1 (min) for a minimum value or F2 (max) for the maximum value.
   (Example: Pressing F1(X) and then F2(max) gives Xmax.)

2. To obtain graph variable names:
   (a) Press the key VARS and press F4(GRPH).
   (b) Select from list by pressing the correct function key. Enter numbers after the letters to indicate various function names.

Procedure G22. **To shade a region of a graph (graphing inequalities).**

1. From the Main Menu, select the Graph icon to enter Graphing mode.

2. In Graph Function Memory, delete or turn off other functions.

3. Press F3 (TYPE), then press F6. Select from $Y>$, $Y<$, $Y\geq$, or $Y\leq$ by pressing the appropriate function key.

4. Enter the function into Graph Function Memory. Use the key X, θ, T for the variable X. Press EXE to store the function.

5. Press F6 (DRAW) to see the graph.

Notes: 1. The inequality sign must be highlighted to be graphed.

2. See notes in Procedure G1.

3. Parametric functions may also be graphed from the Run mode. In Run mode, press the SHIFT key and select F4 (SKTCH). Press F5 (GRPH), then F6. Select from $Y>$, $Y<$, $Y\geq$, or $Y\leq$ by pressing the appropriate function key. After the inequality sign, enter function.

4. A restricted domain may be specified for the graph by entering the minimum x and maximum x in square brackets after the function, separated from the function by a comma. (See Procedure G7) Example: $Y>2x+3$, [-2, 2]

Procedure G23. **To graph in polar coordinates**

1. From the Main Menu, select the Graph icon to enter Graphing mode.

2. In Graph Function Memory, delete or turn off other functions.

3. Press F3 (TYPE), then press F2 ($r=$).

4. Enter the function into Graph Function Memory. Use the key X, θ, T for the variable θ. Press EXE to store the function.

5. Press F6 (DRAW) to see the graph.

Notes: 1. The equal sign must be highlighted to be graphed.

2. See notes in Procedure G1.

3. Polar functions may also be graphed from the Run mode. In Run mode, press the SHIFT key and select F4 (SKTCH). Press F5 (GRPH), then press F2 ($r=$). Enter the function and press EXE.

4. It may be necessary to adjust view window values, particularly the pitch size for θ.

Procedure G24. **To change values for θ.**

With the graph on the screen:

1. Press the key V-WINDOW
The first screen gives the familiar values for x and y.

2. Move the bottom cursor key to scroll down to the next screen, which will give value for T,θ. Now, we can change values for θmin, θmax, and ptch. Ptch is the pitch and determines how far apart the points are plotted.
Standard viewing rectangle values are: θmin = 0, θmax = 6.28(2π) or 360°, and ptch = 0.0628..($2\pi/100$) or 3.6°.

3. To exit, press EXE .

Procedure G25. **Drawing a tangent line at a point.**

1. With the graph on the screen, press the key Sketch.

2. Press F1(Tang). (The normal line may be drawn by pressing F3 (Norm).)

3. Use the cursor keys to move the cursor to a point on the graph where the tangent is to be drawn.

4. Press EXE.

Procedure G26. **Determining the value of the derivative from points on a graph.**

1. Press the key SET UP.

2. Move the highlight to Derivative.

3. Press F1 (On) to turn the derivative on or F2 (Off) to turn the derivative off.

4. Press EXE.

With the graph on the screen, trace the graph. The value of the derivative of he function at any point appears on the bottom of the screen as well as the coordinates of the point.

Procedure G27. **To find the constant of integration for an indefinite integral.**

This procedure assumes that the general antiderivative is known.

1. In Run Mode store a value of 0 for C. ($0 \rightarrow$ C)

2. In Graph Mode store the antiderivative function with C attached for a function name and graph.

3. Given the information that $y = y_0$ when $x = x_0$ and with the graph of the antiderivative (for C = 0) on the screen, evaluate the antiderivative function at $x = x_0$ (Procedure G13). With the cursor on the graph of the antiderivative, read the corresponding y-value, y_a. The constant of integration is the difference between the given y-value y_0 and the value from the graph y_a. That is C = $y_0 - y_a$.

4. To check your answer, store the value obtained for C, graph the antiderivative function, and evaluate at $x = x_0$.

Procedure G28. Function to determine the area under a graph.

With the graph on the screen:
1. Press the key G-Solv, then press F6.
2. Press F3 ($\int$ dx) for integral.
3. Move the cursor to the first x-value (lower limit) and press EXE.
4. Move the cursor to the second x-value (upper limit) and press EXE

The area will be shaded and the value of the definite integral between the two x-values will be given on the bottom of the screen.

Procedure G29. For drawing points and lines.

In Run Mode:
1. Press the key Sketch, then press F6.
2. To plot points:
    (a) Press F1(PLOT)
    (b) Press F2(Pl-On)
    (c) Enter x- and y-coordinates of point separated by a comma.
    Form: **PlotOn** x_1, y_1
    Example: PlotOn $-5,4$
3. To draw a line:
    (a) Press F2 (F-LINE)
    (b) Enter the x- and y-coordinates of the beginning and ending points of the line.
    Form: **F-Line** x_1, y_1, x_2, y_2
    Example: F-Line $-3,-4,4,3$ (Draws a line from $(-3, -4)$ to $(4, 3)$.)

C.4 Programming Procedures

Procedure P1: To enter a new program.

1. From the Main Menu select the icon PRGM.
2. Select F3 (NEW)
3. Enter a name (up to 8 characters) for the program and press EXE.
    (No need to press the alpha, the calculator is already for alpha characters.)
4. The calculator is now in programming mode and ready for the steps of the program.

Notes: 1. The symbol ◢ is used to pause program execution and displays the results of the last calculation. (Pressing EXE at this point in the execution of the program will cause the program to continue.)
2. The question mark and the symbol ◢ are obtained by pressing the key PRGM then pressing F4 (?) or F5 (◢).
3. Enter a programming statement on each line. Press EXE after the statement. Or, several statements may be placed on one line by using a

colon between statements. The colon is obtained by first pressing the key PRGM, then F6, then F5 (:).

Procedure P2. To execute, edit or erase a program.

1. From the Main Menu select the icon PRGM.
2. Highlight the name of the program in the program list.
3. To execute a program: Press F1 (EXE)
4. To edit a program: Press F2 (EDIT)
5. To delete a program: Press F4 (DEL) then, press F1 (YES).
6. To delete all programs: Press F5 (DEL-A) then, press F1 (YES).

Procedure P3. To input a value for a variable.

The question mark (?) is the input command for the Casio.
1. Press the key: PRGM   (while in programming mode)
2. Press F4 to select ? from the PRGM menu.
3. Press the key: → (to store the value)
4. Enter an alpha character for the variable, press EXE
(Example: ?→C)
Note: When the program is executed, the ? will cause a question mark to appear on the screen. Any number entered is stored for the given variable and when the variable is used in a program that value is used in the calculation.

Procedure P4. To evaluate an expression and store that value to a variable.

1. Enter the expression to be evaluated.
2. Press the key: →
3. Enter the variable.
(Example: (9/5)C+32→F)

Procedure P5. To display quantities from a program.

1. The last variable stored, expression calculated, or alpha expression given inside of quotes before the end of the program **or** before the ◢ symbol is displayed when the program is executed.
   Note: To display more than one value, place the ◢ after each quantity to be displayed. If another value is to be displayed, the quantity -Disp- will appear on the screen. Press EXE to see additional values.
   (Example: (9/5)C+32→F◢ prints a value for F.)
2. To display characters as they appear, enter the characters inside of quotes. The quotes are obtained from a menu obtained by pressing the ALPHA key.
   Press the ALPHA key again to return to previous menu.
   (Example: "ENTER C" will cause ENTER C to be displayed.)

Procedure P6. To add or delete line from a program.

1. Use Procedure P2 to bring a program to the screen so that it may be edited.
2. To add a line:

      (a) To insert a line before a given line move the cursor to the first character of the line and to insert a line after a given line move the character to the last character in the given line.

      (b) Press the key: INS then press EXE. This will create a blank line either before or after the given line.

      (c) Make sure the cursor is in the desired line. An instruction may now be entered. When inserting a line before a given line, INS will need to be pressed again.

3. To delete a line

      (a) Move the cursor to the desired line

      (b) Press the key: DEL several times

Procedure P7. Making a decision (If... Then... command)

The form of the statement is

> (program statements before)
> **If** *condition*
> **Then** statement
> (program statements after)

To access the If... and Then.. statements:
While in program edit mode (at a step in the program):

1. Press the key PRGM

2. Press F1 (COM)

3. Select from the menu.

If condition is true, the statement following Then command is executed next. If condition is not true, the program statements after the Then statement are executed next.

The *condition* may involve one of the mathematical symbols $<, \leq, >, \geq, =,$ or $\neq$ to compare two variables and/or constants. These symbols are obtained by pressing the key PRGM, the key F6, then the key F3(REL) and selecting from the menu. The Then... command must be included with the If... command.

Procedure P8. Making a decision with options (If...Then...Else... command)

The form of this combination is

> (program statements before)
> **If** *condition*
> **Then** statement
> (other program statements)
> **Else** statement
> (other program statements)
> **IfEnd**
> (program statements after)

To access the If..., Then..., Else..., and IfEnd... commands:
While in program edit mode (at a step in the program):

1. Press the key PRGM

2. Press F1 (COM)

3. Select from the menu.

If the *condition* is true, the statements following the Then statement are executed. If the *condition* is not true, the statements following the Else statement are executed. After either, the statements after are executed. A statement must follow on the same line as the If... command and also the Else... command.

The *condition* may involve one of the mathematical symbols $<, \le, >, \ge, =,$ or $\ne$ to compare two variables and/or constants. These symbols are obtained by pressing the key PRGM, the key F6, then the key F3(REL) and selecting from the menu. Else is optional and the Else section may be omitted.

Procedure P9. Finding the area under a graph.

1. Store the desired function for a function name (in this case Y1) in the Graph Function Memory.

2. Enter the following program on your calculator to find the area between two values B and C. (It is assumed that the function is stored under the function name f_1.)

```
AREA
"FIRST Z"
?→B
"SECOND Z"
?→C
"AREA IS"
∫(Y1,B,C)
```

The symbol ∫(is obtained by pressing the key OPTN, then F4(CALC), then F4(∫( ).

Procedure P10. Creating a While... loop (While, WhileEnd commands).

The construction of the loop is of the form:

```
... (program statements before loop)
While condition
... (program statements within loop if condition true)
WhileEnd
... (programming statements after loop)
```

To access While... and WhileEnd commands,
While in programming edit mode (at a step in program):

1. Press the key PRGM
2. Press F1 (COM)
3. Press F6, then press F6 again.
4. Select from the menu.

In this case, if *condition* is a true statement the statements following the word While are performed until the condition becomes not true, then the program will transfer to the statements following the word WhileEnd.

The *condition* may involve one of the mathematical symbols $<, \le, >, \ge, =,$ or $\ne$ to compare two variables and/or constants. These symbols are obtained by pressing the key PRGM, the key F6, then the key F3(REL) and selecting from the menu.

Procedure P11. Creating a Do..., LpWhile loop (Do, LpWhile commands).

The construction of the loop is of the form:

> ... (program statements before loop)
> **Do**
> ... (statements within loop repeated until condition is false)
> **LpWhile** *condition*
> ... (programming statements after loop)

To access Do and LpWhile commands,
While in programming edit mode (at a step in program):

1. Press the key PRGM
2. Press F1 (COM)
3. Press F6, then press F6 again.
4. Select from the menu.

In this case, if *condition* is a true statement, the statements following the word Do are performed until the condition becomes false. Then, the program will transfer to the statements following the word LpWhile.

The *condition* may involve one of the mathematical symbols $<, \leq, >, \geq, =,$ or $\neq$ to compare two variables and/or constants. These symbols are obtained by pressing the key PRGM, the key F6, then the key F3(REL) and selecting from the menu.

Procedure P12. Creating a For... loop (For command).

The construction of the loop is of the form:

> ... (program statements before loop)
> **For** *beg* $\rightarrow$ *controlvar* **To** *end* **Step** *inc*)
> ... (statements within loop repeated if *end* not exceeded)
> **Next**
> ... (programming statements after loop)

To access commands,
While in programming edit mode (at a step in program):

1. Press the key PRGM
2. Press F1 (COM)
3. Press F6
4. Select from the menu.

Where *beg* is a beginning value
       controlvar is a variable name (A letter of alphabet)
       end is an ending value
       inc is an increment value. STEP command and *inc* is optional
           (Increment is 1 if this is omitted.)

In this case, the first time through the loop, *controlvar* has the beginning value, and the next time through the loop *controlvar* has the value of *beg* + *inc*. Each time through the loop the value for *controlvar* is increased by *inc*. The loop stops when the value for *controlvar* exceeds the ending value (*end)* and the program will then transfer to the statements following the word End.

Procedure P13. Creating an If...Goto... loop (Lbl, If, Goto commands).

The construction of a loop is:

> ... (program statements before the loop)
> **Lbl** *label*
> ... (program statements within loop)
> **If** *condition*
> **Then Goto** *label*
> **IfEnd**
> ... (program statements after loop)

To access the Lbl and GoTo statements,
While in programming edit mode (at a step in the program).

1. Press the key PRGM
2. Press F3 (JUMP)
3. Select from the menu.

To access If, Then, and IfEnd statements see Procedure P8.

The *condition* may involve one of the mathematical symbols $<, \leq, >, \geq, =,$ or $\neq$ to compare two variables and/or constants. These symbols are obtained by pressing the key PRGM, the key F6, then the key F3(REL) and selecting from the menu. *label* is a digit from 1 to 9.

Procedure P14. To display a graph from within a program.

1. Press the key Sketch
2. Press F5 (GRPH)
3. Press the key F1 (*Y*=)
4. Enter the function after the equal sign.

Procedure P15. Drawing a line from within a program.

1. Press the key Sketch
2. Press F6, then press F2 (LINE)
3. Press the key F1 (*Y*=)
4. Then press F2 (F-Line)
5. Enter the *x* and *y*-coordinates of the end points of the line.
   (See Procedure G30, Step 3)

Procedure P16. To temporarily halt execution of a program.

Use the symbol ◢ within a program to cause a pause in the program.
Press EXE to continue the program.

C.5 Matrix Procedures

Procedure M1. To enter or modify a matrix.

1. Select the MAT icon from the Main Menu.
2. Highlight a matrix name.
3. Enter the number of rows and press EXE.
4. Enter the number of columns and press EXE.
5. A matrix appears on the screen. Enter or modify the element highlighted and press EXE. Elements are entered row by row. The cursor keys may be used to move to different elements.

Note: 26 matrices named from A to Z may be entered into the calculator.

Procedure M2. To view a matrix.

1. Select the MAT icon from the Main Menu.
2. Highlight a matrix name and press EXE.

Procedure M3. Addition, subtraction, scalar multiplication or multiplication of matrices.

1. Select the RUN from the Main Menu.
2. Press the key OPTN
3. Press F2 (MAT)
4. To add, subtract, multiply matrices:
   (a) Select F1(Mat) and enter a matrix name.
   (b) Press the desired operation (+,−, or ×)
   (c) Select F1 (Mat) and enter the second matrix name.
   (d) Press EXE
5. To multiply a matrix by a scalar:
   (a) Enter the scalar
   (b) Select F1 (Mat) and enter a matrix name.
   (c) Press EXE

Many of these operations may be combined into one statement. The result of a matrix calculation is stored under Mat Ans. An error will occur if the operation is not defined for the matrices selected.

Procedure M4. To find the determinant, inverse, and transpose of a matrix.

1. Select the RUN from the Main Menu.
2. Press the key OPTN
3. Press F2 (MAT)
4. To find the determinant:
   (a) Press F3 (Det)
   (b) Press F1 (Mat)
   (c) Enter the name of a matrix.
   (d) Press EXE

5. To find the inverse:
   (a) Press F1((Mat)
   (b) Enter the matrix name.
   (c) Press the key x^{-1}
   (d) Press EXE.
6. To find the transpose:
   (a) Press F4 (Trn)
   (b) Press F1 (Mat)
   (c) Enter the matrix name.
   (d) Press EXE

Procedure M5. To store a matrix.

1. Select the RUN from the Main Menu.
2. Press the key OPTN
3. Press F2 (MAT)
4. Enter the first matrix (the one to be stored):
   (a) Press F1 (Mat)
   (b) Enter the matrix name.
5. Press the key →
6. Enter the other matrix (must be of same size of first matrix).
   (a) Press F1 (Mat)
   (b) Enter the matrix name.
7. Press EXE.   The contents of the second matrix will be replaced by the contents of the first matrix.

Procedure M6. Row operations on a matrix.

1. Obtain the matrix on the screen.
   (a) Select the MAT icon from the Main Menu.
   (b) Highlight a matrix name.
   (c) Press EXE..
2. Press F1 (R-OP)
3. Indicate the row operation:
   (a) To interchange two rows on a matrix:
       (1) Select F1 (SWAP)
       (2) Enter the number of one row (for m) and press EXE.
       (3) Enter the number of the second row (for n) and press EXE.
       The two rows will be interchanged.
   (b) To multiply a row by a constant, select:
       (1) Select F2 (×Rw)
       (2) Enter the constant (for k) and press EXE.
       (3) Enter the number of the row (for m) and press EXE.
       To divide a row by a constant, multiply by the reciprocal.
   (c) To multiply a row by a constant and add to another row, select
       (1) Select F3 (×Rw+)
       (2) Enter the constant (for k) and press EXE.
       (3) Enter the number of the row to be multiplied (for m) and press EXE.
       (4) Enter the number of the row to be added (for n) and press EXE.

C.6 Statistical Procedures

Procedure S1. To enter or change one variable statistical data.

    1. Select the STAT icon from the Main Menu. This brings up a scren showing lists of numbers. Up to six lists may be stored on the Casio calculator.

    2. Enter the one variable data in one list. If both data values and frequencies are to be entered, enter the data in one list and the frequencies in a second list. Make sure the frequency of each data item is in the same position in the list as the data item. Enter each value and press EXE.

    3. The cursor keys may be used to move the cursor to a particular value to change that value.

    Note: When viewing data, only 5 lines of data appear at one time on the screen, but pressing the up or down cursor key will scroll the data on the screen.

Procedure S2. To graph single variable statistical data.

    1. Select the STAT icon from the Main Menu.

    2. Press F1 (GRPH)

    3. Set up the data for any of the three graphs: Graph1, Graph2, or Graph3.:
        (a) Press F6 (SET)
        (b) Highlight each line and select from the menu at the bottom of the screen the desired choice.

            For Graph Type, select either scatter or a *xy*-line graph or Histogram or Box Graph (for these last two press F6 first).
            For Xlist, select the list containing the data.
            For Frequency, select either 1 (if each item occurs once) or the list containing the frequencies of the data.
            For Mark Type, select the type of mark to be used in plotting the data.
            For Graph Color, select the color for the graph.
        Then, press EXIT.

    4. Select from the menu at the bottom of the screen, the desired graph.

    Note: Pressing F4 (SEL) will allow any of the three graphs to be turned on or off.

Procedure S3. To clear the graphics screen.

    The graphics screen should clear itself before each new set of data is plotted.

Procedure S4. To obtain mean, standard deviation and other statistical information.

    After data has been entered:

    1. Select the STAT icon from the Main Menu.

    2. Press F2 (CALC)

    3. Press F6 (SET) to set up calculations.
       Highlight each line and select from the menu at the bottom of the screen the desired choice.
          For 1Var Xlist, select the list containing the data.
          For 1Var Freq, select the list containing the frequencies of the data.
       Press EXIT

4. Press F1 (1VAR). The screen show shows the following:
   The mean $\bar{x}$ of the data.
   The sum of the x's and of the x^2's.
   The σ-standard deviation, $_x\sigma_n$.
   The s-standard deviation, $_x\sigma_{n-1}$.
   The total number of data items n.

Procedure S5. **To clear statistical data from memory.**
   1. Select the STAT icon from the Main Menu.
   2. Press F6, then press F3 (DEL) to delete one data value
      or press F4 (DEL-A) to delete all items in a list and press F1 (YES).

Procedure S6. **To enter or change two variable data.**
   1. Select the STAT icon from the Main Menu. This brings up a scren showing lists of numbers. Up to six lists may be stored on the Casio calculator.
   2. Enter the one set of data (x-values) into one list and the other set of data (y-values) into a second list. If both data values and frequencies are to be entered, enter the frequencies into a third list. Make sure the frequency of each data item is in the same position in the list as the data item. Enter each value and press EXE.
   3. To change a value, move the cursor keys to that particular value and make the desired changes.
   Note: When viewing data, only 5 lines of data appear at one time on the screen, but pressing the up or down cursor key will scroll the data on the screen.

Procedure S7. **To determine a regression equation.**
   1. Select the STAT icon from the Main Menu.
   2. Press F2 (CALC)
   3. Press F6 (SET) to set up the data lists.
      Highlight each line and select from the menu at the bottom of the screen the desired choice.
         For 2Var XList, select the list containing the x-values of the data.
         For 2Var YList, select the list containing the y-values of the data.
         For 2Var Freq, select the list containing the frequencies of the data or select 1 if each item occurs once.
      Press EXIT
   4. Press F3 (REG) and select from the menu at the bottom of the screen:
         Select F1 (x) for linear regression. ($y = ax + b$)
         Select F2 (MED) for linear regression that minimizes extremes. ($y = ax + b$)
         Select F3 (x^2) for quadratic regression. ($y = ax^2 + bx + c$)
         Select F4 (x^3) for cubic regression. ($y = ax^3 + bx^2 + cx + d$)
         Select F5 (x^4) for quartic regression. ($y = ax^4 + bx^3 + cx^2 + dx + e$)
      Or after pressing F6:
         Select F1 (Log) for logarithmic regression. ($y = a + b\ln x$)
         Select F2 (Exp) for exponential regression. ($y = ae^{bx}$)
         Select F3 (Pwr) for power regression. ($y = ax^b$)

Procedure S8. To graph data and the related regression equation.

1. Select the STAT icon from the Main Menu.
2. Press F1 (GRPH)
3. Set up the data for any of the three graphs: Graph1, Graph2, or Graph3.:
    (a) Press F6 (SET) and select the graph number.
    (b) Highlight each line and select from the menu at the bottom of the screen the desired choice.

    For Graph Type, select scatter.
    For XList, select the list containing the x-values.
    For YList, select the list containing the y-values.
    For Frequency, select either 1 (if each item occurs once) or the list containing the frequencies of the data.
    For Mark Type, select the type of mark to be used in plotting the data.
    For Graph Color, select the color for the graph.
    Then, press EXIT.
4. Select from the menu at the bottom of the screen the graph to be drawn.
5. From the menu at the bottom of the screen, select the regression type to see the regression equation.
6. Press F6 (DRAW) to graph the equation on the data plot.

Answers

To Selected Problems

Exercise 1.1

1. 358.7
2. 15.89
3. −13.18
4. 7.47
5. 22
6. 3.15
7. 4.84
8. 11.6
9. 3.0° C
10. 76.3; 78.0
11. 53
12. 285.5
13. 196, 216, 235, 255
14. 50.4, 59.0, 68.0, 77.9, 88.2
15. 0.0, 20.0, 37.0, 37.78, 68.61
16. (a) 14.7   (b) 92.20
17. 129
18. 3.35
19. 192
20. 175.9
21. 60.9%
22. 7.8 cm, 9.6 cm, 9.3 cm
23. 479001600
24. 4
25. 5040
26. $4.0329... \times 10^{26}$
27. 0.66
28. 27
29. 2
30. −2
31. 4
32. −123

Exercise 1.2

1. 5.22×10^7
2. 3.7×10^3
3. 1.61×10^1
4. 1.3×10^3
5. 4.17×10^8
6. -1.46×10^{-11}
7. 1.46×10^2
8. 9.91×10^{11}
9. {110, 68,105, 68}, {−42, 22 19, −22}, {2584, 1035, 2666, 1035}, {0.45, 1.9, 1.4, .51}
10. {38, 47, 130, 100, 210}, {1.8, 2.3, 6.5, 5.1, 10}, {−5, −2, 25, 16, 51}
11. {506, 135, 1060, 4100}, {4.74, 3.41, 5.70, 8.00}
12. {68.0, 32.0, −40.0, 98.6, 212}
13. {283.50, 416.50, 23.75, 198.90}
14. 0.46
15. −.698
16. 0
17. undefined
18. 10.7
19. 19.2
20. (5.34, −2.68)
21. (−7.87, 2.08)
22. 1.8×10^{-6} s
23. 1.07×10^5, 1.68×10^5 ft^3/h

Exercise 2.1

1. 6
3. 0.22, undefined
5. (a) 18.18, undefined
7. (a) undefined, 0.75, −0.54, 0, 0.79
9. (a) 2, not real, 3.32, not real, 4.47
11. 3, 5.48, 3, 2.24, 3.16
13. 0, 0.10, 6.1, 2.2, 3
15. 1, 0, −4, 0, −1
17. (a) 9.7, 5.4, 0.29 (b) 7.7, 3.4, −1.7
19. 12.2 m, 0.190 m
21. (a) 16.0, 17.2, 18.4, 19.6
23. (a) 51.5 ft., 65.5 ft., 13.0 ft.   (b) highest after 1.9 s and hits ground after 4.0 s
25. $w = 5874 - 2.17t$, weights are: 5841 t, 5828 t, 5820 t
27. $C = 465 + 5.14(L - 50) = 208 + 5.14L$, costs are: $542.10, $707.61, $1302.31

Exercise 2.2

	(a)	(b)	(c)
1.	1.8	−6.3	none
3.	4.0	9.7	none
5.	2.6, −2.6	−7.0	(0.0, −7.0) [min]
7.	none	5.0	(1.5, 2.8) [min]
9.	0.0, 2.6, −2.6	0.0	(1.5, −7.1) [min] (−1.5, 7.1) [max]
11.	none	none	none
13.	3.5	1.4	none
15.	2.0	1.4	none

17. Parallel lines are obtained. The lines have y-intercepts of 2, 4, and 6 respectively.
19. (a) straight lines
    (b) line becomes steeper, increasing from left to right
    (c) line becomes steeper, decreasing from left to right
    (d) gives horizontal line
    (e) line moves vertically
    (f) line passes through origin
21. $43,000., 73,500 miles

	x-int.	y-int.	$f(-1)$	$f(1)$
23.	none	c	c	c
25.	0	0	1	1
27.	0	0	−1	1
29.	none	none	−1	1
31.	$0 \le x < 1$	0	−1	1

Exercise 2.3

Answers estimated to two significant digits.

1. (a) $-1.0, 6.0$ (b) -6.0 (c) $(2.5, -12)$ [min]
   (d) D: all x R: $y \geq -12.25$ (e) dec: $x < 2.50$ inc: $x > 2.50$

3. (a) $-16, 16$ (b) 250 (c) $(0.0, 250)$ [max]
   (d) D: all x R: $y \leq 250$ (e) dec: $x > 0$ inc: $x < 0$

5. (a) $-8.1, 8.1$ (b) 0.0 (c) $(4.7, -210)$ [min], $(-4.7, 210)$ [max]
   (d) D: all x R: all y (e) dec: $-4.7 < x < 4.7$ inc: $x < -4.7$ and $x > 4.7$

7. (a) 0.75 (b) -1.0 (c) none
   (d) D: all $x \neq 3$ R: all $y \neq -4$ (e) inc: for all $x \neq 3$

9. (a) -4 (b) -4 (c) none
   (d) D: all $x \neq 4$ R: all $y \neq -8$ (e) dec: for all $x \neq 4$

11. (a) $-3.9, 3.9$ (b) 3.9 (c) $(0.0, 3.9)$ [max]
    (d) D: $-3.9 \leq x \leq 3.9$ R: $0 \leq y \leq 3.9$ (e) dec: $0 < x < 3.9$ inc: $-3.9 < x < 0$

13. (a) $-6.1, 15$ (b) 9.6 (c) $(4.5, 10.6)$ [max]
    (d) D: $-6.1 \leq x \leq 15$ R: $0 \leq y \leq 10.6$ (e) dec: $4.5 < x < 15$ inc: $-6.1 < x < 4.5$

15. (a) none   (b) none   (c) $(1.7, 3.5)$ [min], $(-1.7, -3.5)$ [max]
    (d) D: all $x \neq 0$ R: $y < -3.5, y > 3.5$
    (e) dec: $-1.7 < x < 0, 0 < x < 1.7$ inc: $x < -1.7, x > 1.7$

17. (a) 3.1 (b) 7.5 (c) $(3.1, 0.0)$ [min]
    (d) D: all x R: $y \geq 0$ (e) dec: $x < 3.1$ inc: $x > 3.1$

19. (a) -2.0 (b) 2.0 (c) none   (d) D: all x, R: all y (e) inc: $x < 2, x > 2$

21. 18.37 L/h, 1678 r/min

23. (a) 606 feet, 2360 feet   (b) 50 miles per hour, 24 miles per hour

25. (a) D: all x R: all y (b) inc: all x (c) $-1, 0, 1$

27. (a) D: $x \geq 0$ R: $x \geq 0$ (b) inc: $x > 0$ (c) not a real value, 0, 1

29. (a) D: all x R: all y (b) inc all $x, x \neq 0$ (c) $-1, 0, 1$

31. (a) D: all x R: $y \geq 0$ (b) dec: $x < 0$ inc: $x > 0$ (c) $1, 0, 1$

33. (a) circle
    (b) both have x-intercepts of -5 and 5, for y_1: y-intercept is 5, for y_2: y-intercept is -5
    (c) for y_1 D: $-5 \leq x \leq 5$ R: $0 \leq x \leq 5$ for y_2 D: $-5 \leq x \leq 5$ R: $-5 \leq x \leq 0$

Exercise 2.4

1. (a) $5.42, -4.42$ (b) $(0.500, -24.3)$ [min]   (c) D: all x R: $y > -24.3$

3. (a) 1.38 (b) $(-0.390, 5.591)$ [min] $(0.640, 9.97)$ [max]   (c) D: all x R: all y

5. (a) 1.67 (b) none   (c) D: all $x \neq 4$ R: all $y \neq 3$

7. (a) 4, −4   (b) (4, 0) [max]   (c) D: $-4 < x < 4$, R: $0 < x \le 4$

9. 0.857   11. 5.85, -0.854

13. 2.46   15. No solution

17. 4.84 years, $7500, $5100   19. 4.39 volts, $v > 4.39$, 0.0712, 0.0395

21. $V = 140x - 48x^2 + 4x^3$; 1.92 in

Exercise 2.5

1. Disp "ENTER DEGREES F"   Ans.: 100, 38, 37.0, −18, −40
   Input F
   $5(F-32)/9 \to C$
   Disp "DEGREES C"
   Disp C

3. Prompt x Ans.: −2, −1, 24., 67.4, 38.5
   x^4−x^3+x−2→Y
   Disp "$F(x) = $"
   Disp Y

5. (Function must be stored for Y_1 before running program.)
   Prompt x See answers for Problem 3.
   Disp Y_1

7. Disp "ENTER AMOUNT"   Ans.: $376.13, $377.69, $378.46
   Input P
   Disp "ENTER NO. YEARS"
   Input T
   Disp "ENTER TIMES PER YEAR"
   Input N
   Disp "ENTER RATE (AS DECIMAL)"
   Input R
   $P(1 + R / N)^{\wedge}(N * T) - P \to I$
   Disp "INTEREST IS"
   Disp I

9. Disp "ENTER MILES"   Ans.: 22.5 mi/gal
   Input M
   Disp "ENTER GALLONS"
   Input G
   Disp "MILES PER GALLON"
   Disp M/G

13. Prompt A

    Prompt B

    Prompt C

    $B^2 - 4*A*C \rightarrow D$

    $(-B + \sqrt{D})/(2A) \rightarrow X$

    $(-B - \sqrt{D})/(2A) \rightarrow Y$

    Disp X

    Disp Y

Ans. (a) 0.68, −3.7

(c) −2.4, 4.6

(e) 2.89, −2.89

15. Prompt A

    Prompt B

    Prompt C

    $"A*x^2 + B*x + C" \rightarrow Y_1$ (or on the TI-85 or 86: $y1 = A*x^2 + B*x + C$)

    DispGraph

Exercise 3.1

1. 0.399
2. 0.698
3. 0.132
4. 0.0002172935
5. 1.25
6. 0.0769
7. 0.916
8. 54.56
9. 46.435
10. 1.210
11. 3.3814
12. 0.0001745
13. 57.94°
14. 58.37°
15. 88.6°
16. 51.50°
17. 57.79°
18. 20.37°
19. not a real value
20. 0.03°
21. 33.0°
22. 18.8577°
23. 7.055°
24. not a real value
25. (a) $\sin 23.5° = \cos 66.5° = .399$, $\cos 23.5° = \sin 66.5° = .917$

    (b) $\sin 17.45° = \cos 72.55° = .2999$ $\cos 17.45° = \sin 72.55° = .9540$

    (c) $\sin 77.8° = \cos 12.2° = .977$ $\cos 77.8° = \sin 12.2° = .211$

    Conclusion: $\sin A = \cos (90°-A)$ or $\cos A = \sin (90°-A)$

27. (a) 1   (b) 1   (c) 1

    Conclusion: for any angle A, $(\sin A)^2 + (\cos A)^2 = 1$

29. (a) 1   (b) 1   (c) 1

    Conclusion: for any angle A, $(\csc A)^2 - (\cot A)^2 = 1$

31. $\theta = 35.48°$
32. not a real value
33. $\theta = 33.4°$
34. $\theta = 81.26°$
35. $x = 1.33$
36. $x = 16.61$
37. $x = 24.36$
38. $x = 10.4$
39. $x = 2.070$
40. $y = 210$
41. $B = 38.1°$
42. $C = 53.8°$
43. $A = 72.9°$
45. 370 cm^2
47. (a) $l = 46.49$ ft, (b) a = 11.1 ft.
49. 0.9169
51. 0.86327

Exercise 3.2

1. $B = 66.59°$, $b = 58.32$, $c = 63.55$

3. $A = 44.2°$, $b = 1.61$, $a = 1.57$

5. $b = 295.1$, $A = 33.64°$, $B = 56.36°$

7. $c = .9942$, $A = 8.89°$, $B = 81.11°$

9. Sides of 52.4 and 38.9, angle of 36.6°

11. Angles of 55.4° and 34.6° and side 3.08

13. 0.80°, 0.60°, 0.48°, 0.40°, 0.34°

14. 269 in, 72 in, 19 in

15. 38.0 ft, 33.8 ft, 30.4 ft, 27.6 ft, 25.4 ft

16. The maximum total length is 10π.

17. (a) PROGRAM:RTRI1   (Let x and y be the legs.)
    Prompt X
    Prompt Y
    $\tan^{-1}(X/Y) \to A$
    $90 - A \to B$
    $X / \sin(A) \to H$
    Disp A
    Disp B
    Disp H

Exercise 4.1

1   $x = 1.19$, $y = -0.875$

3. $x = 11.8$, $y = 3.82$

5. $x = 3.7$, $y = 3.2$

7. Same line, dependent system

9. $x = 21.5$, $y = 27.3$

11. No solution, inconsistent system

13. Type A 55.0 cents, type B 76.8 cents

15. 4424 L of first, 7801 L of second

17. $x = 3.37$, $x = 4.03$, $x = 11.0$, $x = -5.37$

19. (a) No profit, loss of $615.50   (b) Profit of $603.50

    (c) 130 items, 116 items   (d) (92.5, 3305), Profit is zero.

21. $F_1 = 9.43$, $F_2 = 8.33$, $F_3 = 1.67$

Exercise 5.1

1. [5.91, 120.9]

3. [−0.118, −0.129]

5. [−12146, 1534.5]

7. [0.00436, 1.25]

9. [30.1 ∠56.8°]

11. [1.485 ∠186.8°]

13. [1279 ∠5.71°]

15. [158 ∠164°]

17. [29.9 ∠62.0°]

19. [63.11 ∠170.49°]

21. [9.58 ∠156.5°]

23. [1223 ∠154.43°]

25. horz: 18.0 lb, vert: 16.4 lb

27. 20.4 miles east, 14.7 miles south

29. 115.7 km east, 88.27 km north

31. 12300 N at an angle of 49.3° above horizontal.

Exercise 5.2

1. $4.5 \angle 39°$

2. $22.3 \angle 236°$

3. $199.4 \angle 128.3°$

4. $67.8 \angle 270°$

5. $0.139 \angle 206°$

6. $2083 \angle 307.7°$

7. $-437 + 714j$

8. $12.4 + 3.49j$

9. $-1.111 - 1.225j$

10. $-0.617 + 3.54j$

11. $-80.0 + 36.3j$

12. $0 - 175.6j$

13. $1.6 + 4.2j$

14. $-9.71 - 8.34j$

15. $-242 + 166j$

16. $-0.18 + 0.17j$

17. $25.6 - 7.60j$

19. $-9654 + 10037j$

21. $0.828 - 0.878j$

23. $27.6 \angle 111.5°$

25. $305 \angle 50.2°$

27. $8.1e^{1.0i}$

29. $9840 \angle 122.8°$

31. (a) $8 \angle 0° = 8 + 0j$, $8 \angle 0° = 8 + 0j$ (b) yes, $-1 + 1.732j$, $-1 - 1.732j$

33. (a) $4 \angle 180 = -4$ (b) yes

35. $300 + 26.0j$

37. $-32.6 + 533j$

39. $2.967 \angle 337.7°$

41. $45.4 \angle 44.7° = 32.3 + 31.9j$

Exercise 6.1

1.

0	$\frac{\pi}{2}$	π	$\frac{3\pi}{2}$	2π
0	3	0	-3	0

3. (a) max: 1; 2; 3 min: -1; -2; -3   (b) 1; 2; 3   (c) changes the amplitude

5. (a) max; 2; 3; 2; 3   min: -2; -3; -2; -3   (b) 2; 3; 2; 3

(c) reflects the graph about the x-axis

7. (a) $1.57 \ (\frac{\pi}{2})$, $4.71 \ (\frac{3\pi}{2})$, $7.85 \ (\frac{5\pi}{2})$, $7.85 - 1.57 = 6.28 \ (2\pi)$

$0.79(\frac{\pi}{4})$, $2.36 \ (\frac{3\pi}{4})$, $3.93 \ (\frac{5\pi}{4})$, $3.93 - 0.79 = 3.14 \ (\pi)$

$0.524 \ (\frac{\pi}{6})$, $1.57 \ (\frac{\pi}{2})$, $2.62 \ (\frac{5\pi}{6})$, $2.62 - .53 = 2.09 \ (\frac{2\pi}{3})$

(b) 4.0, $6.28 \ (2\pi)$; 4, $3.14 \ (\pi)$; 2, $2.09 \ (\frac{2\pi}{3})$

(c) changes the period of the graph, period $= \dfrac{2\pi}{b}$

9. (a) Period appears to be π.   (b) 0.04π or 0.126 (c) Xmin = 0, Xmax = .251

11. Xmin = 0, Xmax = 0.502, Ymin = -5, Ymax = 5; x-intercepts are 0.0, 0.13, 0.25, 0.38, 0.50

13. Xmin = 0, Xmax = 167, Ymin = -0.75, Ymax = 0.75;
    x-intercepts are 20.8, 62.5, 104.2, 145.8

15. Better: Xmin = 0, Xmax = 0.021, Ymin = -20, Ymax = 20; x-intercept is 0.010

17. (a) 15.0, -70.5 (b) 0.0, 0.0083, 0.0167
    (c) 0.0022, 0.0061, 0.0189   (d) 0.0042, 0.0208

Exercise 6.2

1. (a) 2, 6.28 (2π); 2, 6.28 (2π);   2, 6.28 (2π)
   (b) 3.14, 2.36, 1.57

   displacement first to second is -0.78 to left $(-\frac{\pi}{4})$, displacement first to third is -1.57 $(-\frac{\pi}{2})$

   (c) shifts graph horizontally by $-c$ units   (d) amplitude has no effect

3. (a) 2, 3.14 (π);   2, 3.14 (π);   2, 3.14 (π)
   (b) 0.0, 0.78, 0.39

   displacement first to second is $+0.78$ $(\frac{\pi}{4})$, displacement first to third is $+0.39$ $(\frac{\pi}{8})$

   (c) shifts graph horizontally by $-\frac{c}{b}$ units

5. (a) (1) 5   (2) 4π (3) 0   (b) (1) 3   (2) 0.5   (3) $+.25$

   (c) (1) 2.5 (2) 2π (3) $+\frac{\pi}{4}$ (d) (1) 3.5 (2) π (3) 0.25

7. (a) (1) 0.00, 1.00, 2.00, 3.00   (2) max: (0.50, 2.00);   min: (1.50, -2.00)
   (b) (1) 1.28, 2.59   (2) max: (1.94, 3.50);   min: (0.63, -3.50)

9. (a) 38.03, -43.13 (b) 0.88, 1.54;   2.79, 3.63

11. (a) 1400 N, 2000 S, 4500 S   (b) at 29 min, 240 min, 280 min, 490 min

13. (a) 0.0069 s and 0.0153 s   (b) 0.0028 s
    (c) 0.0194 s, difference = 0.0167 = period of graph
    (d) 0.0057 s, 0.0165 s   (e) -100.5 v.

15. (a) 2.42 cm, -0.095 cm   (b) 0.0342 s, 0.0440 s, 0.0787 s, 0.0885 s

17. (a) y_2 is a reflection of y_1 about the x-axis (b) same graph - functions are equivalent
    (c) same graph - functions are equivalent   (d) same graph - functions are equivalent
    (e) graphs differ only by amplitude, x-intercepts are the same

Exercise 6.3

1.

	x-intercepts	local minimum	local maximum
(a) $y = \sin x$	−6.2, −3.1	(−1.6, −1.0)	(1.6, 1.00)
	0, 3.1, 6.2	(4.7, −1.0)	(−4.7, 1.0) no asymptotes
$y = \csc x$	none	(1.6, 1.0)	(−1.6, −1.0)
		(−4.7, 1.0)	(4.7, −1.0)

asymptotes at $x = -3.1$, $x = 0.0$, $x = 3.1$

(b) $y = \tan x$	−3.1, 0.0, 3.1	none	none

asymptotes at $x = -4.7$, $x = -1.6$, $x = 1.6$, $x = 4.7$

$y = \cot x$	−4.7, −1.6,	none	none
	1.6, 4.7		

asymptotes at $x = -3.1$, $x = 0$, $x = 3.1$

(c) $y = \cos x$	−4.7, −1.6,	(−3.1, −1.0)	(0.0, 1.0) no asymptotes
	1.6, 4.7	(3.1, −1.0)	(6.3, 1.0) (-6.3, 1.0)
$y = \sec x$	none	(−6.3, 1.0)	(−3.1, −1.0)
		(0.0, 1.0), (6.3, 1.0)	(3.1, −1.0)

asymptotes at $x = -4.7$, $x = -1.6$, $x = 1.6$, $x = 4.7$

3. As a becomes larger, the graph stretches. Zeros and asymptotes stay same.

5. The graphs have different periods. As b doubles, the period is cut in half.

7. As c changes, the graph is displaced by $-c$ units.

9. The negative sign reflects the graph about the x-axis.

11. (a) 3.57 cm, 3.72 cm   (b) min b = 3.01 cm when $A = 1.57$ radians.
    (c) As A gets closer to π, b becomes very large.

13. max: (0.00, 3.00) and (6.28, 3.00), min: (3.14, 1.00), x-int: none

15. max: (0.00, 1.00) and (3.17, 1.20), min: (1.56, −0.95), x-int: 0.79, 2.30

17. max: (0.37, 1.63), (2.77, 1.63), (4.71, 0.00);
    min: (1.57, −2.00), (3.91, −0.71), (5.52, −0.71)
    x-int: 0.94, 2.20, 3.46, 4.71, 5.97

19. max: (0.65, 5.71), (4.31, 3.59); min: (1.97, −3.59), (5.63, −5.71)
    x-int: 0.00, 1.36, 3.14, 4.92, 6.28

21. max: (0.12, 3.58), (2.20, 0.12), (4.00, 2.88);
    min: (1.28, −1.94), (2.91, −0.90), (5.20, −3.95)
    x-int: 0.79, 2.04, 2.35, 3.30, 4.56, 5.81

23. (a) (2.02, 2.34)   (b) 0.55, 1.50, 1.57, 1.64

25. (a) max is 3.01 when t = 2.20 s   (b) highest − lowest = 6.71 feet

Exercise 6.4

1. (a) Gives line from $(-14, 9)$ to $(4,3)$ (b) no x-int, y-int $= 4.33$ (c) $(-14, 9), (4, 3)$
3. (a) Part of parabola $y = 9x^2$ from $(-15, 25)$ to $(15, 25)$
   (b) x-int: 0.00, y-int: 0.00   (c) $(-15, 25), (15, 25)$
5. (a) Part of hyperbola $y = 1/x$ (b) No intercepts, end points $(-32, -0.50), (32, .50)$
   (c) $(-32, -0.5), (32, 0.5)$
7. Graph is ellipse. x-int: $-3.0, 3.0$ y-int: $-5.0, 5.0$
9. (a) Graph is line from $(-3, -5)$ to $(3, 5)$
   (b) Graph is like figure eight along y-axis with two loops.
   (c) Graph is figure along y-axis with four loops.
   Loops are not closed if A is odd.
11. (a) Graph is a circle of radius 5 with center at $(0,0)$.
   (b) Graph is an ellipse at a 45° angle.
   (c) Graph is a line from $(-5, -5)$ to $(5, 5)$
   As c gets closer to $\pi/2$, graphs approaches a straight line at 45° angle.
   When $c = \pi$, graph is a circle again.
13. x-int: $-4.5, -2.25, 0.0, 2.25, 4.5$ y-int: $-2.5, 2.5$

Exercise 7.1

1. (a) For larger base number, graphs increase slower for $x < 0$ and faster for $x > 0$.
   (b) Inc: all x (c) Asymptote is $y = 0$ for all 3 graphs
   (d) no x-intercept, y-intercept is 1   (e) 2, 3, 4
   (f) domain: all x; range: $y > 0$
3. (a) For larger base number, graphs increase faster for $0 < x < 1$ and slower for $x > 1$.
   (b) Inc: all x (c) Asymptote is $x = 0$
   (d) x-intercept at $x = 1$; no y-intercept   (e) 2, 3, 4
   (f) domain: $x > 0$; range: all y.
5. (a)-(b) $y = 4^x$ continuously increases and $y = 4^{-x}$ decreases for all x.
   (c) In each case asymptote is $y = 0$; They are same.
   (d) In each case y-intercept is $y = 1$; no x-intercept.
   (e) In each case domain: all x; range: $y > 0$
   (f) Yes, reflection about the y-axis.
9. Graphs are reflection about $y = x$. Functions are inverse functions.
11. Graphs are reflection about $y = x$. Functions are inverse functions.
15. $V = 1225(1.0225)^{2t}$ (a) \$1635.91, \$1828.42   (b) 15.6 years
17. (a) 1730 m/s   (b) 3470 m/s   (c) $n = 4270$
19. (a) 54.3 m/s, 73.8 m/s   (b) Asymptote is $y = 95$, limiting velocity is 95 m/s

Exercise 7.2

1. 0.803	2. −2.8608	3. 4.21093
4. 8.88	5. 0.548	6. 12.45430
7. -2.7881	8. 36.597	9. 2.6233
10. −1.728	11. 8.7147	12. −6.4769
13. 133000	14. 0.004847	15. 1.33300
16. 2.32×10^{-13}	17. 4.6	18. 1279
19. 6×10^{-6}	20. 0.37	21. 4.5
22. 0.063	23. 855	24. 1.7

25. a) 0.3713   b) 1.3713
    c) 2.3713   d) 3.3713
    e) 9.3713   All the decimal portions (mantissas) are the same.

27. 3.17   29. 1.8754

31. 3.5   33. 0.234

35. (a) 15.4 decibels   (b) 2.57 W   37. (a) 17.4 yrs (b) 11.6 yrs (c) 8.75 yrs

39. (a) 9.64 min   (b) Will approach 23.5°, but never reach it.

Exercise 8.1

1. $x = 2.79, y = -0.196$; $x = -3.46, y = 3.97$

3. $x = 2.75, y = 4.17$; $x = -2.75, y = 4.17$; $x = 1.78, y = -4.67$; $x = -1.78, y = -4.67$

5. No solution.

7. $x = 4.02, y = -14.1$; $x = -1.84, y = 15.2$

9. $x = 1.86, y = 4.31$; $x = -1.86, y = -4.31$

11. $x = 1.41, y = 0.987$; $x = -0.637, y = -0.595$

13. $x = 1.41, y = 0.148$; $x = 0.0386, y = -1.41$

15. $x = 2.02, y = 2.09$; $x = 0.034, y = -2.00$

17. $x = 4.74, y = 0.748$; $x = 1.03, y = 0.0127$

19. $x = 5.22$

21. The helicopter is 1.694 miles east and 5.083 miles north of the tower.

23. (31.8, 7.52) and (23.8, 13.2);   32.8 ft and 57.6 ft.

25. (a) $x = 1.0, y = 2.0$ $x = -1.0, y = 2.0$
    (b) $x = 1.414, y = 0$ $x = -1.414, y = 0$
    (c) $x = 1.41, y = 0.102$ $x = -1.41, y = 0.102$
    (d) $x = 1.0, y = -2.0$; $x = -1.0, y = -2.0$

Exercise 9.1

1. A is 3×3; B is 3×1; C is 3×3; D is 1×4; E is 1×4

2. A, C

3. −1 is a_{23}, c_{13} and d_{13}

4. 7 is $a_{33}, b_{21}, c_{33}, d_{14}$

5. A and C; D and E

6. AB, AC, CA, CB, BE, BD

7. $\begin{bmatrix} 7 & -5 & -5 \\ 6 & 6 & -6 \\ 8 & -7 & 14 \end{bmatrix}$

9. not possible, not same size

11. $\begin{bmatrix} -8 & 22 & -14 \\ 6 & 0 & 6 \\ 14 & 2 & 14 \end{bmatrix}$

13. $\begin{bmatrix} -25 & 7 & 31 \\ -30 & -26 & 22 \\ -44 & 29 & -70 \end{bmatrix}$

15. $\begin{bmatrix} -14 \\ 18 \\ 62 \end{bmatrix}$

17. $\begin{bmatrix} 0 & 21 & -39 \\ 21 & -8 & -20 \\ 45 & -78 & 54 \end{bmatrix}$

19. $\begin{bmatrix} -165 \\ -132 \\ 66 \end{bmatrix}$

21. $\begin{bmatrix} -18 \\ 11 \\ 54 \end{bmatrix}$

23. Yes, $AI = IA = A$

Exercise 9.2

	transpose	determinant	inverse matrix
1.	$\begin{bmatrix} 2 & -3 \\ -5 & 4 \end{bmatrix}$	−7	$\begin{bmatrix} -.571 & -.714 \\ -.429 & -.286 \end{bmatrix}$
3.	$\begin{bmatrix} 3 & -4 & 5 \\ 2 & 0 & 3 \\ -1 & 3 & -2 \end{bmatrix}$	−1	$\begin{bmatrix} 9.000 & -1.000 & -6.000 \\ -7.000 & 1.000 & 5.000 \\ 12.000 & -1.000 & -8.000 \end{bmatrix}$
5.	$\begin{bmatrix} 1 & -5 & 7 \\ -2 & 3 & -7 \\ -4 & 1 & -9 \end{bmatrix}$	0	No inverse
7.	$\begin{bmatrix} .03 & -.04 & .11 \\ .12 & .45 & .95 \\ .10 & .32 & .04 \end{bmatrix}$	−.013	$\begin{bmatrix} 22.147 & -6.985 & .511 \\ -2.850 & .759 & 1.053 \\ 6.776 & 1.185 & -1.417 \end{bmatrix}$

	determinant	inverse	solution
9.	-13	$\begin{bmatrix} .230 & .154 \\ .154 & -.231 \end{bmatrix}$	$x = 2.231$ $y = -.846$
11.	0	none	no unique solution

	determinant	inverse	solution
13.	-114	$\begin{bmatrix} -.237 & -.447 & .026 \\ .053 & .211 & .105 \\ .272 & .254 & .044 \end{bmatrix}$	$x = -6.789$ $y = 3.842$ $z = 5.351$
15.	6	$\begin{bmatrix} 1.500 & -1.333 & .833 \\ 1.500 & -1.667 & 1.167 \\ .500 & -.667 & .167 \end{bmatrix}$	$x = -10$ $y = -9$ $z = -5$
17.	$-.00905$	$\begin{bmatrix} 4.111 & 1.547 & -1.149 \\ 5.438 & -2.255 & -6.897 \\ 12.467 & -7.405 & -12.356 \end{bmatrix}$	$x = 1024.3$ $y = -392.6$ $z = -402.1$
19.	-11.97	$\begin{bmatrix} .253 & 1.553 & .062 \\ -.165 & 1.036 & .042 \\ -.0025 & 2.757 & -.050 \end{bmatrix}$	$x = .0627$ $y = .0418$ $z = -.0501$

21. $x = 2.5, y = 2.0$ and $x = 2.5, y = -2.0$

23. $\begin{bmatrix} 0.793 & 0.609 & 0 \\ -.609 & .793 & 0 \\ 0 & 0 & 1 \end{bmatrix}$

25. $x = 89.1$ g, $y = 78.1$ g, $z = 32.8$ g

Exercise 9.3

1. $x = 5.23, y = 3.27$ 3. No solution.

5. $x = -4.5, y = 5.7, z = 7.8$

7. Infinite Solutions, $x = 0.333z + 5.227$, $y = 1.333z + 0.867$

9. $x = 11.5, y = -3.7, z = 7.2$ 11. $x = 14.5, y = 0.0, z = 1.5$

13. 312 parts/h, 228 parts/h, 132 parts/h

Exercise 10.1

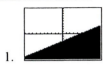 1.    3.   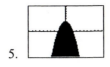 5.

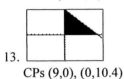 7.
CP (–3.0, 0)   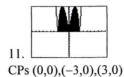 9.
CP (1,0)   11.
CPs (0,0),(–3,0),(3,0)

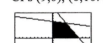 13.
CPs (9,0), (0,10.4)    15.
CPs (3..92, 0), (0, 2.35)   17.
CPs (0,0),(3.8,0),(1.4,3.0),(0,3.0)

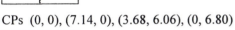 19.
CPs (0, 0), (7.14, 0), (3.68, 6.06), (0, 6.80)   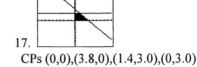 21.
CPs (9.2, 11.2), (13.7, 0.6), (15,0), (65,0)

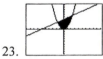 23.
CPs (2.36, 5.57), (–1.69, 2.86)   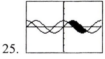 25.
CPs (0.785, 0.707), (3.92, 0.707)

27. $x + y \leq 185, x \geq 0, y \geq 0$; Corner points: (0, 0), (185, 0), (0, 185)

29. $125x + 158y \leq 47500, x \geq 100, y \geq 0$; Corner points: (100, 0), (380, 0), (100, 35000)

31. $x \geq 0, x \leq 322, 175 \leq y \leq 475$; Corner points: (0, 175), (322, 175), (322,475), (0,475)

Exercise 11.1

All points given in polar coordinates with angles in radians.

1. (a) circle, radius = 2   (b) 4   (c) (0,0), (4,0)

3. (a) 4-petal rose   (b) 6   (c) (6, 0), (6, 1.57), (6, 3.14), (6, 4.71), (0, 0)

5. (a) A cardioid   (b) 10   (c) (6, 0), (10, 1.57), (6, 3.14), (2, 4.71)

7. (a) Ellipse, $F(0, 0)$ (b) 0.25   (c) (0.2, 0), (0.17, 1.57), (0.2, 3.14), (.25, 4.71)

9. (a) Spiral   (b) infinite   (c) (0,0), (9.87, 1.57), (39.5, 3.14), (88.8, 4.71)

11. Circles with centers at the origin and with radii of 3, 5, and 7.

13. Circles with center on polar axis and passing through pole. As A changes, radius of circle changes. Radius = $\frac{A}{2}$.

15. Roses of 4 petals, 8 petals, and 12 petals. B determines the number of petals the graph has. If B is even, number of petals = $2B$. Not true if B is odd.

17. Cardioid, vertex at $(2, \frac{\pi}{2})$; Cardioid, vertex at $(0, 0)$; limacon with an inner loop.

If $|A| > |B|$, then cardioid with vertex not at origin; If $|A| = |B|$, then cardioid with vertex at origin; If $|A| < |B|$, then limacon with inner loop.

19. Ellipse, parabola, hyperbola with vertices at $(0, 0)$. If $|B| > |C|$, then ellipse;

If $|B| = |C|$, then parabola; If $|B| < |C|$, then hyperbola.

21. $(0, 0)$, $(3.54, 0.79)$

23. Spiral, $(0.49, 1.57)$, $(1.97, 3.14)$, $(4.44, 4.71)$

25. Min. dist. $= 3.5$, max. dist. $= 13.9$

25. Distance along x-axis $= 6.50$ m; angle is 1 radian (or $57.3°$)

Exercise 12.1

1. (a) x 5 6 7 8 9 10   (e) 7.92   (f) 8   (g) 7, 8, 10   (h) 1.62
f 1 1 3 3 1 3

3. (a) x 75 80 85 90 95 100   (e) 88.2   (f) 90   (g) 90   (h) 8.15
f 1 2 2 3 1 2

5. (a) x 75 80 85 90 95 100   (e) 86.7   (f) 87.5   (g) 75, 90   (h) 9.13
f 3 1 2 3 1 2

7. (a) no. of instructions 18 19 20 21 22 23 24 25
    frequency   2 3 0 2 3 1 3 1
   (e) 21.4   (f) 22   (g) 19, 22, 24   (h) 2.35

9. (a) 22.6   (b) 0.84

11. (a) \$383.90   (b) 65.3

13. mean $= 3.32"$, median $= 3.31"$, mode $= 3.32"$, s $= 0.027"$

15. mean $= 1201$ hr., median $= 1202.5$ hr., mode $= 1205$ hr., s $= 20.1$ hr.

Exercise 12.2

1. (a) 0.6826   (b) 0.4772   (c) 0.0014

3. Area $= 0.6826$ Statement: "$(1/\sqrt{2\pi})e^{\wedge}(-x^2/2)$"$\to Y_1$ **or** $y1 = (1/\sqrt{2\pi})e^{\wedge}(-x^2/2)$

5. (a) 0.4861, 48.61%   (b) 0.9141, 91.41%   (c) 0.0344, 3.44%   (d) 0.9280, 92.80%

7. (a) 81.84%   (b) 31.57%   (c) 56.75%   (d) 24.84%

9. (a) 0.2120   (b) 0.5403   (c) 0.0808   (d) 0.5000

11. (a) 756   (b) 641   (c) 162   (d) 1098

13. (a) 341   (b) 375   (c) 49.4%   (d) 10.6%

15. 10 of the 13 values, 76.9%

17. (a) 0.68   (b) −1.65   (c) −1.65, 1.65   (d) −2.58, 2.58
19. (a) Between 0.4989 and 0.5071   (b) Between 0.4981 and 0.5079
    (c) Between 0.4966 and 0.5094
21. (a) Between 1.186 and 1.224   (b) Between 1.180 and 1.230
    (c) Between 1.171 and 1.239

Exercise 12.3

1. $y = 2.97 + 0.251\,x$, $r = 0.968$, good fit
3. $p = 650.5 - 0.2056x$, $r = -.999$, very good fit
5. $y = 3.069 + 1.970 \ln x$, $r = 0.996$, good fit, $y = 7.61$
7. $y = 10.15\,(2.13)^x$, $r = 0.99988$, very good fit, $y = 209$
9. $y = 0.006625x^2 + 0.7487x + 30.49$, good fit, 91.7 lb/in^2, −0.7°
11. $y = 1442.5x^{-1.01}$, $r = -0.99946$, very good fit, $P = \$2.72$
13. linear, $R = 0.8471 + 0.008240\,T$
    (a) 0.929   (b) 1.012   (c) 18.5°   (d) 20.98°
15. power, $R = 0.004811D^{-2.016}$
    (a) 3.91 ohms   (b) 0.193 ohms   (c) 0.057"
17. exponential, $P = 14.644755(0.999963284)^x$
    (a) 8.6 lb/in^2 (b) 14.6 lb/in^2 (c) about 29300 ft.
18. (c) linear: $y = 0.8737x - 0.3588$ (y is postage, x is years after
    1960)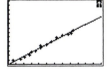
    (e) 1980 calculated = 17, actual = 15
        1995 calculated = 30, actual = 32
    (f) 2017
19. (c) The semi-logarithmic graph gives nearly a straight line for data after 1965, but the data
        prior to 1965 is not near the line.
    (d) exponential, $D = 316.716(1.1094^x)$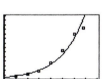
        (D is debt in billions of dollars, x is year after 1965)
    (f) 1995 calculated = 4511, actual = 4921   (in billions of dollars)
        1999 calculated = 6745, actual = 5606
    (g) 2009

Exercise 13.1

	right hand limit	left hand limit	limit
1.	−2	−2	−2
3.	4	4	4
5.	0	0	0

	right hand limit	left hand limit	limit
7.	1	1	1
9.	$-\infty$	∞	does not exist
11.	1	1	1

13. 0.750   15. -3.16

17. 2.72   19. 4

Exercise 13.2

1. PROGRAM: GRDAVE   Answer: Ave is 85.8
   $0 \rightarrow K$ (K is counter)
   $0 \rightarrow S$ (S is accumulator)
   Disp "Enter No. Grades"
   Input N
   While $K<N$
   Disp "Enter Grade"
   Input G
   $1 + K \rightarrow K$
   $G + S \rightarrow S$
   End
   Disp "Ave is"
   Disp S/N

3. PROGRAM:LIMIT   (Note: Function must be stored for $y1$ before execution.)
   $0 \rightarrow K$
   $10 \rightarrow D$
   Prompt A
   Prompt I
   Repeat $K>4$
   $A+I/(D^\wedge K) \rightarrow x$
   Disp $y1$
   $K+1 \rightarrow K$
   End

5. PROGRAM:LIMIT   (Note: Function must be stored for $y1$ before execution.)
   $10 \rightarrow D$
   Prompt A
   Prompt I
   For $(K,1,5)$
   $A+I/(D^\wedge K) \rightarrow x$
   Disp $y1$
   End

7. Limit is -1.

8. Left-hand limit is 0, right hand limit is undefined, limit is undefined.

9. PROGRAM:LIMDER   Answers: 12, 75, 27, 0

```
0→K
10→D
Disp "ENTER X VALUE"
Input x
Disp "ENTER H VALUE"
Input H
While K<5
H/D^K→H
((x+H)^3−x^3)/H→Y          (Note: Change this statement to use another function.)
Disp Y
K+1→K
End
```

Exercise 13.3

Derivative when

x =	−2	−1	0	1	2	
1.	2	2	2	2	2	Derivative is the constant 2.
3.	−4	−2	0	2	4	Derivative is twice the x-value.
5.	−32	−4	0	4	32	Derivative is 4 times the cube of the x-value.
7.	−0.25	−1	undef	−1	0.25	Derivative is negative of reciprocal of x^2.

9. PROGRAM:DERIV2   (Note: Function must be stored for y1 before execution.)

```
0→K                              Answers: −2, 2, 8
Disp "NO OF VALUES"
Input N
While K<N
Disp "ENTER X VALUE"
Input A
nder(y1,x,A)→Y          (For TI-82 and 83 use: NDeriv($Y_1$,X,A)→Y )
Disp Y
1+K→K
End
```

11. For $x < 1$, derivative < 0, f(x) decreasing; for $x > 1$, derivative > 1, f(x) increasing.

13. For $x < −2$, or $x > 2$, derivative < 0, f(x) decreasing;
    for $−2 < x < 2$, derivative > 0 positive, f(x) increasing.

15. For $x < 0$, derivative > 0, f(x) increasing; for $x > 0$, derivative < 0, f(x) decreasing

17. Horizontal tangent line at $x = 2.5$.

19. Horizontal tangent line at $x = 0.0$

21. $s =$ −10   −2.5   5.0   17.5   30.0 s is linear, increasing
    $v =$ 5.0   5.0   5.0   5.0   5.0 v is constant and positive.

23. $s =$ 60   90   90.25   90.24   90 s is increasing, then decreasing
    $v =$ 44   4.0   0.0   −0.8   −4.0 v is positive, then negative

25. $x =$ 35   45   50   55   65   The cost decreases for $x < 50$ and increases for $x > 50$.
    der = −30   −10   0   10   30   Minimum cost of $2000 occurs when $x = 50$.

27. 440, 450, 460, 470

Exercise 13.4

1. $y = 3x + 0.25$ 3. $y = -1.14x + 2.81$
5. $y = -0.67x + 10.82$ 7. $y = 109.2x - 331.1$
9. derivative > 0 for $x < 4$ and $f(x)$ increasing; derivative <0 for $x > 4$, $f(x)$ decreasing.
11. derivative > 0 for $-1/2 < x < 0$, $x > 0$, $f(x)$ increasing;
     derivative < 0 for $x < -1/2$, $0 < x < -1/2$, $f(x)$ decreasing

13.-17. (a) When derivative > 0, function is increasing
         (b) derivative < 0, function decreasing
         (c) At x-intercepts of derivative tangent line to graph is horizontal

13. Graph of derivative is straight line with a negative slope.
     Positive for $x < 0$, negative for $x > 0$, is 0 at $x = 0$.

14. Graph of derivative is a parabola.
     Positive for $x < 0$, negative for $0 < x < 10$, positive for $x > 10$.

15. Derivative is cubic function, has 3 x-intercepts $(-1.7, 0, 1.7)$

16. The derivative is positive for $x < 0$, negative for $x > 0$, and is 0 at $x = 0$.

17. At $x = 5$ and $x = -5$, graph of derivative becomes undefined.
     Derivative is positive for $-5 < x < 0$, negative for $0 < x < 5$, and 0 at $x = 0$.

18. Function and derivative are undefined at $x = 3$.
     Derivative positive for $x < 3$ and negative for $x > 3$.

19. Graph of function has a local maximum at $x = -1$.

20. Graph of function has a local minimum as $x = -2$ and a local maximum at $x = 2$.

21. At $x = 200$ where ball reaches maximum height, derivative > 0 for $x < 200$, ball is rising.

23. Second derivative < 0 for $x < 0$; graph concave down
     Second derivative > 0 for $x > 0$; graph concave up
     Where second derivative is zero, graph changes concavity.

25. Second derivative > 0 for $x < -1.63$, $x > 1.63$; graph concave up
     Second derivative < 0 for $-1.63 < x < 1.63$; graph concave down
     Where second derivative is zero, graph changes concavity.

27. −3.741657, 3.741657   28. 2.410142
29. −2.504978, 2.504978
                                                  30. −1.00, 1.171573, 6.828427

Exercise 14.1

1. $y = \dfrac{3}{2}x^2 - 7x + 4$

3. $y = \dfrac{1}{3}x^3 + x^2 - 3$

5. $y = \dfrac{1}{4}x^4 + 3$

7. $y = (x^2 - 2)^4 - 2$

9. $C = -5$, at $x = 4$, $y = 75$

11. $C = 14$, at $x = -1.2$, $y = 28.8$

13. $C = 6.7$, at $x = 5.2$, $y = 8.3$

15. $i = 2t^2 - 0.2t^3 + 0.7$ at $t = .92$ s, $i = 2.2$ A

Exercise 14.2

1. 24, 28, 30; area = 32

3. 23.6, 21.1, 19.6, 18.8; upper bound

5. 5.1, 11.4, 15.5, 17.8; lower bound

7. 3.67, 2.11, 1.42, 1.12, upper bound

9. Make the following changes in program ARECTG1:
   Change line 8 from $A \to x$ to $A + H \to x$ Change line10 to LINE($x–H$, 0, $x–H$, Y_1)
   Change line 11 to LINE(x, 0, x, Y_1)   Change line 12 to LINE($x–H$, Y_1, x, Y_1)
   In line 9, change $x<B$ to $x<B+H$

13. 40, 36, 34; 30 < area < 34 (exact area = 32)

15. 10.1, 14.3, 16.2, 17.1; 17.1 < area < 18.8 (exact area = 18)

17. 45.6, 31.6, 25.6, 22.9; 17.8 < area < 22.9 (exact area = 20.25)

19. 0.23, 0.39, 0.56, 0.69; 0.69 < area < 1.12 (exact area = 0.875)

21. 21.36, 23.64, 25.52, 25.76; exact value = 26

23. 40.96, 51.84, 61.47, 62.73; exact value = 64

25. −44.6, −42.1, −40.6; graph below x-axis.

27. (a) 84.48, 85.30   (b) 25.92, 5.64   (c) Area from 0 to 12 is algebraic sum of positive area from 0 to 8 and negative area from 8 to 12 (within limits of accuracy).

29. 0.835, 0.919, 0.984, 1.008; exact area = 1.00

Exercise 14.3

1. 32, 32, exact = 32, error = 0.0%, 0.0%

3. 17.7188, 17.9824, exact = 18, error = 1.56%, 0.10%

5. 21.5156, 20.3291, exact = 20.25, error= 6.25%, 3.91%

7. 0.98865, 0.90575, exact = 0.875, error = 12.99%, 3.51%

9. 17.75, 17.1875, 17.04688, 17.01172; errors = 4.41%, 1.10%, 0.276%, 0.0689%; doubling N decreases error by factor of 4.

11. 0.0, 3.99, −3.99; $\displaystyle\int_{-2}^{2}(x^3 - 4x)dx = \int_{-2}^{0}(x^3 - 4x)dx + \int_{0}^{2}(x^3 - 4x)dx$

13. 120.12, –120.12; $\displaystyle\int_1^3 6x^3\,dx = -\int_3^1 6x^3\,dx$

15. PROGRAM: SIMPRULE   (Note: Function must be stored for $y1$ before execution.)

        $0 \to S$
        Prompt A
        Prompt B
        Prompt N
        $(B–A)/N \to H$
        $A \to x$
        $S + y1 \to S$
        For $(K,1,N–1,2)$
        $A + K*H \to x$
        $S + 4y1 \to S$
        End
        (Continued in next column.)

(Continued from previous column.)
For$(K,2,N–2,2)$
$A + K*H \to x$
$S + 2y1 \to S$
End
$B \to x$
$S + y1 \to S$
$S*H/3 \to S$
Disp "Value is"
Disp S

17. 32, 32; no error

19. 18, 18; no error

21. 20.25, 20.25; no error

23. 0.9011, 0.8781; error = 2.98%, .35%

25. 66937 ft-lb

27. $F = 0.1767k$

Exercise 14.4

1. 5.436

3. –0.0986

5. 0.9701

7. 39.27

9. 15.63

11. 18.7

13. 5.33

15. 13.7

17. 0.284

19. 17.7

21. 1067 ft^2

23. 97.8 km

Index

1.5 A Word About Calculator Mode

The **mode** of a calculator provides internal settings for the calculator and gives criteria for doing calculations and graphing of functions.

Procedure C6. Changing the mode of your calculator.

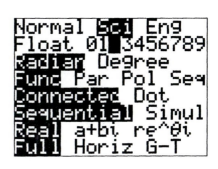

Figure 1.4

Press the key: MODE
You will see several lines. The items that are highlighted in each line set a particular part of the mode. An example of this screen is seen in Figure 1.4.

The first two lines set the way that numbers are displayed on the screen. In the first line, Normal (or Norm) is used for normal mode, Sci for scientific notation, and Eng for engineering notation.

In the second line, Float causes numbers to be displayed in floating point form with a variable number of decimal places. If one of the digits 0 to 9 is highlighted, all calculated numbers would be displayed with that number of decimal places.

(Example: All numbers calculated using the settings in Figure 1.4 will be displayed in scientific notation rounded to 2 decimal places.)

The third line is used for trigonometric calculations and indicates if the angular measurements are in degrees or radians. Other lines on this screen will be considered later.

To change mode:

1. Use the cursor keys to move the cursor to the item desired.
2. Press ENTER. The new item should now be highlighted.
3. To return to the home screen, press the CLEAR or QUIT key.

(Example: To change the mode so that numbers are displayed in scientific notation with two decimal places, highlight Sci in the first line and press ENTER, then highlight 2 in the second line and press ENTER. Then, return to the home screen.)

1.6 Scientific Notation

Often times in engineering work it is necessary to work with numbers whose absolute value is very large or very small. For example the speed of light is approximately 670,000,000 miles per hour and the speed of calculations done by a computer are often measured in nanoseconds where 1 nanosecond is 0.000000001 seconds. It is often more convenient to write numbers such as these in terms of scientific notation. A positive number written in **scientific notation** is written as a product of a number between 1 and 10 times a power of 10. When a number is written in scientific notation, only the significant digits are shown.

Example 1.7: Write the numbers (a) 54,000 and (b) 0.00253 in scientific notation.

Solution: (a) The number 54,000 can be written as $5.4 \times 10,000$ and $10,000 = 10^4$,

   Thus, $54,000 = 5.4 \times 10^4$. The last number is in scientific notation.

(b) The number 0.00253 can be written as $\dfrac{2.53}{1000}$. This is the same as $\dfrac{2.53}{10^3}$.

But $\dfrac{1}{10^3} = 10^{-3}$.   Thus, $0.00253 = 2.53 \times 10^{-3}$.

This last number is in scientific notation.

A short cut way of writing a number in scientific notation is done by counting the number of decimal places the decimal point is moved to place the decimal point behind the first non-zero digit. The number of places the decimal point is to be moved determines the power of 10. If the decimal point is to be moved to the left, the power of 10 is positive and if the decimal point is to be moved to the right, the power of 10 is negative. In the number 54,000 in Example 1.7 it is necessary to move the decimal place 4 places to the left and the power of 10 is 4. For the number 0.00253, it is necessary to move the decimal point 3 units to the right and the power of 10 is –3. To convert numbers from scientific notation to normal notation, the process is reversed.

On a graphing calculator, a number in scientific notation usually appears as a number followed by the letter E and then the appropriate power of 10. The number 670,000,000 would appear as 6.7E8 and the number 0.000000001 would appear as 1E–9. The letter E on the calculator indicates the exponent to the number 10. Thus, E8 means 10^8 and E–9 means 10^{-9}.

In order to enter numbers in scientific notation into the calculator, we make use of the EE key. To enter 6.35×10^4 we would enter 6.35, press the EE key, then enter the number 4. To enter the number 1.250×10^{-3}, we would enter the number 1.250, press the EE key, then enter the number –3 (be sure to use the negative symbol (-) and not the subtraction sign –). The numbers as entered would appear on the screen as 6.35E4 and1.25E–3. After the Enter key is pressed, the numbers would appear as 6.35E4 and1.25E–3 if calculator is in scientific notation mode and as 6350 and .00125 if the calculator were in normal mode.

Example 1.8: Change the calculator into scientific mode and to set the number of decimal places displayed to 2. Then enter and perform the following calculations.
   (a) $63.2(7.25 \times 10^4)$
   (b) $\dfrac{235000 \times 0.0000024}{1.46 \times 10^{-5}}$
   (Before entering numbers, write each in scientific notation.)

Solution: Use procedure C6 to change the calculator into scientific mode and to set the number of decimal places displayed to 2.

(a) $63.2(7.25\times10^4)$ is the same as $63.2\times7.25\times10^4$.

To enter we enter 63.2, press ×, enter 7.25, press the **EE** key, and enter 4. After the problem is entered, press the ENTER key. The result is 4.58E6. In normal notation this is the number 4580000.

(b) First change each number into scientific notation.
$$235000 = 2.35\times10^5$$
$$\text{and } 0.0000024 = 2.4\times10^{-6}.$$

Enter this problem into the calculator using the **EE** key for the numbers in scientific notation. After the problem is entered, press the ENTER key. The result is 3.86E4. Rounded to the proper number of significant digits, the answer is 3.8×10^4 or the number 38000 in normal notation.

Both problems are shown in Figure 1.5.

```
63.2*(7.25E4)
               4.58E6
(2.35E5*2.4E-6)/
1.46E-5
               3.86E4
```

Figure 1.5

1.7 Working with Sets of Numbers

On some calculators it is possible to work with sets of numbers called **lists**. Lists provide a convenient way to handle a set of numbers that are related.

<u>Procedure C7</u>. Storing a set of numbers as a list.

On the TI-82, 83, or 84 Plus:
Up to six lists (of up to 99 values each) may be stored using the names L_1, L_2, L_3, L_4, L_5, and L_6
or
On the TI-83 and 84 Plus additional lists may be stored under a user given name up to 5 characters long and starting with a letter.

On the TI-85 or 86:
Lists (of any length) may be stored using a list name of up to eight characters. The first character of the name a must be a letter.

Method I:

1. Press the key {
2. Enter the numbers to be in the list separated by commas.
3. Press the key }
4. Press the key STO▷
5. Press the key for the list name (or enter the list name) and press ENTER.

Method II:

1. Press the key STAT
2. From the menu, select 1:EDIT
3. Move the highlight to the list and position where numbers are to be entered, and enter the number. The entered value will appear at the bottom of the screen. Press ENTER to place the number in the list.

4. When all values in the list have been entered, press QUIT to exit from the list table.

Note: A value may be changed by moving the highlight in the table to that value, entering the correct value, and pressing ENTER.

Method I:

1. Press the key LIST.
2. Select {
3. Enter the numbers to be in the list separated by commas.
4. Select }
5. Press the key STO▷
6. Enter the list name and press ENTER.

Method II:

1. Press the key LIST
2. Select EDIT
3. To enter a new list, enter the new name after Name= . On the TI-86, the numbers may be stored in a pre-named list (xStat, yStat, fStat) or we can name a new list by going to an unnamed column.

To get a blank column, move the cursor up to the name line and move the cursor to the right to a new column.

4. Enter the values as elements of the list by moving the cursor to each position and entering a value. Press ENTER after each entry. When finished, press QUIT.

Note: A value may be changed by moving the cursor to that value and entering the new value.

Procedure C8.   Selecting a list and viewing a set of numbers.

Select a List:
   On the TI-82, 83, or 84 Plus:
   Press the key for the name of the list,
   or

On the TI-85 or 86:
Enter the name of list and press ENTER.
or

on the TI-83 or 84 Plus:
Press the key LIST and with
NAMES highlighted at the top,
select the name of the list, press
ENTER

View a List:
With name on the screen,
press ENTER

or

1. Press the key STAT
2. Select 1:EDIT
3. Move the highlight to the list.
4. Press QUIT to exit.

1. Press the key LIST
2. Select NAMES, then select the
   list name and press ENTER

or

Select EDIT and move the cursor
to the column desired.
On the TI-85, a list name will need
to be entered or selected.

The arithmetic operations may be performed on lists. When operations are performed on lists the actions are performed on the corresponding elements of the lists. For example, the product {1, 2, 3, 4}{5, 6, 7, 8} will yield {5, 12, 21, 32}. The first numbers in each list will be multiplied, the second numbers, etc. It is also possible to operate on all elements of a list by a constant. For example, the sum {1, 2, 3, 4} + 5 adds 5 to each element of the list and gives {6, 7, 8, 9}and the square root of {1, 2, 3, 4} will yield {1, 1.41, 1.73, 2} (rounded to three significant figures).

Procedure C9. **To perform operations on lists.**

1. To add, subtract, multiply or divide lists, enter the list or the name of a list, press the appropriate key (+, −, ×, or ÷), enter the second list or the list name and press ENTER. Lists must be of the same length in order for these operations to be performed.
2. To perform an operation on all the elements of a list, enter a value, press the appropriate key (+, −, ×, or ÷), enter the list or list name and press ENTER.
3. Functions such as finding the squares or square roots, may be performed on each element of a list by performing the function of the list or list name.

Example 1.9: Find the sum of the set of numbers 12.5, 16.7, 22.1, and 16.3 and the set of numbers 32.1, 22.8, 45.6, 23.1 and then find 11.5% of each number in the result by multiplying the result by 0.115.

Solution: 1. Enter the set of number 12.5, 16.7, 22.1, and 16.3 in one list using the list name *L1* and the set of numbers 32.1, 22.8, 45.6, and 23.1 into a second list using the list name *L2* as in Procedure C7.

2. Find the sum of the two lists by adding the two list names, *L1* + *L2*, and pressing ENTER. This sum is stored under Ans. The result is the list 44.6, 39.5, 67.7, and 39.4.

3. To find 11.5% of the product, multiply Ans by 0.115 (this may be done by first pressing × and then entering .115) and press ENTER. The result, rounded to two decimal places, is the list 5.13, 4.54, 7.79, and 4.53.

Exercise 1.2

In Exercises 1-8, use the graphing calculator to evaluate each of the following. First use Procedure C6 to change the mode of the calculator to scientific notation with three decimal places and perform the following calculations. Round answers to the proper number of significant digits. Leave answers in scientific notation.

1. 362.4×144000

2. $19.172 \div 0.0052$

3. $0.00000237(6.783 \times 10^6)$

4. $\dfrac{8250000}{6.3 \times 10^3}$

5. $3.45 \times 10^8 + 7.2 \times 10^7$

6. $6.8 \times 10^{-12} - 2.41 \times 10^{-11}$

7. $\dfrac{0.02157(6.32 \times 10^8)}{0.125(7.457 \times 10^5)}$

8. $\dfrac{(1.832 \times 10^{12})(4.83 \times 10^{-5})}{8.93 \times 10^{-5}}$

In Exercises 9-13, use lists.

9. Store {34, 45, 62, 23} as one list and {76, 23, 43, 45} as a second list. Find the sum, difference, and product of the two lists. Also, divide the first list by the second.

10. Store {12, 15, 42, 33, 68} as a list. Multiply the list by π, divide the list by 6.5, and subtract 17 from the list.

11. Use lists to find the square and square root of each of the numbers 22.5, 11.6, 32.5, and 64.0.

12. Store {20.0, 0.00, −40.0, 37.0, 100.0} as a list. Multiply the list by (9/5) and add 32 to the result. What to these lists represent?

13. A clerk in the storeroom determines that the number of washers 0.50 inches in diameter is 135 and the number of washers 0.25 inches in diameter is 238, the number of washers 0.125 inches in diameter is 25, and the number of washers 0.375 inches in diameter is 78. The value of each of the 0.50-inch washers is $2.10, of the 0.25-inch washers is $1.75, of the 0.125-inch washers is $0.95 and of the 0.375-inch washers is $2.55. Set up two lists and multiply them to determine the value of the washer of each size in stock.

The slope of a straight line is given by $m = \dfrac{y_2 - y_1}{x_2 - x_1}$ where (x_1, y_1) is one point on a line and (x_2, y_2) is a second point on the line.

14. Find the slope of the line through the points (7.2, –3.1) and (3.11, –5.0).
15. Find the slope of the line through the points (–1.63, 2.50) and (3.72, –1.233).
16. Find the slope of the line through the points (3.12, 4.73) and (5.27, 4.73).
17. Find the slope of the line through the points (–4.89, –2.12) and (–4.89, 3.75).

The distance between two points (x_1, y_1) and (x_2, y_2) is given by the formula $\sqrt{(x_2 - x_1)^2 + (y_2 - y_1)^2}$.

18. Find the distance between the points (2.334, 1.77) and (8.34, –7.12).
19. Find the distance between the points (–12.5, –6.34) and (–3.24, 10.5).

The midpoint between two points (x_1, y_1) and (x_2, y_2) is given by $\left(\dfrac{x_1 + x_2}{2}, \dfrac{y_1 + y_2}{2} \right)$.

20. Find the midpoint between the points (2.334, 1.77) and (8.34, –7.12).
21. Find the midpoint between the points (–12.5, –6.34) and (–3.24, 10.5).

♦ 22. *(Washington, Exercise 1.5, #45)* A computer can do an average multiplication in 5.6×10^{-12} seconds. How long does it take to perform 3.2×10^5 multiplications?

♦ 23. *(Washington, Exercise 1.12, #6)* The flow of water from one stream into a lake is 61500 ft^3 more than the flow of a second stream. In 1 h, 2.75×10^5 ft^3 of water flow into the lake from both streams. What is the flow rate of each stream?

24. Evaluate $\left(1 + \dfrac{1}{n} \right)^n$ for $n = 1$, $n = 2$, $n = 5$, $n = 10$, $n = 100$, $n = 1000$, and $n = 10000$.

   (a) Do these number seem to get close to a particular value as n gets large?
   (b) If so, what does this value appear to be? How large does n need to be in order for the successive values to differ by no more than 0.01 units?

Chapter 2

Functions and Graphing

2.1 Functions and Their Evaluation
(Washington, Sections 3.1 and 3.2)

A real function is a relationship between two sets of real numbers such that for every value from the first set of numbers there is exactly one value of the second set of numbers that relates to it. Variables are used to represent elements in these sets of numbers. A variable representing a number in the first set is referred to as the **independent variable** and a variable representing a number in the second set is referred to as the **dependent variable**. Generally, we will let x represent the independent variable and y represent the corresponding dependent variable. The relationship between the two sets of numbers can often be expressed in the form of an equation relating x and y. For example, $y = x^2 + 2$ is a function stating that each number in the first set is squared and added to 2 to obtain a number in the second set.

In mathematics, functional notation is often used to describe a function. The relationship $y = x^2 + 2$ is expressed in functional notation as $f(x) = x^2 + 2$. The evaluation of a function when $x = k$ is indicated as $f(k)$ and means to substitute the value of k for x in the functional expression and evaluate. For example, if we wish to evaluate the function $y = x^2 + 2$ at $x = 3$, we find $f(3)$. This means to substitute 3 in for x everywhere x appears on the right side of the equation. Thus, $f(3) = 3^2 + 2 = 11$.

We will consider the **domain** of a real function to be that collection of real numbers such that when the function is evaluated at any of these numbers a real number is obtained. The domain refers to values of the independent variable. On a graph, the values in the domain are those values of the independent variable that correspond to points on the graph.

The **range** of a function is that collection of real numbers obtained by evaluating the function for all real numbers in the domain. The range of a function refers to the dependent variable. On a graph, the values in the range of a function are those values of the dependent variable that correspond to points on the graph.

If we consider that the complete graph of a given function appears as in Figure 2.1 then we can estimate the domain and range from the graph. As we look at the graph in Figure 2.1, we see that the graph of the function exists for x-values greater than or equal to −8 and less than or equal to +8. Thus, we would conclude that the domain for this function is those x-values satisfying this condition – that is: $-8 \le x \le 8$. The range would be those y-values that correspond to points on the graph.

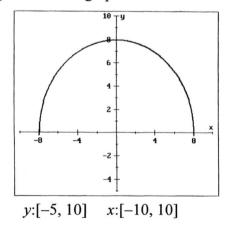

y:[−5, 10]   x:[−10, 10]

Figure 2.1

This appears to be y-values greater than or equal to 0 and less than or equal to 8. We can write the domain as $0 \leq y \leq 8$. When determining the domain or range of a function from a graph, we must be sure we are considering the complete graph.

We can use the graphing calculator to evaluate functions for various numbers in their domain. In order to evaluate a function at different values of x, we first store the function in the calculator's memory.

Procedure C10. To store a function.

On the TI-82 or 83 or 84 Plus:

The TI-82 stores up to eight functions, using function names $Y_1, Y_2, Y_3, \ldots Y_8$. The TI-83 and 84 Plus store up to ten functions.

Use the key X, T, θ for the variable x.

Method I:
1. Enter inside of quotes the function to be stored.
2. Press the key STO▷
3. Press the key Y-VARS
   (On the TI-83 or 84 Plus, press the VARS key and highlight Y-VARS.)
4. Select 1:Function
5. Select a function name (This function will replace any previously stored function.)
6. Press ENTER

Example: $\ ^{"}x^2 + 2^{"} \to Y_1$

Method II:
1. Display the function table by:
   Press the key Y=
2. Move the cursor to the line of the function.
3. Enter the function to the right of a function name.
4. Press QUIT

On the TI-85 or 86:

The TI-85 or 86 may store functions using names that may contain up to eight characters. The first character of a name must be a letter.

Use the key x-VAR for the variable x.

Method I:
1. Enter the name of the function.
   (Names may be up to eight characters long.)
2. Press the key =
3. Enter the function.
4. Press ENTER

To see a stored function, press the key RCL, enter the name of the function, and press ENTER.

Method II:
(Stores functions under the names $y1, y2, y3, \ldots y_{99}$)
1. Display the function table by:
   Pressing the key GRAPH, then selecting y(x)=
2. Enter the function after a function name.
3. Press QUIT

Note: To see stored functions, display the function table by pressing the key Y=. Pressing CLEAR while the cursor is on the same line as the function will erase the function. Be sure to QUIT the function list to return to the home screen.

After storing a function in memory we can use the calculator to evaluate the function at one or more values of the independent variable. On many calculators it is possible to show tables of values for functions. In such a table, a list of selected values of x and corresponding functional values are shown.

Procedure C11. To evaluate a function or create tables.

A. To evaluate a function
Functions may be evaluated at a single value or at several values by entering the values in a list.

On the TI-82 or 83 or 84 Plus:
1. Obtain the function name
   (a) Press the VARS key and highlight Y-VARS at the top of the screen. (On the TI-82 press the Y-VARS key.)
   (b) Select 1:Function...
   (c) Select the name of the function.
2. After the function name enter, in parentheses, the x value, a list of x values, or a list name at which the function is to be evaluated.
   (Example: $Y_1(4)$ or
   $Y_1(\{1,2,3,4\})$ or $Y_1(L_1)$)
3. Press ENTER
The results will be given as a value or a corresponding list.

On the TI-85 or 86:
1. Obtain the function evalF
   (a) Press the key CALC
   (b) Select evalF
2. Enter either the function or the function name then a comma.
3. Enter the variable for which the function is being evaluated followed by a comma.
The function names $y1$, $y2$, etc. may be entered from the keyboard (y must be lower case)
   or
may be selected by:
   (a) Pressing the key VARS
   (b) From menu, select EQU (after pressing MORE).
   (c) Select the variable name
   (may need page↓ or page↑).
4. Enter the value, a list of values, or a list name at which the function is to be evaluated.
(Example: evalF($y1$, x, 4) or
   evalF($y1$,x,$\{1,2,4\}$))
5. Press ENTER

Note: In the evaluation of a function, instead of placing a list of values inside of braces {}, we may enter the list name.
See Procedure C7 for entering lists.

B. To create tables:

The TI-82, 83, 84 Plus, and 86 have the ability to create a table of values for a function by pressing the key TABLE. Parameters for the table are set by pressing the key TblSet on the TI-82, 83, and 84 Plus or by selecting TBLST on the TI-86.

In TblSet:
TblMin or TblStart is the value of the first x value in the table.
ΔTbl is the difference between x-values in the table.
Indpnt: Auto Ask allows the table to be created automatically if Auto is highlighted or by the calculator asking for values of the independent variable if Ask is highlighted.

Pressing (or selecting) TABLE will show a table of values of the independent variable and values for all functions that are turned on in the function table.

By using the cursor keys, we can scroll through the table.

The appearance of an error message when evaluating a function likely indicates that the value of x is not in the domain of the function.

Example 2.1:   (a) Evaluate the function $f(x) = \sqrt{x+3}$ for x values of 6, 2.5, –3, and –4. Are any of these values not in the domain of $f(x)$? Round off answers to 3 significant digits.

(b) Create a table of values for $f(x)$ starting with $x = -3$ and ΔTbl = .5. Then, from the table determine the function value when $x = 1.5$.

Solution: (a) 1. Use Procedure C7 to store the function $\sqrt{x+3}$ and then   Procedure C8 to evaluate at the x values.

2. The value of $f(6)$ is 3.00
The value of $f(2.5)$ is 2.35.
The value of $f(-3)$ is 0.00.
When we try to evaluate the function at –4 we get an error message. This is because –4 is not in the domain of the real function $f(x)$.
(Some calculators may display (0,1) indicating that the complex number, j, is the solution.)

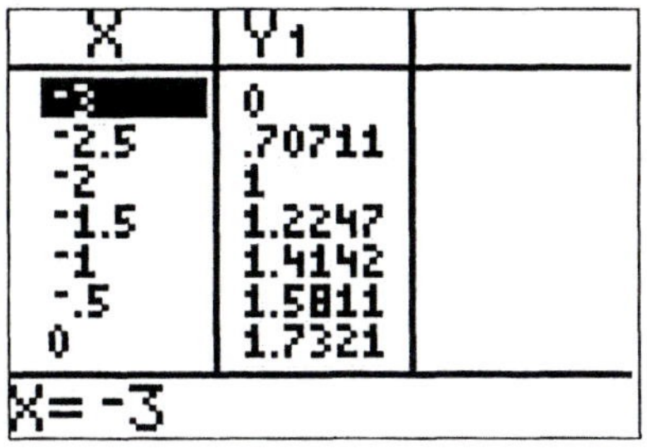

(b) Use Procedure C7 to create a table of values. Set the first value in the table to be –3 and the difference between table values to be 0.5. Press TABLE to see the table. It will appear as in Figure 2.2.
Use the cursor keys to scroll through the table in increasing values of x until we see 1.5. We find that at $x = 1.5$, the value of the function is 2.1213.

X	Y1	
-3	0	
-2.5	.70711	
-2	1	
-1.5	1.2247	
-1	1.4142	
-.5	1.5811	
0	1.7321	

X= -3

Figure 2.2

Sometimes, it is helpful to enter a constant as an alpha character when storing a function in the calculator. The value for the constant is stored for the alpha character and may be changed by storing a new value for the alpha character. For example, to evaluate $\pi x^2 h$ for several values of x when $h = 1.5$ and when $h = 1.8$, we would first store the function $\pi x^2 h$ for a function name, store 1.5 for h and evaluate the function, then store 1.8 for h and again evaluate the function.

Procedure C12. To store a constant for an alpha character.

    1. Enter the value to be stored.
    2. Press the key STO▷
       (STO▷ shows as → on screen.)
    3. Enter the alpha character.
    4. Press ENTER
    (Example: $6.35 \to H$)
    Note: To see the value stored for an alpha character, enter the alpha character and press ENTER.

Example 2.2: (a) Evaluate $F(x) = \pi x^2 h$ for $x = 2.50, 2.60$, and 2.70 cm when $h = 6.35$ cm and when $h = 6.45$ cm.

        (b) Create a table of values for $F(x) = \pi x^2 h$ for $h = 6.35$ and x-values starting at $x = 0$ and with a difference between table values of 0.1. Then, using the table determine the value of x for which the function is nearest 20.

Solution: (a) 1. Use Procedure C7 to store the function $\pi x^2 H$ for a function name.

           2. Use Procedure C9 to store 6.35 for H.

           3. Use Procedure C8 to evaluate the function for $x = 2.50, 2.60$, and 2.70. We find that $F(2.50) = 125$, $F(2.60) = 135$, and $F(2.70) = 145$.

           4. Now, use Procedure C9 to store 6.45 for H.

           5. Evaluate the function again for $x = 2.50, 2.60$, and 2.70. This time we obtain $F(2.50) = 127$, $F(2.60) = 137$, and $F(2.70) = 148$.

        (b) Use Procedure C7 to create a table of values. Set the first value in the table to be 0 and the difference between table values to be 0.1. The table will appear as in Figure 2.3.

Scrolling down through the table, we look for value of $Y1$ near 20. The nearest value is 19.949 when $x = 1.0$.

X	Y1	
0	0	
.1	.19949	
.2	.79796	
.3	1.7954	
.4	3.1919	
.5	4.9873	
.6	7.1817	

X=0

Figure 2.3

Exercise 2.1

Evaluate the functions at the given values of x. Tell which values of x are not in the domain of the function. Round off decimal numbers to 2 decimal places.

1. $f(x) = x^2 + x$ for $x = -3$

2. $g(x) = x^3 - 2x^2 - 4$ for $x = 2.5$

3. $F(x) = \dfrac{1}{x}$ for $x = 4.5$ and for $x = 0.0$

4. $f(x) = \dfrac{x+2}{x}$ for $x = 3.7$ and for $x = -5.1$

5. (a) $g(x) = \dfrac{2}{x-2}$ for $x = 2.11$ and $x = 2$

   (b) Create a table of values for $g(x)$ starting at $x = 0$ with differences of 0.5.

6. (a) $h(x) = \dfrac{3}{x+4}$ for $x = -4.1$ and -4

   (b) Create a table of values for $h(x)$ starting at $x = -5$ with differences of 0.5.

7. (a) $f(x) = \dfrac{x-3}{x^2-4}$ for $x = -2, 0, 2.3, 3, \sqrt{2}$

   (b) Create a table of values for $f(x)$ starting at $x = -4$ with differences of 1.0.

8. (a) $g(x) = \dfrac{x-5}{x^2-2}$ for $x = 5, 2, \sqrt{2}, 3, 0$

   (b) Create a table of values for $g(x)$ starting at $x = -5$ with differences of 1.0.

9. (a) $h(x) = \sqrt{x^2-5}$ for $x = 3, 2, -4, 0, 5$

   (b) Create a table of values for $h(x)$ starting at $x = -3$ with differences of 0.5.

10. (a) $f(x) = \sqrt{x^2-9}$ for $x = 5, -5, 4, -4, \sqrt{3}$

    (b) Create a table of value for $f(x)$ starting at $x = -5$ with differences of 0.5.

11. $g(x) = \sqrt{x^2+5}$ for $x = 2, 5, -2, 0, \sqrt{5}$

12. $h(x) = \sqrt{x^2+\pi}$ for $x = 3, 4, -4, 0, \sqrt{\pi}$

13. $f(x) = |x-3|$ for $x = 3, 3.1, -3.1, 5.2, 0$

14. $F(x) = |x+2|$ for $x = 3.12, -2.15, 2.78, 1.32, -2$

15. $G(x) = [x-1]$ for $x = 2.3, 1.5, -2.34, 1.99, 0.999$

16. $f(x) = [2+x]$ for $x = 0, 2.001, -2.001, -1.999, 1.002$

17. Evaluate $S(x) = h - 9.8x^2$ for $x = 1.0, 1.2$, and 1.4 first when (a) $h = 19.5$, then when (b) $h = 17.5$.

18. Evaluate $g(t) = at^2 - a^2t$ for $t = -0.50, 0.50$, and 1.50 first when (a) $a = -21$, then when (b) $a = 18$.

♦ 19. *(Washington, Exercises 3.1, #41)* A demolition ball is used to tear down a building. Its distance s (in m) above the ground as a function of time t (in s) after it is dropped is $s = 27.25 - 4.90t^2$. Since s = s(t), find s(1.75) and s(2.35).

♦ 20. *(Washington, Exercises 3.1, #43)* The electric power P (in W) dissipated in a resistor of resistance R (in Ω) is given by the function $P = \dfrac{200R}{(100 + R)^2}$.

Since $P = f(R)$, find f(25) and f(45).

21. (a) Evaluate $y = mx + b$ for the x-values 3.00, 3.20, 3.40, and 3.60, when $m = 6.00$ and $b = -2.00$.

(b) Create a table of values for the function in part (a) and from the table determine the value of y at the given values of x.

22. (a) Evaluate $y = a(x - 4.55)^2 - k$ for x values of 2.55, 3.55, 4.55, 5.55, and 6.55 first when
(a) $a = 3.20$ and $k = 1.45$, then when (b) $a = -2.40$ and $k = -1.65$.

(b) Create a table of values for the function in part (a) and from the table determine the value of y at the given values of x.

23. (a) A baseball is thrown upward at a velocity of 62.0 feet per second. The function $h(t) = 62.0t - 16.0t^2 + 5.50$ gives the height of the baseball above ground level as a function of time t in seconds. Evaluate this function to find the height of the ball after 1.0 seconds, 1.875 seconds, and 3.75 seconds.

(b) Create a table of values for the function in part (a) and from the table determine to the nearest tenth of a second the time at which the ball is at its highest point and also the time when the ball hits the ground.

24. (a) The electric power P, in watts, dissipated in a resistor R, in ohms, is given by the function $P = \dfrac{200R}{(100 + R)^2}$. Since $P = f(R)$, find $f(100), f(110)$, and $\underline{f}(120)$.

(b) Create a table of values for the function in part (a) and from the table determine the value of the power R to the nearest 10 units when P is becomes less than 0.40.77

♦ 25. *(Washington, Exercises 3.2, #23)* A rocket burns up at a rate of 2.17 tons per minute after falling out of orbit into the atmosphere. If the rocket weighed 5874 tons before reentry, express its weight w as a function of the time t, in minutes, of reentry. Then evaluate the function to find the weight at 15.0 minutes, 21.0 minutes, and 25.0 minutes.

♦ 26. *(Washington, Exercises 3.2, #24)* A computer part costs $2.87 to produce and distribute. Express the profit p made by selling 129 of these parts as a function of the price of c dollars each. Then, evaluate the function to find the profit if the price is $4.78, $5.24, and $5.98.

♦ 27. *(Washington, Exercises 3.2, #27)* A company installs underground cable at a cost of $465.00 for the first 50.0 feet (or up to 50 feet) and $5.14 for each foot thereafter. Express the cost *C* as a function of the length *L* of underground cable if *L* > 50 feet. Then, evaluate the function to determine the cost if the distance is 65 feet, 97.2 feet, and 212.9 feet.

♦ 28. *(Washington, Exercises 3.2, #33)* A helicopter 117.5 feet from a person takes off vertically. Express the distance d from the person to the helicopter as a function of the height h of the helicopter. Then, evaluate the distance from the person when the helicopter is at a height of 22.3 feet, 52.6 feet, and 117.5 feet.

2.2 Graphing a Function.
(Washington, Section 3.4)

In order to graph a function by hand it is often necessary to evaluate the function at several values of the independent variable, plot these points on a graph, and then connect the points with a smooth curve. The graphing calculator allows us to enter the function and let the calculator do this work for us. The function is plotted on the calculator's screen much quicker than we could graph by hand. The calculator also allows us to graph more than one function at the same time.

Before proceeding, it is helpful to understand the procedure used by a graphing calculator to graph functions. On the graphing calculator, there are only so many dots or pixels that may be plotted. To graph a function the *x*-value of each pixel in the horizontal direction is determined by the calculator and the corresponding *y*-value calculated from an entered function. If the screen has 94 useful pixels in the horizontal direction, the calculator will determine 94 points for the function. The calculator will turn on, or plot, all of these points that lie on the screen and (in connected mode) will turn on points lying between each pair of points. The precision to which points on the screen are found is determined by the difference in values as we move from one pixel to another.

When graphing on the graphing calculator in rectangular coordinates the independent variable will be *x* and the dependent variable will be *y*. To enter a function into the calculator we use a special key for the independent variable *x*.

Procedure G1. To enter and graph functions.

1. Enter the function to be graphed according to Procedure C10.
2. To graph entered functions press the key: GRAPH
3. To return to the home screen while a graph is on the screen, press QUIT or CLEAR.

Notes: 1. Functions entered by this procedure will remain in the calculator's memory until changed or erased or until the memory is cleared.

2. While a graph is on the screen, pressing QUIT or CLEAR will clear the screen and return to the home screen. However, the graph is still in the calculator's memory and can be seen again by pressing GRAPH.

3. When a function has been entered, the equal sign for that function becomes highlighted indicating that the function is turned ON - that is, it is an active function and will be graphed if the GRAPH key is pressed. Only functions that are turned ON will be graphed.

To turn a function ON or OFF, with the function table on the screen:

On the TI-82 or 83 or 84 Plus:

(a) Move the cursor so that it falls on top of the equal sign of the function we wish to turn ON or OFF.

(b) Press the ENTER key. If the function was ON it will be turned OFF and if it was OFF it will be turned ON.

On the TI-85 or 86:

(a) Move the cursor so it is on the same line as the function to be turned ON or OFF.

(b) From the secondary menu, select SELCT. If the function was ON it will be turned OFF and if it was OFF it will be turned ON.

4. On the TI-83 and 84 Plus, different appearances may be obtained for graphs by moving the cursor to the symbol that appears to the left of a function name in the function table. To change this symbol, move the cursor to the symbol and press ENTER until the desired symbol is obtained.

There are times when we may find it is necessary to change or erase a function. This may be done using the cursor, insert and delete keys.

Procedure G2. **To change or erase a function.**

1. Display the function table on the screen (see Method II of Procedure C10).
2. (a) To change a function: Use the arrow keys to move the cursor to the desired location and make the changes by inserting, deleting, or changing the desired characters.
   (b) To erase a function: With the cursor on the same line as the function, press the key: CLEAR
3. Select GRAPH to graph the function or QUIT to exit to the home screen.

The view of a graph that we see on the screen of the graphing calculator is called the **viewing rectangle**. At this point we will use what is called the standard viewing rectangle and defer discussion on other viewing rectangles.

Procedure G3. **To obtain the standard viewing rectangle.**

To obtain the standard viewing rectangle;

On the TI-82 or 83 or 84 Plus:

1. Press the key ZOOM
2. From the menu, select 6:Zstandard

On the TI-85 or 86:

1. Press the key GRAPH
2. From the menu, select ZOOM
3. From the secondary menu, select ZSTD.

For the standard viewing rectangle, the x-value at the left of the screen is -10 and at the right of the screen is 10 and the y-value at the bottom of the screen is -10 and at the top of the screen is 10. Each mark on the axes represents one unit.

We are now ready to graph a function. Generally, we enter functions just as they appear. Be sure to group quantities correctly by using parentheses.

Example 2.3:   Graph the function $y = x^2 - 4$.

Solution: 1. Use Procedure G3 to obtain the standard viewing rectangle.
                 2. Use Procedure G1 to graph the function
                    $y = -4$. The graph is in Figure 2.4.
                    From the screen it appears that the graph
                    crosses the x-axis at -2 and at 2 and crosses
                    the y-axis at -4.

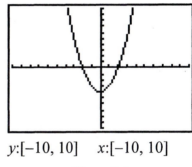

y:[-10, 10] x:[-10, 10]

Figure 2.4

Graphing calculators allow us to graph more than one function at a time. This feature allows us to compare graphs of different functions and to determine the points of intersection of intersecting graphs.

Example 2.4:   Graph the functions $y = 2x$, $y = 2x - 2$, and $y = x - 4$. How do these graphs compare?

Solution: 1. Follow Procedure G1 to graph the three functions. Enter $2x$ for the first
                    function, $2x - 2$ for the second function, and $2x - 4$ for the third function.
                    The resulting graph is in Figure 2.5.
                 2. The graphs are parallel straight lines that
                    intersect the y-axis at different points
                    $(0, -2,$ and $-4)$. We conclude from this
                    that in a function of the form
                    $y = mx + b$, the b value represents where
                    the graph crosses the y-axis.

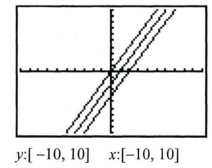

y:[-10, 10] x:[-10, 10]

Figure 2.5

There are certain points on a graph that we shall consider important points. For certain applications these points may have special meanings. We shall consider the x-intercept, the y-intercept, the local maximum point and the local minimum point as important points. The x-value where the graph crosses the x-axis is called the **x-intercept** of the graph of the function. This value is also referred to as the **zero** of the function since when substituted into the function a value of zero is obtained. The y-value where the graph crosses the y-axis is called the **y-intercept** of the graph. If a graph is continuous, that is there are no breaks in the graph, then a point is a **local maximum point** if it is the highest point of all points in the immediate vicinity and if there exists points on either side of the point. Likewise, a point is a **local minimum point** if it is the lowest point of all points in the immediate vicinity and if there exists points on either side of the point. The graph of a function may have several x-intercepts, several local maximum points, and several local minimum points, but only one y-intercept.

The graph in Figure 2.6 has 2 local maximum points (at about $x = -4.5$ and $x = 8$), one local minimum point (at about $x = 1$) and four x-intercepts (at about $x = -7.4$, $x = -1$, and $x = 3$). One of the x-intercepts is off the right of the graph. When giving a local maximum point or a local minimum point, we give the coordinates of the point as an ordered pair. If a local maximum point occurs at $x = -4.5$ and $y = 3$, we state the local maximum point as $(-4.5, 3)$.

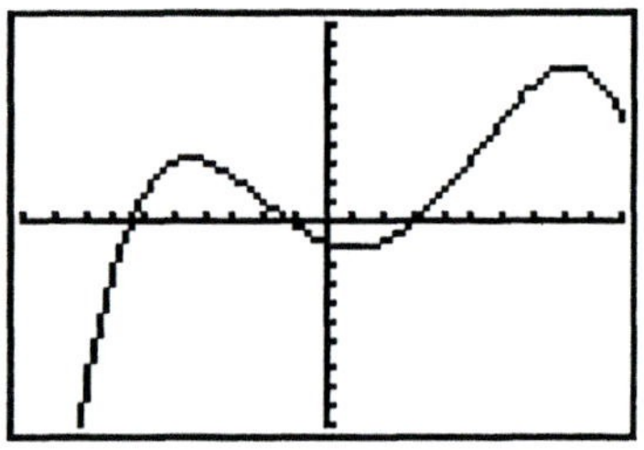

y:[-10, 10] x:[-10, 10}

Figure 2.6

It may be helpful when determining points on a graph to use the trace function on the calculator. With the trace function, the cursor moves along the graph and the x and y coordinates of points on the graph are given at the bottom of the screen. One press of the left or right cursor key moves the cursor one pixel in the horizontal direction. When the trace function is active and if more than one graph is on the screen, one press of the top or bottom cursor key switches the cursor between curves. By moving the cursor to a particular point on the graph, the coordinates of that point are known. Caution is urged in reading the coordinates since the precision of these points depends on the scale of the x- and y-axes.

Procedure G4. To use the trace function and find points on a graph.

1. Select the trace function:
    On the TI-82 or 83 or 84 Plus: On the TI-85 or 86:
    Press the key TRACE   With the graph menu on the screen,
                                             select TRACE.

2. As the right and left arrow keys are pressed, the cursor moves along the graph and, at the same time, the coordinates of the cursor are given on the bottom of the screen.

3. If more than one graph is on the screen, pressing the up or down cursor keys will cause the cursor to jump from one graph to another. The function will appear in the upper right of the screen. On the TI-82, the number of the function will appear at the upper right of the screen.

4. If the graph goes off the top or bottom of the screen, the cursor will continue to give coordinates of points on the graph.

5. If the cursor is moved off the right or left of the screen, the graph scrolls (moves) right or left to keep the cursor on the screen.

Note: If the cursor is moved by using the cursor keys without first pressing the trace key, then the cursor can be moved to any point on the screen. The coordinates of this screen point will be given at the bottom of the screen.

♦ Example 2.5:   *(Washington, Section 3.4, Example 2)* Graph the function $y = 2x^2 - 4$.
Approximate the *x*- and *y*-intercepts and any local maximum or minimum points.
Round answers to one-tenth of a unit.

Solution: 1. Use Procedure G1 to enter and graph the
function $y = 2x^2 - 4$ on the standard viewing
rectangle.

2. Use Procedure G4 to help estimate the
intercepts and maximum point of the graph.
The graph with the cursor at minimum point
is shown in Figure 2.7.

3. Using the trace function, we approximate the
x-intercepts, –1.5 and 1.5,
the *y*-intercept, –4.0, and
the maximum point (0.0, –4.0).

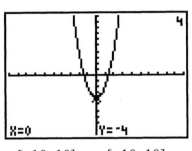

y:[–10, 10] x:[–10, 10]

Figure 2.7

Note: The table function of the graphing calculator may also be useful in
determining important points of a function by observing the y-values as
the x-values increase. Local maximum points occur if y-values increase,
then decrease as x-values increase. Local minimum points occur if y-
values decrease, then increase as x-values increase.

Exercise 2.2

In Exercises 1-16, graph the given function on a graphing calculator using the standard viewing
rectangle. In each case, make a sketch of the graph on your paper and then from the graph, use
the cursor to estimate (a) the *x*-intercepts (zeros of the function), (b) the *y*-intercept, and (c) the
coordinates of any local maximum and local minimum points on the graph. Round answers to
one decimal place.

1. $y = 3.5x - 6.3$ 2. $y = 0.47x + 1.5$

3. $f(x) = -2.4x + 9.7$ 4. $f(x) = -3.5x - 8.9$

5. $y = x^2 - 7$ 6. $y = 9.4 - 0.5x^2$

7. $f(x) = x^2 - 3x + 5$ 8. $f(x) = -2x^2 + 19x - 42$

9. $y = x^3 - 7x$ 10. $y = x^3 - 5x + 3$

11. $y = \dfrac{3}{x}$ 12. $y = \dfrac{4}{x + 4}$

13. $y = \dfrac{2x - 7}{x - 5}$ 14. $y = \dfrac{5}{x^2 - 4}$

15. $y = \sqrt{2 - x}$ 16. $y = \sqrt{18 - x^2}$

17. Graph $y = x - 2$, $y = x - 4$, and $y = x - 6$ on the same axes. What types of graphs are obtained? How do they compare?

18. Graph $y = x + 1$, $y = 2x + 1$ and $y = 4x + 1$ on the same axes. What types of graphs are obtained? How do they compare?

19. Graph several functions of the form $y = mx + b$. To do this replace m and b by several values of your choosing. (b should be between -10 and 10)

    (a) What kind of graph is always obtained?

    (b) What happens for positive values of m as m is increased?

    (c) What happens to the graph for negative values of m as m becomes a larger negative number?

    (d) What happens if $m = 0$?

    (e) What happens as b is change?

    (f) What happens if $b = 0$?

20. Graph several functions of the form $y = |mx + b|$. To do this replace m and b by several values of your choosing. (b should be between -10 and 10)

    (a) What kind of graph is always obtained?

    (b) What happens for positive values of m as m is increased?

    (c) What happens to the graph for various values of b?

    (d) What happens if $m = 0$? If $b = 0$?

◆ 21. *(Washington, Exercises 3.4, #39)* For a certain model of truck, its resale value V (in dollars) as a function of its mileage m is $V = 50000 - 0.34m$. Plot V as a function of m. Use the trace function to estimate the resale value when $m = 20500$ miles and to estimate the number of miles needed for the resale value to be \$25000.

◆ 22. *(Washington, Exercises 3.4, #40)* The resistance R, in ohms, of a resistor as a function of the temperature T, in degrees Celsius, is given by $R = 4.5(1 + 0.082T)$. Plot R as a function of T. Use the trace function to estimate R when $T = 3.65$ degrees and estimate T when $R = 6.55$ ohms.

    There are certain functions that we will call **Basic Functions**. Mental "snapshots" should be taken of these functions. Their shapes, intercepts, symmetry, and relationship to the x- and y-axes should be memorized. In Exercises 23-31 you are asked to graph these Basic Functions. In each case give the x- and y-intercepts, the values of $f(-1)$ and $f(1)$ and take a mental snapshot of the graphs. Estimate all values to one decimal place.

23. Constant function $y = c$. Where c is any constant.

24. Linear function: $y = x$. ($y = x$ is often referred to as the identity function.)

25. Square function: $y = x^2$

26. Square root function: $y = \sqrt{x}$

27. Cube function; $y = x^3$

28. Cube root function: $y = \sqrt[3]{x}$

29. Reciprocal function: $y = \dfrac{1}{x}$

30. Absolute value function: $y = |x|$

31. Greatest integer function: $y = [\, x\,]$
    (On your calculator this is entered as Int(x).)

2.3 Changing Scales: Obtaining a Better Picture.
(Washington, Section 3.4 and 3.5)

Often when graphing a function on a graphing calculator, important sections of the graph are off screen or parts of the graph may be too small to readily tell what the graph looks like. In these cases, it is necessary to adjust the viewing rectangle to allow us to obtain a better view of the graph. In Section 2.2 we used the standard viewing rectangle and in this section we introduce other viewing rectangles.

The **viewing rectangle** is identified by the values for the quantities: Xmin, Xmax, Xscl, Ymin, Ymax, and Yscl. These are defined as:

> Xmin is the *x*-value at the left of the screen.
> Xmax is the *x*-value at the right of the screen.
> Xscl is the scale value on the *x*-axis and indicates the number of units between marks on the *x*-axis.
> Ymin is the *y*-value at the left of the screen.
> Ymax is the *y*-value at the right of the screen.
> Yscl is the scale value on the *y*-axis and indicates the number of units between marks on the *y*-axis.

Procedure G5. **To see or change viewing rectangle.**

1. To see values for the viewing rectangle on the screen.
   On the TI-82 or 83 or 84 Plus: On the TI-85 or 86:
   Press the key WINDOW   With the graph menu on the screen, select RANGE

2. To change viewing rectangle values:
   Make sure the cursor is on the quantity to be changed and enter a new value for that quantity. To keep a value and not change it, either (a) just press the key

ENTER **or** (b) use the cursor keys to move the cursor to a value that is to be changed.

3. To see the graph, after the new values have been entered, select GRAPH
   To return to the home screen, press the key: QUIT

In changing the scales for the axes, Xmin must be smaller than Xmax and Ymin must be smaller than Ymax. If either Xscl or Yscl is 0, no marks will appear on that axis.

Note: The standard viewing rectangle has the following values:

$$Xmin = -10 \qquad Ymin = -10$$
$$Xmax = 10 \qquad Ymax = 10$$
$$Xscl = 1 \qquad Yscl = 1$$

When a function is graphed, all x-values that correspond to points on the graph are in the domain of the function and all y-values that correspond to points on the graph are in the range of the function. The viewing rectangle on the graphing calculator defines a **restricted domain** and a **restricted range** for the function(s) under consideration and generally does not give a view of the whole graph. Since the viewing rectangle does not normally show the whole graph, the question arises as to how much of the graph we should try to view. We will normally try to view that part of the whole graph that contains the important points including the x- and y-intercepts of the graph and all local maximum and local minimum points.

Example 2.6: Graph the function $y = x^3 - 20x + 10$ and adjust the viewing rectangle to obtain a complete graph.

Solution: 1. Use procedure G1 to enter and graph the function $y = x^3 - 20x + 10$ on the standard viewing rectangle.

2. The graph appears in Figure 2.8. This does not give a very good view of the graph. It would be better if the y-scale covered a larger range of values.

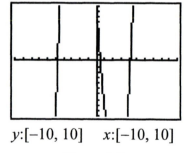

3. Use Procedure G5 to change the viewing rectangle to get a more complete graph on the screen. The trace function can be used to help estimate the minimum and maximum y values. (See Procedure G4).

y:[-10, 10] x:[-10, 10]

Figure 2.8

4. It turns out that a good viewing rectangle has
   x-values from –10 to 10 and y-values from
   –50 to 50 with Yscl = 10. Make these changes.
   The graph in Figure 2.9 gives a better view.

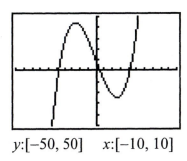

y:[–50, 50] x:[–10, 10]

Figure 2.9

The domain and range may be estimated from the graph
of the function. Each x-value for which there exists a point on the
graph is in the domain of the function and each y-value that
corresponds to a point on the graph is in the range of the function.
Although it must be kept in mind that there are points on this
graph of this function that do not appear in this viewing rectangle.

Caution needs to be exercised when stating which values
are in the domain or range of a function. On the standard viewing rectangle, the graph of the
function

$f(x) = \dfrac{x^2 - 1}{x - 1}$ is an apparent straight line. However, we cannot evaluate this function at $x = 1$,

since division by zero is not allowed. Thus, $x = 1$ is not in the domain and a point cannot exist
on the graph of the function at $x = 1$. If the function is graphed using the decimal viewing
rectangle (see Procedure G6), a hole appears in the graph at $x = 1$. This function is said to be
discontinuous at $x = 1$. The graph of a function does not exist at points where the function is
undefined (such as where division by zero occurs) or where the value of the function is not a real
number. At such points the graph is said to be **discontinuous**.

It is also important to be able to determine the intervals on which a function is increasing
or decreasing. If, on an interval of a graph, the y-values always get larger as x-values increase,
then the function is said to be **increasing** on that interval. If on an interval, the y-values always
become smaller as the x-values increase, then the function is said to be **decreasing** on that
interval. In an intuitive sense, this means that if the graph goes uphill as we look across the
graph from left to right, the function is increasing. If the graph goes downhill, the graph is
decreasing.

On occasion it is helpful to return to a standard viewing rectangle or to obtain other
special viewing rectangles. Since the screen is a rectangle (width = 1.5 times height) and not a
square, the length of one unit on the x-axis and the length of one unit on the y-axis are not the
same when the minimum and maximum values of x and y are the same. If a circle were graphed
using the standard viewing rectangle, it would not appear as a circle but as an ellipse. To remedy
this, we can use the **square viewing rectangle** on which the x- and y-axes have the same scale
distance for each unit. There are also other special viewing rectangles that are useful in different
situations.

Procedure G6: To obtain preset viewing rectangles.

1. Bring the zoom menu to the screen.
   On the TI-82 or 83 or 84 Plus: On the TI-85 or 86:
   Press the key ZOOM   With the graph menu on the screen,
                                           select ZOOM

2. Select the desired preset viewing rectangle. (The left-hand column gives the selection from the ZOOM menu for the TI-82 or 83 or 84 Plus and the right hand column for the TI-85 or 86.)

(a) For the **standard viewing rectangle**:
     Xmin = –10, Xmax = 10, Xscl = 1,
     Ymin = –10, Ymax = 10, Yscl = 1
     Select: 6:ZStandard   ZSTD

(b) For the **square viewing rectangle**:
     The square viewing rectangle keeps the y-scale the same and adjusts the x-scale so that one unit on x-axis equals one unit on y-axis. For a square viewing rectangle the ratio of y-axis to x-axis is about 2:3.
     Select: 5:ZSquare   ZSQR (after pressing MORE)

(c) For the **decimal viewing rectangle**:
     The decimal viewing rectangle makes each movement of the cursor (one pixel) equivalent to one-tenth of a unit
     Select: 4:ZDecimal   ZDECM (after pressing MORE)

(d) For the **integer-viewing rectangle**:
     The integer-viewing rectangle makes each movement of the cursor (one pixel) equivalent to one unit. After selecting the integer-viewing rectangle, move the cursor to the point that is to be located at the center of the screen.
     Select: 8:ZInteger   ZINT (after pressing MORE)
     Then press ENTER.

     Note: To have one movement of the cursor in the x-direction differ by k units set:

     Xmin = –94k/2   Xmin = –126k/2
     Xmax = 94k/2   Xmax = 126k/2

(e) For the **trigonometric viewing rectangle**:
     The trigonometric viewing rectangle sets the x-axis up in terms of π or degrees (depending on mode) and sets Xscl = $\dfrac{\pi}{2}$ (or 90°).
     Select: 7:ZTrig   ZTRIG

In Example 2.6, we did not see enough of the graph using the standard viewing rectangle. It is also sometimes necessary to enlarge a portion of the screen to better view the graph.

♦ <u>Example 2.7</u>: *(Washington, Section 3.4, Example 3)* Graph the function $y = x - x^2$, then from the graph
(a) estimate the x-intercepts of the graph to one decimal place,
(b) determine the range and domain of this function, and
(c) determine the intervals on which the function is increasing and the intervals on which the function is decreasing.

Solution: 1. Graph the function $y = x - x^2$ on the standard
                 viewing rectangle. The graph is in Figure 2.10.
                 From this graph it is difficult to estimate where
                 the graph crosses the x-axis.

              2. Change the scale to obtain a better view of the
                 graph. Use Procedure G5 to set the scale
                 values of Xmin $= -1$, Xmax $= 3$, Xscl $= 1$,
                 Ymin $= -1$, Ymax $= 1$,
                 Yscl $= 1$. This will give a restricted domain of
                 -1 to 3 and a restricted range of -1 to 1.

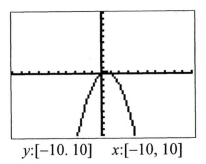

y:[−10. 10] x:[−10, 10]

Figure 2.10

              3. Graph the function again. This graph appears
                 in Figure 2.11.
                 We now answer the questions.

                 (a) We estimate the x-intercepts as being about
                     0.0 and 1.0. The trace function may be
                     used as an aid to determine these values.

                 (b) From the graph of the function, it appears
                     that there are no places where the graph is
                     not continuous and each x-value
                     corresponds to a point on the graph. We
                     conclude that the domain of this function is
                     all real numbers. The highest point on the
                     graph occurs at $y = 0.25$ and all y-values

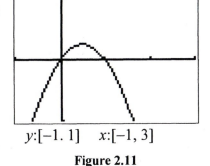

y:[−1. 1] x:[−1, 3]

Figure 2.11

                     below 0.25 appear to correspond to points on the graph. We conclude that
                     the range is all real numbers y such that $y \le 0.25$.

                 (c) From the graph we determine that the function decreases for $x > 0.5$ and
                     increases for $x < 0.5$.

     A type of function that we graph somewhat differently is a piecewise function.
Piecewise functions are functions that are expressed in terms of two or more functions, each
defined on specified intervals of the domain. To graph these functions on the graphing calculator
it is necessary to place the specified interval in parentheses following the given function. The
interval is identified using the relation symbols: $=, \ne, >, \ge, <$, and $\le$.

Procedure C13. **To obtain relation symbols ($=, \ne, >, \ge, <$, and $\le$).**

     1. Press the key: TEST
     2. Select the desired symbol from the menu.

Procedure G7. To graph functions on an interval.

1. To graph a function on the interval $x < a$ or on the interval $x \le a$ for some constant a:

   In the function table after the equal sign, enter the function $f(x)$ followed by either $(x<a)$ or $(x \le a)$.

   (Example: To graph $y = x^2$ on the interval $x<2$, enter for the function: $x^2(x<2)$)

2. To graph a function on the interval $a < x < b$ or on the interval $a \le x \le b$ for some constants a and b:

   In the function table after the equal sign, enter the function $f(x)$ followed by either $(x>a)(x<b)$ or $(x \ge a)(x \le b)$.

   (Example: To graph $y = x^2$ on the interval $-3 \le x \le 2$, enter for the function: $x^2(x \ge -3)(x \le 2)$)

3. To graph a function on the interval $x > a$ or on the interval $x \ge a$ for some constant a:

   In the function table after the equal sign, enter the function $f(x)$ followed by either $(x>a)$ or $(x \ge a)$.

   (Example: To graph $y = x - 5$ on the interval $x>2$, enter for the function: $(x-5)(x>2)$)

   Note: The above forms may be combined by writing their sum.
   Example: $(x+5)(x<2) + (7-x^2)(x \ge 2)$
   Where the function contains more than one term, enclose the function in parentheses.

When considering functional values or the graph of a piecewise function, it is important to watch the intervals on which the function is defined. The graphs of each part of the function only exist on the interval on which that part is defined. **Care should also be taken when graphing with the calculator in connected mode since lines may be drawn connecting end points that are not part of the actual graph.**

◆ <u>Example 2.8</u>: *(Washington, Section 3.4, Example 7)* For the function, $f(x)$, (a) graph the function, (b) give the x- and y-intercepts and (c) give the domain and range.

$$f(x) = \begin{cases} 2x+1 & \text{if } x \le 1 \\ 6-x^2 & \text{if } 1 < x \end{cases}$$

Solution: 1. Writing the $f(x)$ this way indicates that $y = 2x + 1$ for values in the domain that are less than or equal to 1 and $y = 6 - x^2$ for domain values that are greater than 1.

2. Use Procedure G7 to graph this function. Enter the function as:
$$(2x+1)(x \leq 1) + (6 - x^2)(x > 1)$$
On the standard viewing rectangle the graph appears as in Figure 2.12.

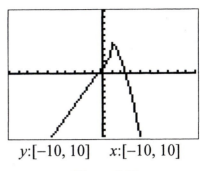

$y{:}[-10, 10] \quad x{:}[-10, 10]$

Figure 2.12

3. Answering the questions:

   (b) There are two x-intercepts, $x = -0.5$ and $x = 2.5$. The y-intercept is 1.0. When using the trace function make sure that the cursor is on the correct curve. Notice that the vertical line connecting the curves at $x = 1$ is not part of the graph. If we were to look at a table of values for this function, we would notice that at $x = 1$, $y = 3$, while at $x = 1.1$, $y = 4.79$. The graph jumps up for x-values just slightly greater than 1 since the function now is $6 - x^2$.

   (c) The graph exists for all x-values and the domain is all real numbers. From the graph we see that only y-values less than 5 correspond to points on the graph. The range is all real numbers y such that $y < 5$.

When graphing discontinuous functions, that is functions which have breaks in the graph, lines sometimes appear on the graph that are not really part of the graph. In Example 2.8 we note that a line appears to go from the point (1,3) to the x-axis and from the point (1,5) to the x-axis. These lines are not really part of the graph. Sometimes it can be helpful with discontinuous functions to graph in dot mode. When the calculator is in **dot mode** individual points of the graph are plotted where each point corresponds to each pixel on the x-axis and the points are not connected.

Procedure G8. Changing graphing modes.

1. To change modes:

   On the TI-82 or 83 or 84 Plus: On the TI-85 or 86:

   Press the key MODE   With the graph menu on the screen, select FORMT (after pressing MORE)

   Then, use the cursor keys to highlight the desired option.

2. To draw graphs connected or unconnected (with dots)

   Highlight Connected to connect plotted points with lines.   Highlight DrawLine to connect plotted points with lines.

   Highlight Dot to just plot individual points.   Highlight DrawDot to just plot individual points.

3. To draw graphs simultaneously or sequentially

   Highlight Sequential to graph each function in sequence.   Highlight SeqG to graph each function in sequence.

   Highlight Simul to graph all selected functions at the same time.   Highlight SimulG to graph all selected functions at the same time..

4. Highlight the desired option in each line then press ENTER.
5. To return to the home screen, press CLEAR or QUIT.

Exercise 2.3

On all exercises, obtain a graph showing all important points and draw a sketch of the graph on your paper indicating values for Xmin, Xmax, Ymin, and Ymax at the ends of the axes.

In Exercises 1-20, from the graph estimate (a) the x-intercepts, (b) the y-intercepts, (c) the coordinates of the local maximum and local minimum points, (d) give the domain and range of each function, and (e) give the values of x for which the function is increasing and for which it is decreasing. Estimate answers to two significant digits - do not try to obtain a high degree of precision.

1. $y = x^2 - 5x - 6$

2. $y = 11 - 12x^2 - x^3$

3. $y = 252 - x^2$

4. $y = x^3 - 1200$

5. $y = x^3 - 66x$

6. $y = x^3 - 5x^2 + 6.25x$

7. $y = \dfrac{4x - 3}{3 - x}$

8. $y = \dfrac{x^2 - 5}{3x - 7}$

9. $y = \dfrac{x^2 - 16}{4 - x}$

10. $y = \dfrac{2x^2 + 5x - 3}{x + 3}$

11. $y = \sqrt{15 - x^2}$

12. $y = -\sqrt{20 - 14x - x^2}$

13. $y = \sqrt{95x^3 - x^2 + 9x - 2}$

14. $y = 4x^3 - 44x^2 + 100x - 64$

15. $y = x + \dfrac{3}{x}$

16. $y = x^2 + \dfrac{3}{x^2}$

17. $y = |2.4x - 7.5|$

18. $y = |x^2 - 3x - 2|$

19. $f(x) = \begin{cases} 2 + x \ if \ x \leq 2 \\ (x - 2)^2 \ if \ x > 2 \end{cases}$

20. $g(x) = \begin{cases} \dfrac{1}{x - 1} \ if \ x < 1 \\ \sqrt{x - 1} \ if \ x \geq 1 \end{cases}$

Do exercises 21-24 as indicated. Estimate values to two significant digits.

♦ 21. *(Washington, Exercises 3.4, #38)* The consumption c of fuel, in liters per hour (L/h), of a certain engine is determined as a function of the number r, in revolutions per minute (r/min) of the engine to be $c = 0.0132r + 3.85$. This formula is valid from 500 r/min to 2000 r/min. Plot c as a function of r. Use the trace function to estimate the consumption when $r = 1100$ r/min and to estimate the revolutions per minute when $c = 26$ L/h.

♦ 22. *(Washington, Exercises 3.4, #47)* A formula used to determine the number N of board feet of lumber that can be cut from a 4-ft section of a log of diameter d (in inches) is given by $N = 0.22d^2 - 0.71d$.
   (a) Graph N as a function of d for d from 5 in. to 50 in.
   (b) Use the trace function to estimate the number of board feet of lumber when $d = 20$ in. and when $d = 40$ in.
   (c) A logging company only wants trees that will produce at least 100 board feet of lumber.   What is the smallest diameter tree the company will use?

♦ 23. *(Washington, Exercises 3.4, #43)* The maximum speed v, in miles per hour, at which a car can safely travel around a circular turn of radius r, in feet, is given by $r = 0.42\,v^2$. Graph r as a function of v. Allow for speed up to 75 miles per hour.
   (a) What needs to be the radius for a car traveling 38 miles per hour? For a car traveling 75 miles per hour?
   (b) At what speed can a car travel around a curve of radius 1075 feet? Around a curve of radius 249 feet?

♦ 24. *(Washington, Exercises 3.4, #44)* The height h, in meters, of a rocket as a function of the time t, in seconds, is given by the function $h = 1449t - 4.9t^2$. Graph h as a function of t. Assume level terrain.
   (a) What is the greatest height obtained by the rocket?
   (b) For what values of time is the rocket going up (the height increasing)?
   (c) After what period of time does the rocket hit the ground?

In Exercises 25-30, graph each of the basic functions, then give (a) the domain and range of each function, (b) the intervals on which each function is increasing or decreasing, and (c) the y-values for $x = -1, 0$, and 1.

25. $y = x$ 26. $y = x^2$

27. $y = \sqrt{x}$ 28. $y = \dfrac{1}{x}$

29. $y = x^3$ 30. $y = \sqrt[3]{x}$

31. $y = |x|$ 32. $y = [\, x \,]$

In Exercises 33 and 34, first graph the functions on the same standard viewing rectangle, then on the same square viewing rectangle. Connect the graph (where points are missing). (a) What type of graph is obtained by combining the two functions? (b) What are the x- and y-intercepts? (c) What are the domains and ranges of each of the given functions?

33. $y_1 = \sqrt{25 - x^2}$ and $y_2 = -\sqrt{25 - x^2}$

34. $y_1 = \sqrt{50 - x^2}$ and $y_2 = -\sqrt{50 - x^2}$

2.4 Zoom: Finding Special Points / Solving Equations
(Washington, Section 3.5)

In Section 2.3, we changed the viewing rectangle to obtain a better view of the graph. In this section, we introduce an efficient method of changing the viewing rectangle by making use of the zoom function of the graphing calculator. The **zoom function** allows us to zoom in or to zoom out on the graph in the region of interest.

Procedure G9. To zoom in or out on a section of a graph.

1. With a graph on the screen,

   On the TI-82 or 83 or 84 Plus: On the TI-85or 86:
   Press the key ZOOM   With the graph and graph menu on the screen, select ZOOM

2. From the menu:

   To zoom in, select 2:Zoom In   To zoom in, select ZIN
   or   **or**
   To zoom out, select 3:Zoom Out   To zoom out, select ZOUT

3. The cursor keys may be used to move the cursor near the point of interest. After zooming, the point at the location of the cursor will be near the center of the screen.

4. Press ENTER

5. Repeated zooms may be made (if no other key is pressed in the meantime) by moving the cursor near the point of interest and pressing ENTER.

   Note: It is a good idea to use the TRACE function together with ZOOM.

The magnification determined of the zoom operation is determined by the **zoom factors**. These factors are preset on new calculators but may be changed to allow a greater magnification. If the zoom factors are set too large, it is possible that the portion of the graph that is of interest may not appear and only a blank screen will be obtained.

Procedure G10. To change zoom factors

1. Obtain the zoom menu:
   On the TI-82 or 83 or 84 Plus: On the TI-85 or 86:
   Press the key ZOOM   With the graph menu on the screen,
                                          select ZOOM

2. Then from the menu:
   Highlight MEMORY in the top row   Select ZFACT
   and select 4:SetFactors.   (after pressing MORE twice).

3. Change the factors as desired.
   There are factors for magnification in both the x- and y-directions. The preset
   factors are 4 in both directions. Make changes to XFact for the magnification
   in the x-direction and to YFact for magnification in the y-direction.
   Caution: Do not set factors too large.

4. To exit this screen press the key: QUIT

The zoom function may be used to zoom in and find the coordinates of the point to any
desired accuracy (limited only by the capability of the calculator). Example 2.9 uses the zoom
function to find the x-intercepts.

Example 2.9:   Graph the function $y = x^2 - 3x - 5$ and locate the x-intercepts to two decimal
places.

Solution: 1. Start by graphing the function, $y = x^2 - 3x - 5$ in the standard viewing
                  rectangle. The graph is a parabola that crosses the x-axis near -1 and near $+4$.
                  We will first find the x-intercept near -1.

               2. Use Procedure G10 to check the zoom
                  factors. Set both zoom factors to 4.

               3. Return to the graph and use the trace
                  function to move the cursor to a point on the
                  graph near the negative x-intercept. On the
                  bottom of the screen are the x- and y-values
                  for the location of the cursor on the graph.
                  With the cursor near $x = -1$, the screen
                  appears as in Figure 2.13. (Your x- and y-
                  values may differ.)

               4. Use Procedure G9 to zoom in.

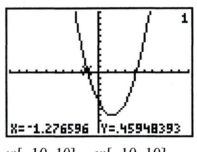

X= -1.276596   Y=.45948393

y:[$-10, 10$] x:[$-10, 10$]

Figure 2.13

5. Use the trace function and move the cursor near where the graph crosses the *x*-axis. The graph is given in Figure 2.14. At one point the cursor will be below the *x*-axis and at the next point to the left will be above the *x*-axis. This can also be determined by noting that for one point the *y*-value is negative and for another point the *y*-value is positive. The *x*-intercept must be between the two points. In this case, this means the solution must be between approximately $x = -1.22$ and $x = -1.17$.

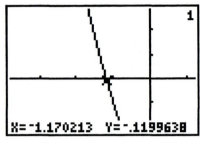

y:[−2.18, 2.82] x:[−3.78, 1.22]

Figure 2.14

6. Move the cursor to a point near the intersection point of the graph and the *x*-axis (where $x = -1.27$) and zoom in again (Procedure G9).

7. Repeat steps 5 and 6 until the *x*-values immediately on either side of the intersection point are the same rounded off to two decimal places. The *x*-intercept is −1.19.

8. To determine the other *x*-intercept, first return to the standard viewing rectangle and obtain the original graph.

9. Move the cursor to the positive *x*-intercept and repeat the procedure until the zero near $x = 4$ can be found to two decimal places. The *x*-intercept is 4.19.

The *x*-intercepts of $y = x^2 - 3x - 5$ are −1.19 and 4.19.

The location of the *x*-intercept in Example 2.9 involved repeating the three steps:

1. Move the cursor near the desired point.
2. Zoom in on the point.
3. Use the TRACE function to help determine the intercept.

Note: Continue these steps until the desired degree of precision is obtained.

In Example 2.9, we zoomed in on a section of the graph. In some cases, it may be desirable to zoom out in order to see more of the graph. To do so, use Procedure G9 to zoom out on a section of the graph and repeat as necessary to obtain the desired results

An approximate solution to an equation can be found relatively quickly and to a high precision by using a graphing calculator. In application problems, it is often not necessary to obtain an exact answer, but to find an answer to a desired precision. To solve an equation of the form $f(x) = 0$, where $f(x)$ represents a function of *x*, first let $y = f(x)$ and graph the function. The solutions to the equation $f(x) = 0$ are the *x*-intercepts of the graph. Solutions to an equation are often referred to as **roots** of the equation.

It is important to note that the solutions (or roots) to the equation $f(x) = 0$, the zeros of the function $f(x)$, and the *x*-intercepts of the graph of $y = f(x)$ are all the same numerical values.

Procedure G11. **Solving an equation in one variable.**

1. Write the equation so that it is in the form $f(x) = 0$, let $y = f(x)$ and graph the function. Use a viewing rectangle so that the x-intercept of interest appears on the screen.

2. Determine a solution by:

   (a) Using Procedure G9 to zoom in on the x-intercept.

   or

   (b) Using the built in procedure:

   On the TI-82 or 83 or 84 Plus:

   (1) Press the key CALC

   (2) From the menu, select 2:zero (2:root on the TI-82)

      The words Left Bound? (or Lower Bound?) appear.

   (3) Move the cursor near, but to the left of the intercept and press ENTER. The words Right Bound? (or Upper Bound?) appear.

   (4) Move the cursor near, but to the right of the intercept and press ENTER.
      The word Guess? appears.

      Move the cursor near the desired point. (It is usually sufficient to leave the cursor at its last position.)

   (5) Press ENTER again. The root (x-intercept) will appear at the bottom of the screen.

   On the TI-85 or 86:

   (1) From the graph menu, select MATH (after pressing MORE).

   (2) From the secondary menu, select ROOT.

   (3) On the TI-86 the words LeftBound? Appear. Follow steps (3), (4), and (5) for The TI-82 and 83.

   On the TI-85, move the cursor near the intercept and press ENTER. The root (x-intercept) will appear at the bottom of the screen.

♦ <u>Example 2.10</u>: *(Washington, Section 3.5, Example 3)* Solve the equation $x^2 - 2x = 1$. Find the answers to five decimal places.

Solution:

1. First place the equation in the form $f(x) = 0$ by subtracting 1 from both sides. This gives $x^2 - 2x - 1 = 0$. Let $y = x^2 - 2x - 1$.

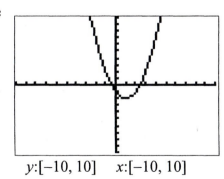

y:[−10, 10] x:[−10, 10]

Figure 2.15

2. Graph the function $y = x^2 - 2x - 1$ in the standard viewing rectangle. Figure 2.15 shows the graph. From the screen, it appears this graph crosses the x-axis just to the left of zero and between 2 and 3.

3. Use Procedure G11 to determine the negative root between −1 and 0, then determine the root between 2 and 3. The roots are found to be $x = -0.41421$ and $x = 2.41421$.

We may use a similar method to find the coordinates of local maximum and local minimum points to a high degree of precision.

Procedure G12. Finding maximum and minimum points.

1. Graph the function and adjust the viewing rectangle so that the desired local maximum or local minimum point is on the screen.
2. Determine the local maximum or local minimum point by:
    (a) Using Procedure G9 to zoom in on the point
    or
    (b) Use the built-in procedure:

On the TI-82 or 83 or 83 Plus:

(1) Press the key CALC
(2) From the menu, select 3:minimum or 4:maximum. The words Left Bound? (or Lower Bound?) appear.
(3) Move the cursor near, but to the left of the desired point and press ENTER. The words Right Bound? (or Upper Bound?) appear.
(4) Move the cursor near, but to the right of the desired point and press ENTER. The word Guess? appears.

Move the cursor near the desired point. (It is usually sufficient to leave the cursor at the last position.)

(5) Press ENTER again. The coordinates of the point will appear at the bottom of the screen.

On the TI-85 or 86:

(1) Adjust the viewing rectangle so that the point of interest is the highest point or lowest point on the screen.
(2) From the graph menu, select MATH (after pressing MORE).
(3) From the secondary menu, select FMIN for a local minimum or FMAX for a local maximum point (after pressing MORE).
(4) On the TI-86 the words LeftBound? appear. Follow steps (3), (4), and (5) for the TI-82 and 83.

On the TI-85 move the cursor near the desired point and press ENTER. The coordinates of the point will appear at the bottom of the screen.

Note: On the TI-85 LOWER and UPPER may be used to select lower and upper bounds on which the maximum and minimum values are to be found.

Example 2.11: Find the local maximum and local minimum points on the graph of $y = 3x^2 - 7x - 1$ to four decimal places, give the x-values for which the function is increasing, and give the domain and range of the function.

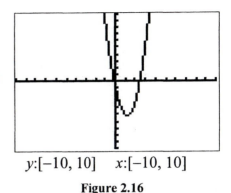

y:[−10, 10] x:[−10, 10]

Figure 2.16

Solution:

1. Graph the function $y = 3x^2 - 7x - 1$ on the standard viewing rectangle. The graph is shown in Figure 2.16. From the graph we see that the function gives a parabola with a local minimum point.

2. Use Procedure G12 to determine the coordinates of the local minimum point. The minimum point is located at (1.1667, −5.0833).

3. From the graph we see that the graph is increasing to the right of the minimum point. Thus, the function increases for $x > 1.1667$.

4. From the graph and the function, we see that every x-value corresponds to a point on the graph. Thus, the domain is all real numbers. Since the graph has a minimum point, and there are no points on the graph below this point, the range must be the y-values corresponding to this point or points above this point. Thus, the range is $y \geq -5.0833$.

On occasion it is helpful to be able to evaluate a function to a high precision at a given value of x.

Procedure G13. Find the value of a function at a given value of x.

1. Graph the function and adjust the viewing rectangle so that the given x-value is within the restricted domain of the screen.

2. Determine the value of the function by:

   (a) Using Procedure G9 to zoom in on the point

   or

   (b) Use the built-in procedure:

   On the TI-82 or 83 or 84 Plus:
   (1) Press the key CALC
   (2) From the menu, select 1:value. $X=$ appears on the screen.
   (3) Enter the given x-value and press ENTER. The value of the function will appear at the bottom of the screen.

   On the TI-85 or 86:
   (1) From the graph menu, select EVAL (after pressing MORE twice).
   (2) The words Eval $x =$ appear on the screen.
   (3) Enter the x-value and press ENTER. The value of the function will appear at the bottom of the screen.

Another method that can be used to zoom in on a portion of a graph is to create a box containing the area to be enlarged and have the calculator redraw the region inside the box to fill the screen. This method can be used in combination with other zooming methods.

Procedure G14. To zoom in using a box.

1. Obtain the zoom menu:
   On the TI-82 or 83 or 84 Plus: On the TI-85 or 86:
   Press the key ZOOM   With the graph menu on the screen, select ZOOM

2. Then
   Select 1:Zbox   From the secondary menu
                                           select BOX

3. Use the cursor keys to move the cursor to a location where one corner of the box is to be placed and press ENTER.

4. Use the cursor keys to move the cursor to the location for the opposite corner of the box and press ENTER. As the cursor keys are moved a box will be drawn and when ENTER is pressed, the area in the box is enlarged to fill the screen.

Exercise 2.4

Graph each of the functions in exercises 1-8 on your calculator. Find (a) the x-intercepts, (b) the coordinates of local maximum and local minimum points, and (c) the domain and range of each function. Find answers to three significant digits.

1. $y = x^2 - x - 24$ 2. $y = x^3 + 8x$

3. $y = 7 + 6x + 3x^2 - 8x^3$ 4. $y = 4 - x^3$

5. $y = \dfrac{3x - 5}{x - 4}$ 6. $y = \dfrac{3x + 1}{5 - 2x}$

7. $y = \sqrt{16 - x^2}$ 8. $y = \sqrt{x^2 - 4}$

In Exercises 9-16 solve each of the equations by graphing on the graphing calculator. Evaluate solutions to three significant digits.

9. $7x - 6 = 0$ 10. $x^2 - 37 = 0$

11. $x^2 - 5x = 5$ 12. $x^3 - 15x = 0$

13. $5x - 3x^2 + x^3 = 9$ 14. $x^3 + 2x^2 = 6$

15. $\dfrac{8}{x^2+6}=0$ 16. $2\sqrt{x}+5=2x$

Do Exercises 17-22 as indicated.

17. A small business depreciates some equipment according to the equation $V=15500-3200t$ were V is the value in dollars after t years. According to this formula, when will the equipment be fully depreciated (have a value of zero)? Determine the value when $t=2.5$ years and when $t=3.25$ years.

18. The height, h, (in feet) of a baseball thrown upward from a cliff is a function of time, t, (in seconds) according to the equation $h=150+70t-16t^2$. At what time will the height be zero? At what time will the ball be at its highest point? For what times will the ball be descending? How high will the ball be when $t=2.5$ s and when $t=4.75$ s.?

♦ 19. *(Washington, Exercises 3.5, #37)* In an electric circuit, the current i, in amperes, as a function of the voltage v is given by $i=0.0127v-0.0558$. Find v when $i=0$. For what values of v is the current positive? Determine the current when $v=10.0$ and $v=7.5$?

♦ 20. *(Washington, Exercises 3.5, #40)* The height (in ft.) of a rocket as a function of time t (in s) of flight is given by $h=125+245t-16.0t^2$. Determine the maximum height of the rocket and the value of t when this occurs and then determine the value of t when the rocket hits the ground.

♦ 21. *(Washington, Exercises 3.5, #44)* A rectangular bin is to be made from a sheet of sheet metal, which measures 14.0 inches by 10.0 inches, by cutting out equal squares of side x from each corner and bending up the sides. Find the value of x that gives the maximum volume for the box. (Hint: Find a function for the volume of the box, then find a local maximum point on the graph.)

♦ 22. *(Washington, Exercises 3.5, #44)* For the box of problem 17 determine the value of x that will give a volume of 95 cubic inches.

2.5 Introduction to Programming the Graphing Calculator

The graphing calculator has the ability to be programmed in much the same way that one would program a computer to perform a specific task. A **program** is a set of individual instructions to tell the calculator what tasks are to be performed. In a program, each step must be explicitly stated. The symbols and words used to state these instructions form what is called a programming language. Each step of the program is called a statement and is a single instruction for the program.

The language used for programming the graphing calculator is different from other languages used for programming computers with which the reader may be familiar. To program the graphing calculator, we make use of the various keys and other instructions that are accessed by selection of menu items on the calculator. Essentially any calculation or instruction that may be done directly may also be programmed into the calculator. The best way to see how to enter and execute (run) a program is to follow the steps in Example 2.12. First we list the basic procedures needed for programming. To enter a new program or to change an existing program, it is first necessary to enter programming mode as described in Step 2 of Procedure P1. **Programming mode** allows the instructions of the program to be entered or modified.

Procedure P1. To enter a new program.

1. Press the key: PRGM
2. Enter programming mode:

   On the TI-82 or 83 or 84 Plus: On the TI-85 or 86:

   Highlight NEW in the top row and   Select EDIT.
   select 1:Create New

3. After Name=, enter a name for the program. The name may be up to eight characters long. (Notice that the calculator is ready for alpha characters.) Then, press ENTER.
4. The word PROGRAM appears on the top row followed by the name of the program and the cursor is on the second row preceded by a colon (:). The calculator is now in programming mode and is ready for the first statement in the program.
5. Enter the steps of the program one line at a time, pressing ENTER after each step. (On some calculators, more than one instruction may be placed on the same line if the instructions are separated by a colon (:). Care needs to be exercised in doing this.)
6. When finished entering the program, press QUIT to return to the home screen.

Once a program has been entered it may be executed, edited or erased. To **execute** a program means that each of the instructions in the program is carried out in order, one at a time. To **edit** a program means to make changes in the statements of the program.

Procedure P2. To execute, edit, or erase a program.

1. Press the key: PRGM
2. To **execute** a program:

   On the TI-82 or 83 or 84 Plus: On the TI-85 or 86:

   With EXEC highlighted in the top   Select NAMES
   row, select the name of the program
   to be executed.   From the secondary menu, select the
   Then, press ENTER.   name of the program.
                                         Then, press ENTER.

   A program may be executed additional times by pressing ENTER if no other keys have been pressed in the meantime.
3. To **edit** a program:

On the TI-82 or 83 or 84 Plus:

Highlight EDIT in the top row and select the name of the program to be edited.

On the TI-85 or 86:

Select EDIT.

From the secondary menu, select the name of the program to be edited.

Make changes in the program as desired. The cursor keys may be used to move the cursor to any place in the program. The calculator is now in programming mode.

4. To **erase** a program:

On the TI-82 or 83 or 84 Plus:

Press the key MEM.

On the TI-82 or 83:

  Select 2:Delete...

  Select 7:Prgm (or 6:prgm on the 82)

On the TI-83 Plus or 84 Plus

  Select 2:Mem Mgmt/Del...

  Then select 7:Prgm

On the TI-85 or 86:

Press the key MEM

Select DELET

Select PRGM

(after pressing MORE)

Move the marker on the left until it is next to the name of the program to be erased and press DEL. (On the TI-82 or TI-83, press ENTER.) The program will be erased from memory. (Caution: Before erasing a program, make sure that you really want to erase it.)

As mentioned previously, a program consists of a series of explicit steps that tell the calculator what operations to perform. Quite often it is desirable to enter a value from the keyboard for a variable. This is referred to as "inputting" a value. The value entered is then stored for the specified variable. When the variable is later used in a calculation, the value, which has been stored, replaces the variable in the calculation.

Procedure P3. To input a value for a variable.

Entering a value for a variable uses the keywords: Input or Prompt.
The form of the command is: **Input** *v* or **Prompt** *v*
(*v* represents the variable name)

With the calculator in programming mode:

1. To access the word Input:

On the TI-82 or 83 or 84 Plus:

Press the key PRGM

Highlight I/O in the top row.

Select 1:Input **or** 2:Prompt

On the TI-85 or 86:

From the menu, select I/O

From the secondary menu,
Select Input **or** Promp

2. Enter the variable name and press ENTER.
(Example: Input *C* **or** Prompt *C*)

When the program is executed the word Input followed by a variable will cause a question mark to appear on the screen while the word Prompt will cause the name of the variable followed by the question mark to appear. Any number entered is stored for that variable and when the variable is later used in the program that value is used in the calculation.

Once a value for a variable has been entered, it is likely that it will be used to evaluate an expression. This is accomplished by entering a statement, which gives the expression and stores the result to another variable.

Procedure P4. To evaluate an expression and store that value to a variable.

With the calculator in programming mode:

On the TI-82 or 83 or 84 Plus:

Enter the expression, press the key STO▷, and then, enter a variable name.

On the TI-85 or 86:

Enter the expression, press the key STO▷ and then, enter a variable name.

or

Enter a variable name, press the key =, and then enter the expression.

The expression on the left will be evaluated at values that have been stored for the variables used on the left and then, stored for the variable name on the right.   (Example: $(9/5)C + 32 \rightarrow F$)

After an expression has been evaluated and the results stored for a variable, the results need to be displayed. It is also possible to display words or string expressions on the screen. A **string expression** is a set of characters enclosed in quotes.

Procedure P5. To display quantities from a program.

Either the value for a variable or an alpha expression may be displayed by using the keyword Disp.

The form of the command is: **Disp v**   (v represents the variable name)

The quantity v may be a variable name, a number, or a set of characters enclosed inside of quotes. If v is a quantity enclosed in quotes, the quantity will be printed exactly as it appears.

With the calculator in programming mode:
1. To access the word Disp:

On the TI-82 or 83 or 84 Plus:

Press the key PRGM
Highlight I/O in the top row.
Select 3:Disp.

On the TI-85 or 86:

From the menu, select I/O
From the secondary menu, select Disp

2. Enter the variable name and press ENTER.
   (Examples: Disp *F* or Disp "ANSWER IS")

When the program is executed, the value stored for the variable or the expression will be displayed on the screen.

Note: Quotes (") are located on the keyboard of the TI-82, 83 and 84 Plus.
        On the TI-85 and 86, they are located under the I/O menu when in
        programming mode.

Now, we illustrate how a program is entered and executed. When programming, it a good idea to enter the basic steps needed, then, when we are certain the program works, go back and make improvements.

Example 2.12: Write a program to calculate the temperature in degrees Fahrenheit when a temperature in degrees Celsius is entered. Use the formula
$F = (9/5)C + 32$. Then, evaluate *F* if *C* = 100 and if *C* = −40.

Solution: 1. Obtain the correct mode for entering a new program by using Procedure P1.

2. For the program name enter: CTOF

3. Use Procedure P3 to enter on the second line a statement to input a value for *C*. (Input *C*)

4. On the next line, use Procedure P4 to enter the expression $\dfrac{9}{5}C + 32$ and store the resulting value for *F*. ($(9/5)C + 32 \rightarrow F$)

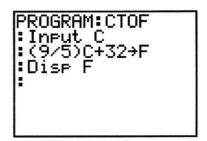

Figure 2.17

5. On the next line use Procedure P5 to display the value for *F* (Disp *F*).
This constitutes a brief, but complete program. The program appears in Figure 2.17.

6. Exit from the programming mode, then execute this program (see Procedure P2). The input command will cause a question mark to appear on the screen. The program is requesting a numerical value to be entered for the variable *C*. Enter the number 100 and press ENTER. The corresponding value for *F* should appear at the right of the screen. The result is 212.

7. The program may be executed several more times by pressing the key ENTER if no other keys are pressed in the meantime. If another key has been pressed, use Procedure P2 again. Execute the program a second time and enter −40. The result should be −40. This is the temperature where Celsius and Fahrenheit temperatures are the same.

If errors occur when entering or running a program or if it is desired to change the program, the lines of the program may be edited by moving the cursor to the desired position and making the changes. (See Procedure P2.)

One difficulty with the program in Example 2.12 is that when the question mark appears on the screen, we may not be aware of the value desired. Likewise, when the value is printed out we may not be aware of its meaning. This is particularly true in large programs where several variables are used. This problem may be overcome by displaying a word or two prior to the question mark or output. Example 2.13 illustrates how we handle this. A word or two that appears before the question mark is called a **prompt command**. We could also have used the command Prompt to print the variable when showing the question mark. Before considering the example, we consider how to add or delete instructions from a program.

Procedure P6. To add or delete line from a program.

With the calculator in programming mode, select EDIT, then:
1. To add a line:
    (a) To insert a line before a given line move the cursor to the first character of the line and to insert a line after a given line move the character to the last character in the given line.
    (b) Press the key INS, then press ENTER This will create a blank line either before or after the given line
    (c) Place the cursor in the desired line. An instruction may now be entered.
2. To delete a line
    (a) Move the cursor to the desired line
    (b) Press the key CLEAR, then press DEL.

Example 2.13: For the program converting Celsius temperatures to Fahrenheit, have a prompt displayed before the question mark and identify the output.

Solution: 1. Use Procedure P2 to edit the program CTOF.

2. Use Procedure P6 to insert a blank line before the line Input C.

3. To display a message that will indicate what value is to be entered, use Procedure P5 to insert the instruction to display the string "Enter C" (Use: Disp "ENTER C")

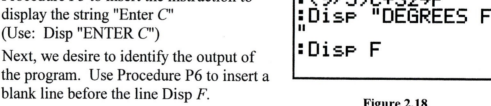

```
PROGRAM:CTOF
:Disp "ENTER C"
:Input C
:(9/5)C+32→F
:Disp "DEGREES F
"
:Disp F
```

4. Next, we desire to identify the output of the program. Use Procedure P6 to insert a blank line before the line Disp F.

Figure 2.18

5. Now, as in step 3 enter the instruction to display "Degrees F".
(Disp "DEGREES F ")

6. The complete program appears in Figure 2.18. Execute the program. Try several values for C.

Often when writing a program it is desirable to make decisions in a program and in some cases to transfer to other sections of the program. We next look at how these are accomplished within a program.

Making a decision in a program involves looking at a *condition* and determining if that *condition* is true or false. If the *condition* is true, a given statement or statements are executed if the *condition* is false, another statement or statements are executed. The *condition* usually involves one of the mathematical symbols $<, \leq, >, \geq, =,$ or $\neq$ to compare two variables and/or constants. (For example: $x < 5$ or $y \geq 0$) A *condition* that is true has a value of 1 and a *condition* that is false has a value of 0.

If a decision is to be made whether only one statement is executed or not, we may use the If... command.

Procedure P7. Making a decision (If... command).

The form of the If command is **If** *condition*

To access If...
While in programming edit mode (at a step in program):

On the TI-82 or 83 or 84 Plus:	On the TI-85 or 86:
To select the If command:	To select the If command
(a) Press PRGM	(a) Select CTL
(b) With CTL highlighted,	(b) Select If
select 1:If	

Following the word *If* enter the *condition*. (Examples: $x < 0$ or $T + 5 \geq 20$)

If *condition* is true, the statement following the If command is executed next. If *condition* is not true, the second statement following the If command is the next statement executed.

The *condition* may involve one of the mathematical symbols $<, \leq, >, \geq, =,$ or $\neq$ to compare two variables and/or constants. To obtain these symbols, press the key TEST and select from the menu.

Example: Input x
    If $x < 0$
    Disp "NEGATIVE"
    If $x \geq 0$
    Disp "NON NEGATIVE"

In many cases the decision is more complex. In this case we make use of the If...Then... or If...Then...Else... commands.

Procedure P8. Making a decision with options (If...Then...Else... commands).

The form of this combination is

(program statements before)
If *condition*
Then
(program statements)
Else
(program statements)
End
(program statements after)

To access If... Then... Else...
While in programming edit mode (at a step in program):

On the TI-82 or 83 or 84 Plus:	On the TI-85 or 86:
1. Press PRGM and with CTL highlighted	1. Select CTL

2. (a) For the If command
        Select 1:If
	2. (a) For the If command
	        Select If

(b) For the Then command
        Select 2:Then
	(b) For the Then command
	        Select Then

(c) For the Else command
        Select 3:Else
	(c) For the Else command
	        Select Else

(d) For the End command
        Select 7:End
	(d) For the End command
	        Select End

If the *condition* is true, the statements following the Then command are executed. If the *condition* is not true, the statements following the Else command are executed. After either set of statements is executed, the statements after the END statement are executed.

The *condition* involves one of the mathematical symbols $<, \leq, >, \geq, =,$ or $\neq$ to compare two variables and/or constants. To obtain these symbols, press the key TEST and select from the menu.
Else is optional (the Else section may be omitted).

Example: Input x
If x < 0
Then
Disp "NEGATIVE"
Else
Disp "NON NEGATIVE"
End

Exercises 2.5

Write programs to do the following:

1. Modify the equation $F = \dfrac{9}{5}C + 32$ to find values for C, given values for F. Write a program to perform this calculation. Evaluate for $F = 212, 100, 98.6, 0, -40$.

2. Write a program to evaluate $A = (1+r)^5$ for values of r to be input. Evaluate for $r = 0.05, 0.055, 0.06, 0.065$

3. Write a program to evaluate the function $f(x) = x^4 - x^3 + x - 2$ for entered values of x. Evaluate for $x = 0, 1, 2.5, 3.14, -2.34$.

4. Write a program to evaluate the function $g(x) = x^5 - x^3 + 2$ for entered values of x. Evaluate for $x = 1, 2.5, -3.1, 1.998$.

♦ 5. *(Washington, Exercises 3.1, #13-28)* Write a program to evaluate a function that has been stored for the function name Y_1 for values of x to input from the keyboard. Repeat Exercises 3 and 4 by first storing the function for Y_1, then using this program.

6. Write a program to evaluate the volume of a cylindrical tank for various values of r and h. The formula is $V = \pi r^2 h$. Evaluate for $r = 123.4$ and $h = 25.6$ and for $r = 127.4$ and $h = 22.5$.

7. Write a program to calculate the interest earned on P dollars deposited for t years at an annual interest rate of r percent per year (entered as a decimal) compounded n times per year. The formula is $P\left(1+\dfrac{r}{n}\right)^{nt} - P$. Find the interest earned on \$1500.00 deposited for 5 years at an interest rate of 4.5% compounded 4 times per year, at 12 times a year, and at 365 times a year.

8. Write a program to calculate the monthly payment M on a loan of P dollars for t years at an interest rate of r percent per year (entered as a decimal). The formula is $M = \dfrac{P}{K}$ where $K = \dfrac{1-(1+i)^{-12t}}{i}$ and $i = \dfrac{r}{12}$. Find the monthly payment on a loan of \$60000.00 for a period of 20 years at an interest rate of 9.0% per year, then, at an interest rate is 10.0%.

9. Write a program to calculate miles per gallon. The number of miles driven and the number of gallons of gas required should be used as input. Evaluate for the case where it took 15.3 gallons to drive 345 miles.

♦ 10. *(Washington, Exercises 5.2, #5-12)* Write and enter a program that accepts the coordinates of two points and gives the slope of the line passing through the two points. Then, find the slope of the line through the following pairs of points:

(a) $(1.2, 2.3), (4.5, 9.2)$ (b) $(-1.2, 2.4), (-3.5, 4.5)$
(c) $(5.63, -2.25), (-2.68, -5.14)$ (d) $(3.45, 5.18), (-2.63, 7.26)$

♦ 11. *(Washington, Exercises 5.2, #13-20)* Enter the following program to graph straight lines with slopes of m and y-intercepts of b:

> Disp "$M=$"
> Input M
> Disp "$B=$"
> Input B
> "$MX+B$"→Y_1
> DispGraph   (or DispG)

DispGraph is selected from the I/O menu when in programming mode. All other functions in the function table should be turned off.

Then, graph the following lines:

(a) $m=2.5$, $b=2.5$ (b) $m=-1.5$, $b=3.5$

(c) $m=-3$, $b=1.4$ (d) $m=0.5$, $b=-6.0$

♦ 12. *(Washington, Exercises 7.3)* Write and enter a program to evaluate the discriminant (b^2-4ac) of a quadratic equation and determine what type of roots the equation has. That is, real and unequal, real and equal, or not real. Apply the program to determine the type of roots in the following equations.

.(a) $3x^2-2x+5=0$ (b) $4x^2+9=12x$

(c) $14x^2+26x=35$ (d) $2.4x^2=75$

♦ 13. *(Washington, Exercises 7.3, #5-32)* Write and enter a program to solve a quadratic equation with real solutions. Then, use this program to solve the following quadratic equations:

(a) $x^2+3.0x-2.5=0$ (b) $4.3x^2-11.2x=6.8$
(c) $22+4.5x=2.0x^2$ (d) $6.72x=4.5x^2$
(e) $4.20x^2=35.2$ (f) $12.8x^2=15.8-13.7x$

14. Write an enter a program that will solve a quadratic equation and that will print out the solutions if they are real and print out the words "Not Real" if they are not real. Use the program to solve the following equations:

(a) $x^2+x+1=0$ (b) $x^2+3x=-5$

(c) $2x^2=2x-5$ (d) $3x-x^2=7$

15. Write and enter a program for graphing a quadratic function $y=ax^2+bx+c$ when values of a, b, and c are input. Use the program to graph the function $y=2x^2+3x-5$.

♦ 16. *(Washington, Exercises 7.4, #9-16).* Write and enter a program for graphing the quadratic function $y=a(x+h)^2+k$. Run the program for various values of a, h and k and note the differences in the graphs.

Chapter 3

Trigonometric Calculations

3.1 Basic Trigonometric Calculations
(Washington, Section 4.3)

Calculations involving trigonometric functions are done on the graphing calculator in much the same way as with standard scientific calculators except that expressions are generally entered as they appear. For example, to find the sine of an angle, press the key: SIN, then enter the angle, and press ENTER. Care must be taken to be sure that the calculator is in the correct mode, degrees or radians. At this point we will be working with angles in degrees and thus we should make sure the calculator is in degree mode.

Procedure C14. To change from radian mode to degree mode and vice versa.

> 1. Press the key: MODE
> 2. Move the cursor to either Radian or Degree on the third line depending on which mode is desired. (The mode, which is highlighted, is the active mode.) Press ENTER.
> 3. Press CLEAR or QUIT to return to the home screen.

We can use the graphing calculator to evaluate any of the six trigonometric functions sine, cosine, tangent, secant, cosecant, and cotangent. Since only the first three of these have calculator keys, we need to know the relationship between these functions and the other trigonometric functions. These relationships are given by

$$\sec \theta = 1/\cos \theta \quad \text{or} \quad \sec \theta = (\cos \theta)^{-1}$$
$$\csc \theta = 1/\sin \theta \quad \text{or} \quad \csc \theta = (\sin \theta)^{-1}$$
$$\cot \theta = 1/\tan \theta \quad \text{or} \quad \cot \theta = (\tan \theta)^{-1}$$

We are now able to find the secant, cosecant, and cotangent by using the keys for the sine, cosine, and tangent. For example, in order to find the secant of an angle, find the cosine of the angle then take the reciprocal. The reciprocal of a value is found by dividing the number 1 by that value or we can use the x^{-1} key on the calculator. The -1 exponent here means find the reciprocal of the number just as it does in algebra. **Warning when using the x^{-1} key, care must be taken to use correct sets of parentheses.**

Example 3.1: Find: (a) cos 27.32° and (b) cot 91.2°

Solution: 1. To find cos 27.32°, we enter the function just as it appears. Press the key for COS, then enter 27.32, and press ENTER. We find

$$\cos 27.32° = 0.8885$$

Rules for rounding of trigonometric values are given in Section 1.4.

2. To find cot 89.2°, we use the relationship that $\cot \theta = 1/\tan \theta = (\tan \theta)^{-1}$

$$\cot 89.2° = 1/ \tan 89.2°$$
$$\text{or} \quad \cot 89.2° = (\tan \theta)^{-1}$$

```
cos(27.32)
          .8884570796
(tan(89.2))-1
          .0139635414
```

Figure 3.1

Either enter 1/ TAN 89.2° or enter (TAN (89.2°)) and press the x^{-1} key. Either way we find that

$$\cot 89.2° = 2.18$$

A second type of problem that we need to be able to do is to find the angle if we know the value of one of the trigonometric functions. This involves use of what are called the inverse trigonometric functions. The inverse sine function is written $\sin^{-1} x$, the inverse cosine function $\cos^{-1} x$, and the inverse tangent function $\tan^{-1} x$. Likewise, there is an inverse secant function, $\sec^{-1} x$, the inverse cosecant function, $\csc^{-1} x$, and the inverse cotangent function, $\cot^{-1} x$.

Note: The −1 exponent has two different meanings. **When used with functions, the −1 exponent often means the inverse function like it does here with the inverse trigonometric functions, while when used in algebra or with a numerical value it means the reciprocal of the number.** For example: The inverse function that we call the inverse sine of x is written as $\sin^{-1} x$ and the reciprocal of $\sin x$ is written as $(\sin x)^{-1}$.

For positive values of the trigonometric functions and for positive angles less than 90 degrees, we define the inverse trigonometric functions as follows:

$$\theta = \sin^{-1} x \text{ if and only if } x = \sin \theta$$
$$\theta = \cos^{-1} x \text{ if and only if } x = \cos \theta$$
$$\theta = \tan^{-1} x \text{ if and only if } x = \tan \theta$$
$$\theta = \sec^{-1} x \text{ if and only if } x = \sec \theta$$
$$\theta = \csc^{-1} x \text{ if and only if } x = \csc \theta$$
$$\theta = \cot^{-1} x \text{ if and only if } x = \cot \theta$$

By this definition, $\sin^{-1} x$ equals an angle θ if and only if the sine of θ is x. Thus, if we know that the sine of an angle θ is a number n ($\sin \theta = n$), then it follows that $\theta = \sin^{-1} n$. The other inverse trigonometric functions may be thought of in a similar manner.

The inverse functions may be found by pressing the appropriate keys on the calculator. If we know the sine of an angle and want to find the angle, we enter SIN^{-1} followed by the sine of the angle, and then press ENTER.

Example 3.2: Find (a) $\sin^{-1} 0.725$ and (b) $\sec^{-1} 1.263$

Solution:
1. Finding $\sin^{-1} 0.725$ is equivalent to finding θ where $\theta = \sin^{-1} 0.725$. In other words we desire to find the angle whose sine is 0.725. We can use the SIN^{-1} key on the calculator. Press the SIN^{-1} key, enter 0.725, and press ENTER. We find
$$\sin^{-1} 0.725 = 46.5°$$

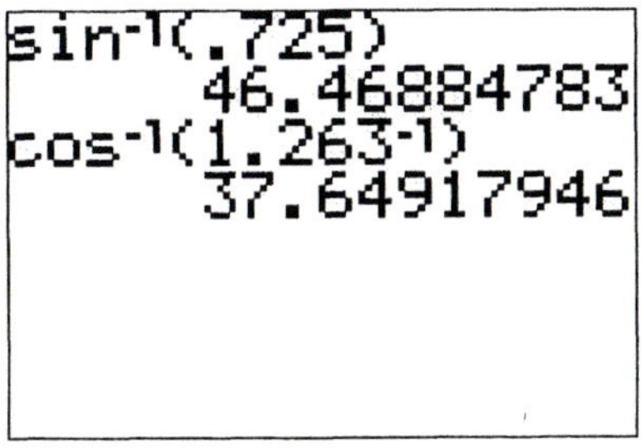

2. In part (b), we use the definition of the inverse trigonometric functions to note that $\theta = \sec^{-1} 1.263$, thus $\sec \theta = 1.263$.
From the relationships, this is the same as
$\dfrac{1}{\cos \theta} = 1.263$. Solving this last equation for

$\cos \theta$ we obtain $\cos \theta = \dfrac{1}{1.263}$.

Again using the definition of inverse functions we see that

$$\theta = \cos^{-1} \frac{1}{1.263} \quad \text{or} \quad \theta = \cos^{-1} 1.263^{-1}.$$

```
sin⁻¹(.725)
         46.46884783
cos⁻¹(1.263⁻¹)
         37.64917946
```

Figure 3.2

To do this calculation, we can first press COS^{-1} key then enter (1/1.263) and press ENTER or we can press COS^{-1} key, enter 1.263 and press the x^{-1} key, then press ENTER. Either way we obtain

$$\theta = 37.65°$$

Exercise 3.1

In exercises 1-24, evaluate the indicated quantity. All angles are in degrees.
Be sure calculator is in degree mode.

1. $\sin 23.5°$ 2. $\cos 45.7°$

3. $\tan 7.5°$ 4. $\sin 0.01245°$

5. $\sec 36.8°$ 6. $\cot 85.6°$

7. $\cos 23.7°$ 8. $\tan 88.95°$

9. $\csc 1.234°$ 10. $\sec 34.27°$

11. $\cot 16.475°$ 12. $\sin 0.01°$

13. $\sin^{-1} 0.8475$ 14. $\cos^{-1} 0.5245$

15. $\cos^{-1} 0.0251$ 16. $\tan^{-1} 1.257$

17. $\sec^{-1} 1.876$ 18. $\csc^{-1} 2.873$

19. $\cos^{-1} 1.256$ 20. $\cot^{-1} 0.5621$

21. $\cot^{-1} 1.54$ 22. $\sec^{-1} 1.05672$

23. $\tan^{-1} 0.12376$ 24. $\sin^{-1} 1.1295$

25. Find the $\sin A$ and $\cos B$ for each of the following:
     (a) $A = 23.5°$ and $B = 66.5°$
     (b) $A = 17.45°$ and $B = 72.55°$
     (c) $A = 77.8°$ and $B = 12.2°$
     In each case, note that $B = 90° - A$. What conclusion can we arrive at about the sine and cosine of complementary angles?

26. Find the $\tan A$ and $\cot B$ for each of the following:
     (a) $A = 45.5°$ and $B = 44.5°$
     (b) $A = 26.54°$ and $B = 63.46°$
     (c) $A = 87.2°$ and $B = 2.8°$
     In each case, note that $B = 90° - A$. What conclusion can we arrive at about the tangent and cotangent of complementary angles?

27. Evaluate $(\sin A)^2 + (\cos A)^2$ of each of the following:
     (a) $A = 37.9°$
     (b) $A = 2.23°$
     (c) $A = 88.52°$
     What conclusion can we arrive at about this expression for any given angle A? Why is it necessary to enter $(\sin A)^2$ and $(\cos A)^2$ rather than $\sin A^2$ and $\cos A^2$?

28. Evaluate $(\sec A)^2 - (\tan A)^2$ for each of the following:
    (a) $A = 52.6°$
    (b) $A = 1.4°$
    (c) $A = 86.23°$
    What conclusion can we arrive at about this expression for any given angle A?

29. Evaluate $(\csc A)^2 - (\cot A)^2$ for each of the following:
    (a) $A = 33.3°$
    (b) $A = 0.565°$
    (c) $A = 83.28°$
    What conclusion can we arrive at about this expression for any given angle A?

30. Evaluate $\sin^{-1} x + \cos^{-1} x$ for each of the following:
    (a) $x = .25$
    (b) $x = .7895$
    (c) $x = 0.0125$
    What conclusion can we arrive at about this expression for any given value x?

Exercises 31-44 contain equations than may be encountered in solving problems in trigonometry. Evaluate the unknown in each equation.

31. $\sin \theta = \dfrac{2.456}{4.231}$

32. $\cos \theta = \dfrac{123.45}{63.88}$

33. $\tan \theta = \dfrac{0.612}{0.928}$

34. $\cot \theta = \dfrac{1.265}{8.231}$

35. $\sin 34.6° = \dfrac{x}{2.345}$

36. $\tan 45.65° = \dfrac{x}{16.24}$

37. $\cos 1.278° = \dfrac{24.35}{x}$

38. $\cot 23.7° = \dfrac{23.64}{x}$

39. $x^2 = 1.245^2 + 3.125^2 - 2(1.245)(3.125)(\cos 25.40°)$

40. $y^2 = 234^2 + 125^2 - 2(234)(125)(\cos 63.5°)$

41. $\sin B = \dfrac{25.2 \sin 23.2°}{16.1}$

42. $\sin C = \dfrac{6.43 \sin 65.5°}{7.25}$

43. $\cos A = \dfrac{16.2^2 + 12.5^2 - 17.3^2}{2(16.2)(12.5)}$

44. $\cos A = \dfrac{7.23^2 + 8.15^2 - 6.24^2}{2(7.23)(8.15)}$

45. In any triangle if b and c represent the lengths of two sides and θ the angle between these two sides, the area of the triangle can be found from the equation $A = \dfrac{1}{2}bc \sin \theta$. Use this formula to find the area of a triangle that has two sides of length 25.6 cm and 34.7 cm if the angle between these two sides is 56.5°.

♦ 46. *(Washington, Exercises 4.3, #53)* the sound produced by a jet engine was measured at a distance of 100 m in all directions. The loudness d of the sound (in decibels) was found to be $d = 62.5 + 28.0 \cos \theta$, where the 0° line was directed in front of the engine.
   (a) Find d if $\theta = 62.5°$.
   (b) Find θ is $d = 50.0$ d.

♦ 47. *(Washington, Exercises 4.3, #54)* A brace is used in the structure and the length of the brace is given by $l = a(\sec \theta + \csc \theta)$.
   (a) Find l if $a = 15.82$ ft and $\theta = 35.95°$.
   (b) Find a if $l = 106$ ft and $\theta = 6.7°$.

48. Find the sine of the angle whose cosine is 0.2387. (That is find: $\sin(\cos^{-1} 0.2387)$)

49. Find the tangent of the angle whose sine is 0.6758. (That is find: $\tan(\sin^{-1} 0.6758)$)

50. Find the secant of the angle whose tangent is 1.35.

51. Find the cosine of the angle whose cosecant is 1.9812

3.2 Right Triangles
(Washington, Sections 4.4 and 4.5)

One of the important uses of trigonometric functions is to find the lengths of sides or the size of angles in a right triangle if a sufficient information about the triangle is known. The following right triangle definitions are used in doing these calculations.

$$\sin A = \frac{\text{side opposite } A}{\text{hypotenuse}}$$

$$\cos A = \frac{\text{side adjacent } A}{\text{hypotenuse}}$$

$$\tan A = \frac{\text{side opposite } A}{\text{side adjacent } A}$$

These definitions allow us to find the missing parts of a right triangle if we know one side and an angle or if we know two sides of the triangle.

Example 3.3: One leg of a right triangle is 6.5 units long. The angle adjacent to this leg is 37.5 degrees. Find the hypotenuse and other leg of the triangle.

Solution:
1. Let x represent the length of the other leg and h represent the length of the hypotenuse.

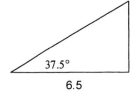

2. To find the other leg we make use of the fact that we know the side adjacent to the given angle and want to find the opposite leg. Since the tangent function involves these three quantities, we use the tangent function for our calculations and we have

$$\tan 37.5° = \frac{x}{6.5} \quad \text{or} \quad x = 6.5 \tan 37.5°$$

Make sure your calculator is in degree mode, then enter 6.5 tan 37.5 and press ENTER. The result is 4.9876 which when rounded to the correct number of decimal places is 5.0.

3. To find the hypotenuse we make use of the fact that we know the side adjacent to the given angle and want to find the hypotenuse. Since the cosine function involves these three quantities, we obtain

$$\cos 37.5° = \frac{6.5}{r} \quad \text{or} \quad r = \frac{6.5}{\cos 37.5°}$$

Enter 6.5/cos 37.5 into your calculator and press ENTER. The result is 8.193 or 8.2 when rounded to the correct number of decimal places.
Thus, the leg is 5.0 units long and the hypotenuse is 8.2 units long.

Example 3.4 shows how to find the missing parts of a right triangle when two sides are known. We first find one angle using trigonometry, the second angle by subtracting the angle we found from 90°, and then find the other side by using a trigonometric function (or we could use the Pythagorean Theorem).

Example 3.4: One leg of a right triangle is 102 feet and the hypotenuse is 152 feet. Find the missing angles and the other side.

Solution:
1. Let θ represent the angle opposite the 102 ft. side.
2. Since we know the hypotenuse and the side opposite an angle and want to find the angle we select the definition of the trigonometric function that make use of these three quantities. In this case it is the sine function.

$$\text{Thus, } \sin\theta = \frac{102}{152}$$

$$\text{This means that } \theta = \sin^{-1}\frac{102}{152}.$$

Enter $\sin^{-1}(102/152)$ into your calculator and press ENTER. The result is about 42.148. Rounding this to the correct number of decimal places we find that $\theta = 42.1°$.

3. Let the missing angle be represented by α. To find α we just need to subtract 42.1° from 90°

$$\alpha = 90° - \theta = 90° - 42.1° = 47.9°$$

4. Let x be the unknown side. This side can be found by using the Pythagorean theorem or by using the tangent or cosine function. Using the tangent function we have

$$\tan\theta = \frac{102}{x} \text{ or } \tan 42.148 = \frac{102}{x}$$

$$\text{Thus, } x = \frac{102}{\tan 42.148} = 112.696$$

or x = 113 rounded to correct number of decimal places. Note that the rounded angle was not used in the calculation, but that the value was rounded after the calculation was completed. The unknown parts of the triangle are the angles 42.1° and 47.9° and the third side 113.

Exercise 3.2

♦ *(Washington, Exercises 4.4, #9-28)* In exercises 1-8 solve the right triangle *ACB* for the missing parts. Angle *C* is the right angle.

1. $A = 23.41°$, a = 25.25
2. $B = 45.67°$, a = 14.76
3. $B = 45.8°$. c = 2.25
4. b = 45.67, c = 56.78
5. a = 163.5, c = 245.7
6. $A = 12.5°$, c = 123
7. a = 0.1537, b = 0.9823

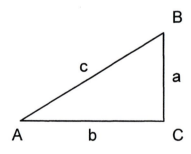

8. $b = 15400$, $a = 25930$

9. Find the missing parts of a right triangle if the hypotenuse is 65.3 inches and one angle is 53.4°

10. Find the missing parts of a right triangle if one angle is 27.95° and the leg opposite this angle is 112.5 meters.

11. Find the missing parts of a right triangle if one leg is 2.54 km and the other leg is 1.75 km.

12. Find the missing parts of a right triangle if one leg is 14.357 inches and the hypotenuse is 22.761 inches.

Hint for problems 13-17: Set up a formula and then evaluate the formula. Using a list of numbers may be helpful.

13. The center of a car's headlight is 25.0 inches above the surface of a highway. Determine the angle of depression needed for the center of the headlight beam to hit the highway 150.0, 200.0, 250.0, 300.0, and 350.0 feet in front of the car.

14. A person 72.0 inches high stands outside in the sunlight every odd numbered hour from 6:00 A.M. till Noon. Determine the length of the shadow of this person at each odd hour between 6:00 AM and Noon. Assume the sun rises at 6:00 A.M. and is directly overhead at noon and that the angle of elevation of the sun is directly proportional to the hour after sunrise.

♦ 15. *(Washington, Exercises 4.5, #11)* In designing a new building, the base of a doorway is 2.65 feet above the ground. A ramp for the disabled is to be built from to the base of the doorway along level ground. Determine the length along the ramp if the ramp makes angles with the ground of 4.0°, 4.5°, 5.0°, 5.5°, and 6.0°

♦ 16. *(Washington, Exercises 4.5, #16)* A guardrail is to be constructed around the top of a circular observation tower. The diameter of the observation area is 10.00 m. The guardrail is to be constructed in equal straight sections. Find the length of each section and also the total length of all sections if 5, 10, 15, 20, 25, 30, 35, or 40 sections are used. What would be the maximum total length of guardrail needed no matter how many sections?

17. Write and test a program to solve a right triangle if:
    (a) Two legs are given.
    (b) One leg and the hypotenuse is given.
    (c) One angle and an adjacent leg is given.
    (d) One angle and an opposite leg is given.
    (e) One angle and the hypotenuse is given.

Chapter 4

Systems of Linear Equations

4.1 Graphical Solutions of Systems of Linear Equations
(Washington, Section 5.3)

A linear equation is an equation that may be written in the form $ax + by = c$ where a, b, and c are real numbers. A graph of such an equation is a straight line. A system of two such equations has a solution if there exists a pair of numbers (for x and y) that satisfy both equations. In this section we will see how a solution may be found by graphical means. We may use graphs to readily solve some application problems that are difficult to solve by algebraic means.

Example 4.1:   Consider the system of equations:

$$3x + 2y = 6$$
$$x - 3y = 13$$

Determine 3 points on the graph of the first equation and 3 points on the graph of the second equation. By trial and error find a point that is on the graph of both equations.

Solution:
1. For the first equation, if we let x be 0, 2, and −2 and solve for y, we obtain pairs of numbers that satisfy the first equation and thus give three points on its graph: (0, 3), (2, 0), and (−2, 6).

2. Likewise in the second equation if we let x be 0, 2, and −2, we obtain the pairs of numbers (0, 13), (2, −11/3), and (−2, −5) that satisfy the second equation and give three points on its graph.

3. If the search for points is continued through enough x values, eventually (perhaps an eternity) it would be found that a pair of numbers (4, −3) satisfy both equations. This would correspond to the point where the two graphs intersect. There is only one such point that will satisfy both these equations since the graphs are not parallel and do not give the same line. The graphs are shown in Figure 4.1.

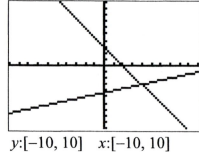

y:[−10, 10] x:[−10, 10]

Figure 4.1

Because of the nature of the equations in Example 4.1, it was relatively easy to find a pair of numbers that satisfied both equations. For most systems of equations such a guessing game would be a ridiculous way of determining the desired values.

In order to solve a system of equations, we desire to find the set of numbers, if such numbers exist, that satisfy all the equations. A system of two linear equations in two unknowns may have no solutions, one solution, or an infinite number of solutions as shown in the following table:

A system of two linear equations in two unknowns:

(1) That gives intersecting lines on a graph, has one solution (a pair of numbers).

(2) That gives parallel lines on a graph, has no solution and is called **inconsistent**.

(3) That gives the same line on a graph, has an infinite number of solutions and is called **dependent**.

To find the solution of a system of two linear equations that has one solution, we need to graph both equations on the same graph and find the point of intersection. To graph a linear equation, solve the equation for the dependent variable (y) and then follow Procedure G1. After both equations are graphed, solve the system by using Procedure G15.

Procedure G15. Finding an intersection point of two graphs.

and adjust the viewing rectangle so the region of the graph that contains the intersection point is on the screen. Then,

1. Enter the functions into the calculator as two different functions (Y_1 and Y_2).

2. Graph the functions on the same screen and adjust the viewing rectangle so that the region containing the intersection point is on the screen.

3. Determine the point of intersection by:

(a) Using Procedure G9 to zoom in on the point. To obtain the desired degree of accuracy, using the trace function move the cursor just to the left and then just to the right of the intersection point after each zoom. A solution to the desired accuracy is obtained when the x-values, on either side of the intersection point, rounded off to the desired significant digits are equal.

or

(b) Use the built-in procedure:

(1) Select the correct mode to find the point of intersection.

On the TI-82, 83 or 84 Plus:	On the TI-85 or 86:
Press the key CALC	With the graph menu on the screen, select MATH (after pressing MORE).
Select 5:intersect	Then, select ISECT (after pressing MORE).

(2) Move the cursor near the point of intersection.

(3) If the cursor is on one of the graphs, continue. Otherwise, use the up and down cursor keys to move the cursor to the correct graph. (On the TI-82, 83 or 84 Plus, and 86 the words "First curve?" appear on the screen.)

(4) Press ENTER. The cursor will jump to another graph.

(5) If the cursor is on the correct second graph, continue. Otherwise, use the up and down cursor keys to change the cursor to the correct graph. (On the TI-82, 83, 84 Plus and 86 the words "Second curve?" appear.)

(6) Press ENTER.

(7) Obtain the coordinates of the intersection point.

On the TI-82, 83, 84 Plus or 86:	On the TI-85:
The word "Guess?" appears on the screen. Make sure the cursor is near the point of intersection and press ENTER.	The word ISECT appears..

The word Intersection appears.
The coordinates of the intersection point appear on the screen.

One way to determine if the cursor is on opposite sides of the point of intersection is to note the y-values on each graph at a particular x on one side then to note the y-values on each graph at a particular x on the other side. The cursor can be made to jump from one graph to the other by pressing the up and down cursor keys. On one side of the intersection point one graph will have the larger y-value, but the same graph will have the smaller y-value on the other side.

Example 4.2: Solve the following system of equations. Find the x- and y-values to three decimal places.

$$5x + 2y = 7$$
$$3x = y + 3$$

Solution: 1. First solve each equation for y as a function of x.

$$
\begin{array}{ll}
5x + 2y = 7 \qquad \text{and} & 3x = y + 3 \\
2y = -5x + 7 & 3x - 3 = y \\
y = (-5/2)x + 7/2 & y = 3x - 3
\end{array}
$$

2. On the standard viewing rectangle, use Procedure G1 to graph
   $y = (-5/2)x + 7/2$ and $y = 3x - 3$
   on the same graph.
   The graph is given in Figure 4.2.

3. Use the trace function (Procedure G4) to move the cursor near the point of intersection. For one value of x the graph will be on one side of the intersection point and for another value, on the other side. The intersection point must be between these two points. In this case for approximately $x = 1.1579$ the cursor is to the left of the intersection point and for $x = 1.3684$ it is on the right hand side. Leave the cursor on the point where $x = 1.3684$. (Your values may differ slightly.)

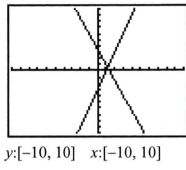

$y:[-10, 10] \quad x:[-10, 10]$

Figure 4.2

4. Use Procedure G15 to find the point of intersection. The result, to a precision of three decimal places, is $x = 1.182$, $y = 0.545$.

It may happen that, when trying to find the solution to a system of equations on a graphing calculator, the intersection point falls outside the viewing rectangle. In this case it is necessary to change the viewing rectangle so that the intersection point appears on the screen. Three different methods may be used to do this:

1. Change the viewing rectangle as described in Procedure G5. This may be particularly useful if the x or y values are very large or very small.

or

2. Use zoom out as described in Procedure G9.

or

3. Use the trace function to move the cursor near the point of intersection until the point of intersection lies on the screen. This works only if the point of intersection lies off the right or off the left of the screen. In this case the screen will scroll (move over) in the direction the cursor is moving.

Example 4.3: Solve the following system. Find answer to two decimal places.

$$x + 10y = 50$$
$$x - 2y = 8$$

Solution: 1. Solve each equation for y as a function of x.

$$x + 10y = 50 \quad \text{and} \quad x - 2y = 8$$
$$10y = -x + 50 \qquad -2y = -x + 8$$
$$y = \frac{-1}{10}x + 5 \qquad y = \frac{1}{2}x - 4$$

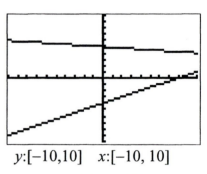

2. Graph $y = (-1/10)x + 5$ and $y = (1/2)x - 4$ on the same graph using the standard viewing rectangle. The graph is in Figure 4.3.

y:[−10,10] x:[−10, 10]

Figure 4.3

3. From the graph it does appear that the lines do intersect, but off the screen to the right. In this case we use the trace function and move the cursor along a graph until the point of intersection is on the screen. After the point of intersection is on the screen, the graph should appear as in Figure 4.4.

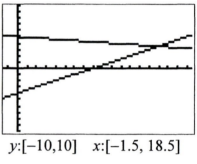

4. Use Procedure G15 to find the point of intersection. The solution is $x = 15.00$ and $y = 3.50$.

y:[−10,10] x:[−1.5, 18.5]

Figure 4.4

Sometimes the values of the variables in the equations are such that when the standard viewing rectangle is used either the point of intersection is not apparent or one or both graphs do not even appear on the screen. In such a situation a better viewing rectangle must be used based on estimates of the values of the variables. The viewing rectangle is then adjusted as seems appropriate. One method of estimating the maximum or minimum y-values needed for the viewing rectangle is to use the trace function. The trace function gives coordinates for points on the graph even when the points do not appear on the screen.

♦ Example 4.4:   *(Washington, Section 5.3, Example 4)* A driver traveled for 1.5 hours at a constant speed along a highway. Then, through a construction zone, the driver reduced the car's speed by 20.0 miles per hour for 30.0 minutes. If 100.0 miles were covered in the 2.0 hours, what were the two speeds.

Solution: 1. We recall that distance equals rate times time and let x = the car's highway speed and y = the car's speed in the construction zone. Then, $1.5x$ represents the distance traveled on the highway and $0.50x$ represents the distance traveled in the construction zone and the sum gives the total number of miles traveled.

$$1.5x + 0.50y = 100.$$

Since the difference in speeds is 20 miles per hour: $y = x - 20$

2. Solve each equation for y, enter each equation into the calculator and graph on the standard viewing rectangle. In this case, nothing appears on the screen since both graphs are located outside the viewing rectangle. Using the trace function and observing the x- and y-intercepts of the functions, we determine that a reasonable viewing rectangle is

y:[−30, 220] x:[−10, 80]

Figure 4.5

$$Xmin = -10, Xmax = 80$$
$$Ymin = -30, Ymax = 220$$

This gives the graph in Figure 4.5.

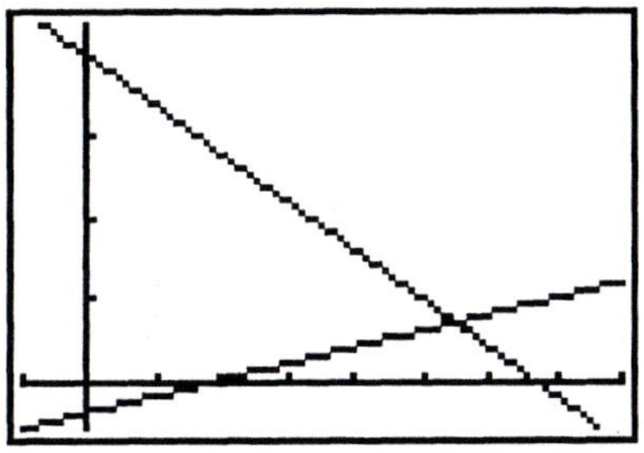

3. Use Procedure G15 to find the point of intersection. Based on the accuracy of the values given in the problem, we round off the answers to two significant digits. The answer is $x = 55, y = 35$. Thus, the highway speed is 55 miles per hour and the construction zone speed is 35 miles per hour.

To check answers to a system of equations, use Procedure C11, to evaluate the functions obtained from each equation at the value of x found in the solution. If the corresponding y-values are essentially the same as the y-value of the solution, the answer is correct.

When we solve a system of two linear equations by graphing and parallel lines are obtained, there are no points of intersection and thus, no solutions, and the system is an inconsistent system. If when graphing the two linear equations only one line is obtained the system is dependent and there are infinite solutions – every point on the line is a solution. The lines can be verified as being parallel or as being the same line by using the trace function.

To verify that two graphs are parallel or the same line:
1. Use the trace function and the right or left cursor to move to a point on one of the graphs and note the y-value.
2. Use the up or down cursor keys to jump the cursor to the other graph and note the y-value for the same x.
3. Note the difference between the y-values.
4. Repeat the process for several x-values – each time noting the difference between the y-values.

If the difference in each pair of y-values is the same the graphs are parallel, the system is inconsistent and there are no solutions.

If the difference in each pair of y-values is 0, the two graphs are the same, the system is dependent and every point on the line is a solution.

The method of finding the intersection of two graphs may also be used to determine an x-value that gives a particular y-value for the function. To do this, graph the given function and also the function $y = k$ where k is the particular y-value. Then, find the intersection point of these two graphs.

Exercise 4.1

In Exercises 1-10 *(Washington, Exercises 5.3, 3-32)*, solve each of the following systems of linear equations. Find each answer to 3 significant digits. Identify any systems that are inconsistent or dependent.

1. $2x - 3y = 5$
    $4x + 2y = 3$

2. $3x + 4y = 4$
    $2x - 5y = 7$

3. $2x + 3y = 35$
    $4y = 3x - 20$

4. $1.3x = 5.4y + 6.3$
    $2.6y + 0.60x + 19.7 = 0$

5. $6x + 1.5y = 27$
    $y = 3.2$

6. $y = -40$
    $5x - 9y = 160$

7. $2.40x + 1.30y = 7.80$
    $1.68x + 0.91y = 5.46$

8. $11x + 25y = 95$
    $2x = 3y - 49$

9. $5.24x - 3.18y = 25.67$
    $3.72x = 4.29y - 37.35$

10. $7.6x - 4.4y = 6.30$
    $5.7x = 3.3y + 8.63$

11. $127x + 245y = 670$
    $25.4x + 49.0y = 402$

12. $60.6x + 50.2y = 233.7$
    $23.7y = 27.2 - 24.8x$

13. A company buys two types of disks for personal computers on two different orders and they do not know the cost of each disk of each type. They do know that in one order they received 425 of type A and 575 of type B and the cost was \$675. In the second order they received 750 of type A and 225 of type B and the cost was \$585. Determine the cost of each type A and each type B disk.

♦ 14. *(Washington, Exercises 5.3, # 34)* Two currents, i_1 and i_2, in a circuit are related by the two equations: $2.1i_1 + 5.8(i_1 + i_2) = 11.8$ and $3.7i_2 + 5.8(i_1 + i_2) = 11.8$. Solve this system of equations to find the two currents.

♦ 15. *(Washington, Exercises 5.4, # 43)* Two grades of gasoline are mixed to make a gasohol blend. The first grade contains 4.650% alcohol and the second grade contains 12.25% alcohol. It is desired to mix x liters of the first grade with y liters of the second to obtain 12,225 liters of a mixture containing 9.500% alcohol. The equations are:

$$0.04650x + .1225y = .09500(12225)$$
$$x + y = 12225$$

Find how many liters of each mixture are needed.

16. Given the two equations

$$-11.5x + 13.8y = 18.4$$
$$22.08y = 18.4x + 29.44$$

Graph these equations. What can you tell about the solution? Which, if any, of the following are solutions to this system?

$$x = 20, y = 18; \ x = 32, y = 28; \ x = 56, y = 48; \ x = 68, y = 58; \ x = 86, y = 73.$$

17. In order to determine the x-value that gives a particular y-value, k, in an equation, we can graph the given equation and the equation $y = k$ and find the point of intersection. For example: To determine the x-value for when $y = 2.7$ in the equation $3.5x - 2y = 6.4$, graph the system of equations $y = (3.5x - 6.4)/2$ and $y = 2.7$ and find the point of intersection. Use this method to find the x-value when $y = 2.7$, $y = 3.85$, $y = 16.0$, and $y = -12.5$

18. Using the procedure outlined in Exercise 15, find the x values that give y-values of 22.7 and 33.5 in the equation $12.5x + 3.2y = 22.4$.

19. A cost function for a business is given by $y = 1225 + 22.5x$. The income is given by the revenue function $y = 35.75x$. In both cases, y is in dollars and x is the number of items. Graph these functions on the same graph then answer these questions.
    (a) Determine the cost and revenue when $x = 46$ items by finding the difference in y-values. (Hint: Jump the cursor between the curves.) If the business produces and sells 46 items is the business making any profit?
    (b) Determine the cost and revenue when $x = 138$ items by finding the difference in y-values. If the business produces and sells 138 items is it making any profit? If so, how much?
    (c) For how many items would the cost be $4150? Would the revenue be $4150?
    (d) Determine the point at which the cost equals the revenue. This is called the break-even point. What is the profit at this point?

20. The cost function for a business is given by $y = 18725 + 125x$ and the income is given by the revenue function $y = 175x$ where y is in dollars and x is the number of items. Graph these functions on the same graph then answer these questions.

   (a) Determine the cost and revenue when $x = 200$ items by finding the difference in y-values. If the business produces and sells 200 items is the business making any profit?

   (b) Determine the cost and revenue when $x = 500$ items by finding the difference in y-values. If the business produces and sells 500 items, is it making any profit? If so, how much?

   (c) For how many items would the cost be $22600? Would the revenue be $22600?

   (d) Determine the point at which the cost equals the revenue. This is called the break-even point. What is the profit at this point?

In certain cases the calculator may be used to graphically solve systems of three equations. One case in which this is true is when one of the equations contains only two variables and the missing variable may be easily eliminated from the other equations.

◆ 21. *(Washington, Exercises 5.6, # 21)* To solve the system of equations:

$$0.707F_1 - 0.800F_2 \qquad\quad = 0$$
$$0.707F_1 + 0.600F_2 - F_3 \quad = 10.0$$
$$3.00F_2 - 3.00F_3 = 20.0$$

Subtract the first equation from the second to obtain the set of equations:

$$1.400F_2 - F_3 = 10.0$$
$$3.00F_2 - 3.00F_3 = 20.0$$

Solve these equations on the graphing calculator. Once values for F_2 and F_3 have been found, F_1 may be readily found by substituting in the first equation.

◆ 22. *(Washington, Exercises 5.6, # 22)* In applying Kirchhoff's laws to an electric circuit, a student comes up with the following equations for currents I_A, I_B, and I_C (in amperes):

$$I_A + I_B + I_C = 0$$
$$3.5I_A - 8.8I_B \qquad\quad = 4.77$$
$$- 8.8I_B + 4.8I_C = 5.58$$

Use the method described in Exercise 17 to solve this system of equations.

Chapter 5

Vectors and Complex Numbers

5.1 Working with Vectors.
(Washington, Section 9-2, 9-3 and 9-4)

Many quantities, such as length or area, may be described by giving a single number corresponding to the magnitude of that quantity. However, other quantities such as force, velocity, and acceleration are best described giving both the magnitude and the direction. In mathematics, **vectors** are quantities that are described by giving both magnitude and direction and are represented as directed line segments. To show a vector we draw a line of a certain length, and place an arrow on the end of the line to show its direction. To identify a particular vector, we need to give a number indicating the magnitude of the vector and an angle indicating the direction of the vector.

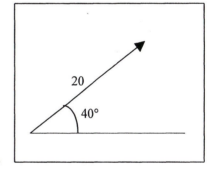

For example, suppose we wish to show a vector **F** describing a force of 20.0 pounds acting in a direction of 40 degrees above the horizontal. We would first select a scale (perhaps 1 cm = 5 lb), and draw a line segment of the appropriate length and at an angle of 40 degrees as in Figure 5.1. An arrow is drawn at the end of the line segment to show the direction of the vector.

Figure 5.1

Vectors are often named using letters printed in bold type to denote that it is a vector (for example, *A*). When writing vectors by hand the name of the vector should be written with an arrow above it (for example, $\vec{A}$). Writing the name of the vector without the arrow or printing the name in regular type represents the magnitude of the vector. The Greek letter θ is often used to represent the angle to the vector. The name of the vector is written as a subscript to θ to denote the angle to a particular vector (for example θ_A). For the vector *F*, $F = 20.0$ and $\theta_F = 40°$.

It is frequently helpful to break down the vector into horizontal and vertical components. These components are vectors in the horizontal and vertical direction whose sum is the given vector.

It is also possible to determine the magnitude and direction of a vector from its horizontal and vertical components. Suppose that 12 minutes after first observing a plane on a radar screen, the plane is 75 miles east and 63 miles north of its original position as shown in Figure 5.2. The distance, *d*, the plane has traveled in the 12 minutes can be found by using the Pythagorean Theorem:

$$d = \sqrt{75^2 + 63^2} = 97.9 \text{ or } 98 \text{ miles}$$

Since the plane traveled these 98 miles in 12 minutes ($\frac{1}{5}$ of an

hour), it is traveling a rate of 5 times 98 or 490 miles per hour.
The angle of direction of the plane can be found by determining

the angle whose tangent is $\frac{63}{75}$:

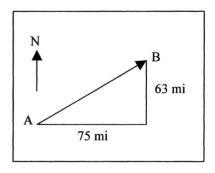

$$\theta = \tan^{-1}\frac{63}{75} = 40.0 \text{ degrees north of east.}$$

Figure 5.2

Thus, the vector that describes the plane has a magnitude of 490 miles per hour at an angle of
40 degrees north of east.

We could also have determined that the eastward component of the vector representing
the plane is 5 times 75 or 375 miles per hour (call this V_x) and the northward component (call
this V_y) is 5 times 63 miles or 315 per hour. Then, the magnitude, V, and direction, θ_V, of such a
vector could be determined in a manner similar to the above method by using the equations

$$V = \sqrt{V_x + V_y}\qquad\qquad \text{[Equation 1]}$$

$$\theta = \tan^{-1}\frac{V_y}{V_x}$$

In the example given above this becomes

$$V = \sqrt{375^2 + 315^2} = 490 \text{ miles per hour}$$

$$\theta_V = \tan^{-1}\frac{315}{375} = 40.0°$$

It is particularly advantageous to consider vectors relative to the x- and y-coordinate
system. Vectors are placed so that their starting point is at the origin and the direction of the
vector is the angle to the vector measured from the positive x-axis. Such a vector is said to be in
standard position. The horizontal component is then the x-component and the vertical
component is the y-component. The x-component is a vector that starts at the origin and is
directed along the x-axis and the y-component starts at the origin and is directed along the y-axis.
For a vector V, the magnitude of the x-component is designated as V_x the magnitude of the
y-component is designated as V_y, and the magnitude of the vector itself is V. If we let θ
represent the angle to the vector from the positive x-axis, then these components are given by

$$V_x = V \cos \theta \qquad\qquad \text{[Equation 2]}$$
$$V_y = V \sin \theta$$

Also, we note that Equation 1 previously given also hold true.

For the vector **F**, shown in Figure 5.1, the magnitude of the horizontal (or *x*-) component is given by 20.0 cos 40° = 15.3 and of the vertical (or *y*-) component is given by 20.0 sin 40° = 12.9.

In order to have a convenient method of working with vector components on a graphing calculator, we next give a brief introduction to the location of points on a graph using polar coordinates. (Polar coordinates and graphing of equations in polar coordinates will be covered in more detail in Chapter 11.) In rectangular coordinates, a point is identified by giving the *x*- and *y*-coordinates of the point. In polar coordinates, a point is identified by giving the distance, *r*, of the point from the origin and the angle, θ, from the positive *x*-axis to the line going from the origin to the point. These quantities are shown in Figure 5.3. Polar coordinates are expressed in the form (*r*, θ). Any point can be identified by giving either the rectangular coordinates of that point or the polar coordinates of the point. In order to convert from one set of coordinates to the other, we use the relationships

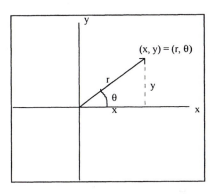

Figure 5.3

$$x = r\cos\theta \quad \text{and} \quad r = \sqrt{x^2 + y^2} \qquad \text{[Equation 3]}$$

$$y = r\sin\theta \qquad \theta = \tan^{-1}\frac{y}{x}$$

If **V** is the vector that begins at the origin and terminates at the point (*x*, *y*) and the point (*x*, *y*) has the polar coordinates (*r*, θ), then the vector can be expressed as the vector [x, y] or by the form [*r* $\angle\theta$]. The magnitude of the vector equals the r-value and the direction of the vector is given by the angle θ. Finding the components of the vector from its magnitude and direction is equivalent to converting polar coordinates to rectangular coordinates. Likewise, finding the magnitude and direction of a vector from the components is equivalent to converting rectangular coordinates to polar coordinates.

Vector (**V**)	Polar coordinates	Rectangular coordinates
magnitude (*V*)	*r*	$\sqrt{x^2 + y^2}$
direction (θ_V)	θ	$\tan^{-1}\frac{y}{x}$
x-component (V_x)	$r\cos\theta$	*x*
y-component (V_y)	$r\sin\theta$	*y*

For our purposes, a <u>vector</u> in terms of its *x*- and *y*-components is in **rectangular form** and is expressed as [*x*, *y*]. If the vector is in terms of its magnitude and direction, it will be considered to be in **polar form** and expressed as [*r* $\angle\theta$].

Many graphing calculators contain a provision for converting from polar to rectangular coordinates and from rectangular to polar coordinates. This feature may be used to find the components of a vector given its magnitude and direction and also to find the magnitude and direction given the components given the vector.

Procedure C15. Conversion of Coordinates

Make sure calculator is in correct mode -- radians or degrees.
A. To change from **polar to rectangular**:

On the TI-82 or 83 or 84 Plus:
1. Press the key ANGLE
2. To find x:
    Select 7:P▷Rx(
    To find y:
    Select 8:P▷Ry(
3. Enter the values of r and θ separated by commas and press ENTER.
Example: P▷Rx(2.0, 45)
    gives 1.4 [in degree mode]

On the TI-85 or 86:
1. In square bracket [],
    Enter the magnitude and angle separated by $\angle$.
2. Press the key VECTR
3. Select OPS
4. Select ▷Rec (after pressing MORE)
5. Press ENTER
Example: [2.0∠45]▷Rec
    gives [1.4, 1.4]
    in degree mode.

B. To change from **rectangular to polar**:

On the TI-82 or 83 or 84 Plus:
1. Press the key ANGLE
2. To find r:
    Select 5:R▷Pr(
    To find θ:
    Select 6:R▷Pθ(
3. Enter the values of x and y separated by commas and press ENTER.

Example: R▷Pθ(1.0, 1.0)
    gives 45 [in degree mode]
Note: On the TI-82, it may be just as easy to use the conversion formulas (Equation 3) directly -- especially when converting from polar to rectangular.

On the TI-85 or 86:
1. Enter the values of x and y inside of [ ] separated by a comma.
2. Press the key VECTR
3. Select OPS
4. Select ▷Pol
5. Press ENTER

Example: [1.0, 1.0] ▷Pol
    gives [1.4 ∠45]
    in degree mode.
Note: On the TI-85 or 86, vectors are entered or displayed as a pair of numbers inside of square brackets, [ ]. When the numbers are separated by a comma they are in rectangular form and when separated by $\angle$ are in polar form.

After conversion from one form to the other, a visual check should be made to make sure the two forms represent the same vector. When converting from rectangular to polar, be careful to obtain the correct angle. When the angle is displayed as a negative angle it is necessary to add 360° to obtain the correct positive angle.

Example 5.1: (a) Find the *x*- and *y*-components of the vector with a magnitude of 3.25 at an angle of 150°.

           (b) Find the magnitude and angle for a vector that has an *x*-component of 2.14 and a *y*-component of –3.56.

Solution: 1. Finding the *x*- and *y*-components in part (a) is equivalent to converting from polar coordinates to rectangular coordinates. The *x*-component may be found using 3.25 cos 150° and the *y*-component may be found using 3.25 sin 150° or use Procedure C15 to perform this conversion. The *x*-component of this vector is –2.81 and the *y*-component is 1.63. We write the answer in rectangular form as [–2,81, 1.63].

            2. Finding the magnitude and direction in part (b) is equivalent to converting from rectangular coordinates to polar coordinates. Use Procedure C15 and perform this conversion. The magnitude is 4.15 and the angle is given as –59.0, which is the same as 301.0° (after adding 360°). The answer in polar form is [4.15 ∠301°]

The sum of two or more vectors is referred to as the **resultant** of the vectors. In order to add vectors in polar form, we first find the *x*- and *y*-components of each vector. Then add all the *x*-components and all the *y*-components to obtain the *x*-component and *y*-component of the resultant vector. The magnitude and direction of the resultant vector are obtained from its components. On some calculators it may be possible to simply enter the vectors in polar form and add the vectors without finding the components.

Procedure C16. Finding the sum of vectors in polar form.

On the TI-82 or 83 or 84 Plus:
1. Find the *x*- and *y*-components of each vector.
(see Procedure C15)
2. Find the sum of the *x*-components and the sum of the *y*-components.
3. Find the magnitude and direction of the resultant vector.
(see Procedure C15)

    or

On the TI-85 or 86:
Vectors may be entered directly in polar form and then added together if the calculator is in correct mode. Press the key MODE and be sure that in the third line, Degree is highlighted and in the seventh line, CylV is highlighted.

Find the sum
$$A\cos\theta_A + B\cos\theta_B + ...$$
and store this for x
Find
$$A\sin\theta_A + B\sin\theta_B + ...$$
and store this for y,
then convert $[x, y]$ to polar form.

or

On some calculators it may be more convenient to treat vectors as complex numbers.

Example:

$$[12 \angle 120] + [15 \angle 75]$$
gives $[25 \angle 95]$ when rounded.

Note: On the TI-85 or 86, vectors may be stored under variable names by entering the vector, pressing the key STO▷, entering the variable name and pressing ENTER. Vectors may also be entered or modified by first pressing the key VECTR, then selecting EDIT. The sum of vectors may be found by finding the sum of the variables.

The form of the output of a vector on the screen may be changed by pressing the key MODE and in the seventh line highlighting
RectV for rectangular form
CylV for polar form

Example 5.2:   Find the sum of the vectors A and B where A has a magnitude of 22.5 in the direction $\theta_A = 75.8°$ and B has a magnitude of 57.2 in the direction $\theta_B = 53.2°$.

Solution: Use Procedure C16 to find the sum of the two vectors.
If we first find the x- and y-components for the resultant vector R:

$$R_x = 22.5\cos 75.8 + 57.2\cos 153.2 = 39.8$$
$$R_y = 22.5\sin 75.8 + 57.2\sin 53.2 = 67.6$$

Then, converting these to polar coordinates we obtain $R = 78.4$ and $\theta_R = 59.5°$.
We can express this as $[22.5 \angle 75.8] + [57.2 \angle 53.2] = [78.4 \angle 59.5]$
Thus, the sum has a magnitude of 78.4 and is in the direction of 59.5°.

The student should be aware that in some cases it is certainly possible for the x- and/or y-components to be negative depending on which quadrant the angle is in. For example the x-component of the vector $[25 \angle 150]$ is 25 cos 150 or -21.6.

Exercise 5.1

In Exercise 1-8, each vector is given in polar form, find the x- and y-components of the vector. The given vectors are in standard position.

1. magnitude = 121, angle = 87.2°   2. magnitude = 35.7, angle = 125.6° 3. magnitude = 0.175, angle = 227.6°   4. magnitude = 1273, angle = 312.5°

5. [12243 ∠172.8°]   6. [65.3 ∠270.0°]

7. [1.25 ∠89.8°]   8. [789 ∠157.5°]

In Exercises 9-16, each vector is given in rectangular form, find the magnitude and direction of the vector. The given vectors are in standard position.

9. x-component is 16.5, y-component is 25.2
10. x-component is 4275, y-component is –3672
11. x-component is –1.475, y-component is –0.1758
12. x-component is –17.3, y-component is 37.4
13. [1273, 127.3]   14. [–6.72, 0.00]
15. [–152, 43.5]   16. [16530, –87634]

In Exercises 17-24, find the sum, $\vec{A}+\vec{B}$ or the sum $\vec{A}+\vec{B}+\vec{C}$ of the given vectors.

17. $A = 24.7$, $\theta_A = 37.8°$ and $B = 12.5$, $\theta_B = 115.9°$
18. $A = 125.2$, $\theta_A = 85.64°$ and $B = 312.8$, $\theta_B = 212.35°$
19. $\vec{A} = [52.34 ∠134.34°]$ and $\vec{B} = [37.25 ∠226.46°]$
20. $\vec{A} = [2.17 ∠12.2°]$ and $\vec{B} = [3.48 ∠294.6°]$
21. $A = 17.8$, $\theta_A = 35.4°$, $B = 25.2$, $\theta_B = 157.6°$, and $C = 16.1$, $\theta_C = 270.0°$
22. $A = 1.752$, $\theta_A = 112.2°$, $B = 2.175$, $\theta_B = 180.0°$, and $C = 5.143$, $\theta_C = 345.6°$
23. $\vec{A} = [243.1 ∠16.35°]$, $\vec{B} = [1724.4 ∠85.63°]$, and $\vec{C} = [1934.6 ∠220.64°]$
24. $\vec{A} = [12370 ∠184.3°]$, $\vec{B} = [11540 ∠269.5°]$, and $\vec{C} = [24310 ∠169.3°]$
25. The force acting on an object is 24.3 pounds at an angle of 42.3° above the horizontal. Find the horizontal and vertical components.
26. The force on a cable is 5674 pounds at an angle of 152.5°. Find the horizontal and vertical components.
27. A ship is sailing in a direction 35.6 degrees south of east at a speed of 25.2 miles per hour. Find the east and south components of the vector representing the ship.

28. A car is driving on a straight interstate highway in Texas. In 15.0 minutes after passing point A, the car is 14.2 miles north and 10.8 miles west of point A. What is the speed of the car in miles per hour and what is its direction?

♦ 29. *(Washington, Exercises 9-2, #20)* A jet is 145.5 km at a position 37.35° north of east of Tahiti. What are the components of the jet's displacement from Tahiti?

♦ 30. *(Washington, Exercises 9-2, #18)* Water is flowing downhill at 17.75 ft/s through a pipe that is at an angle of 65.68° with the horizontal. What are the horizontal and vertical components of the velocity?

♦ 31. *(Washington, Exercises 9-4, #5)* In lifting a heavy piece of equipment from the mud, a cable exerts a vertical force of 7550 N, and a cable from a truck exerts a force of 8230 N at 12.5° above the horizontal. Find the resultant of these forces.

♦ 32. *(Washington, Exercises 9-4, #17)* A plane flies at 535 km/h into a head wind of 62.0 km/h at an angle of 76.3° with the direction of the plane. Find the resultant velocity of the plane with respect to the ground.

5.2 Complex Numbers
(Washington, Sections 12-2, 12-4, and 12-6)

If we were attempting to solve an equation of the form $x^2 = -4$ we would not be able to find any real number as a solution since real numbers squared are always positive. However, there are many applications in which numbers of the type that would solve this equation are valuable. Thus, it is necessary to define numbers that involve the square roots of negative numbers. In mathematics and many areas the lowercase letter i is set by definition to be equal to $\sqrt{-1}$. In areas such as electronics where i is used for current, it is better to set $j = \sqrt{-1}$. In this manual we will generally use j. Numbers of the form $a + bj$ where a and b are real numbers are called **complex numbers**.

Since a complex number is represented by two real values, each complex number corresponds to a point on the rectangular coordinate system where values on the x-axis correspond to the real part of the complex number and values on the y-axis correspond to the imaginary part of the complex number. For the complex number $3 + 2j$, the real part, 3, corresponds to the x-coordinate 3 and the real number, 2, in the imaginary part, corresponds to the y-coordinate 2. Thus, the complex number $3 + 2j$ can be represented by a vector from the origin to the point (3, 2). A coordinate system set up in this way for complex numbers is called the **complex plane** and each complex number corresponds to a vector in the plane. In the complex plane, the x-axis is the called the **real axis** and the y-axis is called the **imaginary axis**.

Considering the complex number $x + yj$ as a vector from the origin to the point (x, y) with a magnitude r and a direction θ, we have the familiar relationships or x, y, r and θ given in Equation 3. Using $x = r \cos \theta$ and $y = r \sin \theta$, the complex number $x + yj$ may also be written in the form $r \cos \theta + (r \sin \theta) j$ or the form $r(\cos \theta + j \sin \theta)$. The form $x + yj$ is called the

rectangular form of the complex number and the form $r(\cos\theta + j\sin\theta)$ is called the **polar form**. Another notation for $r(\cos\theta + j\sin\theta)$ is $r\angle\theta$. From mathematics it can also be shown that $r(\cos\theta + j\sin\theta) = re^{\theta j}$. (On a calculator $re^{\theta j}$ may be displayed as $re\wedge\theta i$.) The form $re^{\theta i}$ is a polar form of a complex number and when using this form, the angle θ, must be in terms of radians. We note here that some calculators may use i and some may use j with complex numbers.

A vector given by its x- and y-components corresponds to the rectangular form of a complex number $x + yj$. A vector identified by its magnitude and direction corresponds to the polar form of a complex number $r\angle\theta$ or $re^{\theta i}$. Finding the sum of two complex numbers is similar to finding the sum of two vectors (see Procedure C16).

All of the familiar operations of addition, subtraction, multiplication, and division may be performed on complex numbers. For complex numbers in rectangular form, these operations are similar to algebraic operations as shown in the following examples. Recall that since $j = \sqrt{-1}$, then $j^2 = -1$.

Additon: $(2 + 3j) + (3 - 4j) = 5 - j$

Subtraction: $(2 + 3j) - (3 - 4j) = -1 + 7j$

Multiplication: $(2 + 3j)(3 - 4j) = 6 - 8j + 9j - 12j^2 = 18 + j$

For division, multiply the numerator and denominator by the conjugate of the denominator. The **conjugate** of the number $a + bj$ is $a - bj$.

$$\frac{2+3j}{3-4j} = \frac{(2+3j)(3+4j)}{(3-4j)(3+4j)} = \frac{6+8j+9j+12j^2}{9-16j^2} = \frac{-6+17j}{25} = -\frac{6}{25} + \frac{17}{25}j$$

The **absolute value of a complex number** is the magnitude of the corresponding vector:

Absolute value of $2 + 3j = |2 + 3j| = \sqrt{2^2 + 3^2} = \sqrt{13}$

If the complex numbers are given in polar form, then the operations may be performed by first converting the complex number to rectangular form, then proceeding to do the operations in rectangular form. The result can then be converted back to polar form. Some calculators perform operations on complex numbers directly regardless of the form of the complex numbers. On calculators where the form $re^{\theta j}$ is used, remember θ must be in radians. A summary of the relationship of vectors and complex numbers is given below:

	Vectors	complex numbers
rectangular form	$[x, y]$	$x + yj$ or (x, y)
polar form	$[r\angle\theta]$	$r\angle\theta$ or $(r\angle\theta)$ or $re^{\theta j}$

Procedure C17.   Conversion of complex numbers.

Make sure calculator is in correct mode -- radians or degrees.

A. To change from **polar to rectangular** form:

On the TI-82 or 83or 84 Plus:

Think of the complex number $r \angle \theta$ as the vector $[r \angle \theta]$ and use Procedure C15.

or

On the TI-83 or 84 Plus, the mode can be set to an output of the form $a + bi$ (under MODE key) and the number entered in the form $re^{\wedge}\theta i$ using the key for $e^{\wedge}$ and the key for i, (not the letter I) and then press ENTER. In this case it is first necessary to make sure that θ is in radians.

On the TI-85 or 86:

1. Enter the magnitude and angle separated by $\angle$ in ( ).
2. Press the key CPLX
3. Select ▷Rec (after pressing MORE) and press ENTER

Example: (2.0 $\angle$45)▷Rec gives (1.4, 1.4) in degree mode.

B. To change from **rectangular to polar** form:

On the TI-82 or 83 or 84 Plus:

Think of the complex number $x + yj$ as the vector $[x, y]$ and use Procedure C15.

or

On the TI-83 or 84 Plus, the mode can be set to an output of the form $re^{\wedge}\theta i$ (under MODE key) and the number entered in the form $a + bi$ and then press ENTER. Whether θ is in radians or degrees will depend on whether the calculator is set to radians or degrees.

On the TI-85 or 86:

1. Enter the values of x and y inside of ( ) separated by a comma.
2. Press the key CPLX
3. Select ▷Pol
4. Press ENTER

Example: (1.0, 1.0) ▷Pol gives (1.4 $\angle$45) in degree mode.

Note: On the TI-85, a complex number is expressed as a pair of numbers inside of parentheses ( ). When the numbers are separated by a comma they are in rectangular form and when separated by $\angle$ they are in polar form.

Example 5.3: (a) Convert $12.4(\cos 25.8° + j \sin 25.8°)$ to rectangular form.

(b) Convert the complex number $3.5 - 2.7j$ into polar form.

Solution: 1. In part (a), the complex number $12.4(\cos 25.8° + j \sin 25.8°) = 12.4 \angle 25.8°$. [This is also $12.4e^{\wedge}0.45i$ with the angle in radians.] Use Procedure C17 to convert this to rectangular form and obtain (11.2, 5.39) or $11.2 + 5.39j$.

2. For part (b) use Procedure C17 to convert the complex number
    $3.5 - 2.7j$ or $(3.5, -2.7)$ to polar form.
    The result is $4.4 \angle 322°$ or $4.4(\cos 322° + j \sin 322°)$
    [or $4.4e^{\wedge}5.62i$ in radians.]

Procedure C18. **Operations on complex numbers.**

On the TI-82 or 83 or 84 Plus:

1. If not already in rectangular form, change the complex number into rectangular form as when changing polar coordinates to rectangular coordinates (see Procedure C15).

    On the TI-83 and 84 Plus, complex numbers may be entered in the form $a + bi$ or $re^{\wedge}\theta i$.

2. Perform the operation on the rectangular forms of the complex numbers.

3. If the answer is to be in polar form, convert the result back to polar form as when changing rectangular coordinates to polar coordinates (see Procedure C15).

    Note: On the TI-83 and 84 Plus the operations may be done leaving the complex numbers in the form $re^{\wedge}\theta i$ if the values for θ are in radians.

On the TI-85 or 86:

Operations are performed by entering the complex numbers in either rectangular or polar form using the operation symbols $+$, $-$, $\times$, or $\div$ between the numbers. (See note on Procedure C17.)

Note: Powers and roots of complex number may be found by raising the complex number to the desired power using $\wedge$.

The form of output of the complex number on the screen may be changed by pressing the key MODE and in the fourth line highlighting
    RectC for rectangular mode
    PolarC for polar mode

Example 5.4: Find the following and give answers in polar form:

(a) $2.24 \angle 135.4° + 3.18 \angle 275.7°$

(b) $(2.24 \angle 135.4°)(3.18 \angle 275.7°)$

(c) $(5.7e^{1.2j}) \div (3.1e^{2.5j})$ (θ in radians)

Solution: 1. In part (a), use Procedure C18 to find the sum of the two complex numbers.
    The result is $2.04 \angle 231.2°$.
    (The rectangular form of $2.24 \angle 135.4°$ is $-1.59 + 1.57j$ and the rectangular form of $3.18 \angle 275.7°$ is $0.316 - 3.16j$. We must add the two rectangular forms then convert back to polar form.)

2. In part (b), use Procedure C18 to find the products of the two complex numbers. The result is 7.12 $\angle$51.1°.

3. In part (c), use Procedure C18.

The result is $1.8e^{-1.3j}$. Note that this is simply performing algebra on the two quantities: dividing 5.7 by 3.1 and subtracting exponents. If we convert this to degrees we find that the answer is 1.8$\angle$285°.

Exercise 5.2

Except where noted all angles are given in degrees.
In Exercises 1-6, convert the complex number in rectangular form into a complex number in polar form.

1. $3.5 + 2.8j$

2. $-12.4 - 18.6j$

3. $-123.7 + 156.4j$

4. $-67.8j$

5. $-0.125 - 0.0615j$

6. $1275 - 1647j$

In Exercise 7-16, convert the complex number in polar form into a complex number in the rectangular form $a + bj$.

7. $837(\cos 121.5° + j \sin 121.5°)$

8. $-12.9(\cos 195.7° + j \sin 195.7°)$

9. $1.654 \angle 227.8°$

10. $3.59 \angle 99.9°$

11. $-87.9 \angle 335.6°$

12. $175.6 \angle 270.0°$

13. $4.5e^{1.2i}$ (θ in radians)

14. $-12.8e^{0.71i}$ (θ in radians)

15. $293e^{2.54i}$ (θ in radians)

16. $0.25e^{2.4i}$ (θ in radians)

In Exercises 17-30, perform the indicated operation on the complex numbers. The answers should be in the same form as the complex numbers in the problem.

17. $(12.5 + 17.8j) + (13.1 - 25.4j)$

18. $(6.134 - 5.234j) - (7.154 + 8.451j)$

19. $(-112.5 + 54.6j)(104.5 - 38.5j)$

20. $(0.165 - 1.285j)(-0.341 - 0.897j)$

21. $\dfrac{63.8 - 12.4j}{43.8 + 31.4j}$

22. $\dfrac{2.384 + 1.783j}{1.879 - 3.183j}$

23. $(25.4 \angle 147°) - (16.3 \angle 226.5°)$

24. $(1.45 \angle 235.9°) + (4.17 \angle 100.4°)$

25. $(-177.2 \angle 178.6°) + (239.4 \angle 85.7°)$

26. $(16.88 \angle 181.32°)(55.32 \angle 289.45°)$

27. $5.0e^{0.75i} + 3.5e^{1.37i}$ (θ in radians)

28. $(0.25e^{1.35i})(1.40e^{1.15i})$ (θ in radians)

29. $110.2 (\cos 12.4° + j \sin 12.4°) \times 89.3 (\cos 110.4° - j \sin 110.4°)$

30. $72.3 (\cos 127.8° - j \sin 127.8°) + (-55.8)(\cos 226.8° + j \sin 226.8°)$

31. (a) Find $(2 \angle 120)^3$ and $(2 \angle 240)^3$.

    (b) Are $2 \angle 120$ and $2 \angle 240$ cube roots of 8? Explain. Write each in rectangular form.

32. (a) Find $(1 \angle 90)^4$ and $(1 \angle 270)^4$.

    (b) Are $1 \angle 90$ and $1 \angle 270$ fourth roots of 1? Explain. Write each in rectangular form.

33. (a) Find $(2 \angle 90)^2$. Write answer in polar and in rectangular form.

    (b) Is $2 \angle 90$ the square root of -4? Explain.

34. Verify by substitution that $-1 + 2j$ is the solution to $x^2 + 2x + 5 = 0$

Exercises 35-42 refer to alternating current circuits. In an alternating current circuit the voltage E is given by $E = IZ$ where I is the current in amperes and Z is the impedance in ohms.

♦ 35. *(Washington, Exercises 12-2, #45)* Find the complex number representation of E if $I = 0.867 - 0.524j$ and $Z = 240 + 175j$.

♦ 36. *(Washington, Exercises 12-2, #46)* Find the complex number representation of Z if $E = 6.24 - 4.17j$ and $I = 0.427 - 0.573j$.

37. Find the complex number representation of E and write in rectangular form if $·I = 6.25 \angle 58.0°$ and $Z = 85.5 \angle 35.5°$

38. Find the complex number representation of Z and write in rectangular form if $E = 105 \angle 30.5°$ and $I = 4.24 \angle 15.0°$

♦ 39. *(Washington, Exercises 12-4, #37)* The voltage of a certain generator is represented by $2.745 - 1.127j$ kV. Write this voltage in polar form.

♦ 40. *(Washington, Exercises 12-4, #40)* The current in a microprocessor circuit is represented by $3.572 \angle 15.25°$ μA. Write this in rectangular form.

♦ 41. *(Washington, Exercises 12-6, #53)* The power P in watts (W) supplied to an element in a circuit is the product of the voltage E and the current I (in A). That is $P = EI$. Find the expression for the power supplied if $E = 7.15 \angle 59.5°$ volts and $I = 6.35 \angle -14.8°$ amperes.

♦ 42. *(Washington, Exercises 12-6, #54)* The displacement d of a weight suspended on a system of two springs is $d = 6.24 \angle 23.4° + 2.88 \angle 75.5°$ inches. Perform the addition and express the answer in polar form.

Chapter 6

Graphs of Trigonometric Functions

6.1 Graphing of Basic Sine and Cosine Functions
(Washington, Sections 10.1 and 10.2)

The graphing of trigonometric functions on the calculator is done in much the same way as the graphing of algebraic functions. However, we need to be aware of two things. First, trigonometric functions should be graphed with the calculator set in radians (See Procedure C6). Second, there is a special trigonometric viewing rectangle for trigonometric functions (see Procedure G6) that marks the x-axis at multiples of $\dfrac{\pi}{2}$. We also need to remember when we read points off the graph that the angular type values will be in terms of radians.

Example 6.1: On the graphing calculator graph $y = \sin x$ and $y = \cos x$ using the trigonometric viewing rectangle. Estimate the local maximum and local minimum points to one decimal place and estimate the x-intercepts to one decimal place for $-2\pi \le x \le 2\pi$.

Solution:
1. Graph $y = \sin x$ on the trigonometric viewing rectangle. (See procedure G6).
   For the standard trigonometric viewing rectangle:

   Xmin = −6.28... (2π),
   Xmax = 6.28... (2π),
   Xscl = 1.57... $(\pi/2)$,
   Ymin = −3 or −4, Ymax = 3 or 4,
   Yscl = 0.25 or 1

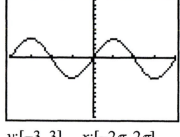

y:[−3, 3] x:[−2π, 2π]

Figure 6.1

2. The graph of the $y = \sin x$ appears in Figure 6.1.

3. Using the trace function we estimate that the local maximum points are (−4.7, 1) and (1.5, 1) and the local minimum points are (−1.5, −1) and (4.7, −1).

   These correspond to points $(-\dfrac{3\pi}{2}, 1)$, $(\dfrac{\pi}{2}, 1)$, $(-\dfrac{\pi}{2}, -1)$ and $(\dfrac{3\pi}{2}, -1)$.

The x-intercepts are $-6.3, -3.1, 0, 3.1, 6.3,$ and these values correspond to $-2\pi, -\pi, 0, \pi, 2\pi$.

4. Graph $y = \cos x$ on the standard trigonometric viewing rectangle. The graph is in Figure 6.2

5. The local maximum points are $(-6.3, 1)$, $(0, 1)$ and $(6.3, 1)$ and the local minimum points are $(-3.1, -1)$ and $(3.1, -1)$. These correspond to $(-2\pi, 1)$, $(0, 1)$, $(2\pi, 1)$, $(-\pi, -1)$, and $(\pi, -1)$. The x-intercepts are $-4.7, -1.5, 1.5, 4.7$ which correspond to $-\dfrac{3\pi}{2}, -\dfrac{\pi}{2}, \dfrac{\pi}{2},$ and $\dfrac{3\pi}{2}$.

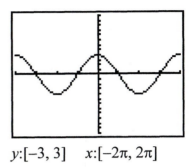

$y{:}[-3, 3] \quad x{:}[-2\pi, 2\pi]$

Figure 6.2

A mental snapshot should be taken of the graphs of $\sin x$ and $\cos x$ (i.e. commit to memory). It is important to know what these graphs look like, the location of their local maximum and local minimum points (in terms of π), and the location of the x-intercepts (in terms of π). Graphs of other sine and cosine functions may be obtained from these basic graphs by modifying the graph. The exercises for this section will explore this more. The graphs shown in Examples 6.1 and 6.2 are typical of graphs of trigonometric functions in that a portion of the curve repeats along the x-axis. The repeating portion of the curve is called a **cycle**. The **amplitude** of the graph of sine or cosine functions is one-half the difference between the minimum and the maximum y-values. The **period** is the length of one complete cycle of the graph. If P is the period of a function, then the function evaluated at x and at $x + P$ gives the same value for the function. That is, $f(x) = f(x + P)$. The amplitude for both $y = \sin x$ and $y = \cos x$ of Example 6.1 is 1 and the period in each case is 2π. Thus, $\sin x = \sin (x + 2\pi)$ and $\cos x = \cos (x + 2\pi)$.

Example 6.2: Graph $y = \cos 0.5x$ and $y = 3 \cos 0.5x$ and determine the difference between the two graphs.

Solution: 1. Make sure the calculator is in radian mode, then graph $y = \cos (0.5x)$ and $y = 3 \cos (0.5x)$ on the trigonometric viewing rectangle. The parentheses may or may not be needed, depending on the calculator. It is always a good idea to include the quantity following a trigonometric function within parentheses.

2. The resulting graph is in Figure 6.3. Note that the differences in the two graphs are the heights or amplitudes. The amplitude of the first graph is 1 and of the second graph is 3. Both curves have the same zeros, which are the same points where they cross the x-axis. These points are $(\pi, 0)$ and $(-\pi, 0)$. We note that the graph of $y = 3\cos(0.5x)$ may be obtained from the graph of $y = \cos(0.5x)$ by stretching the graph in the y-direction.

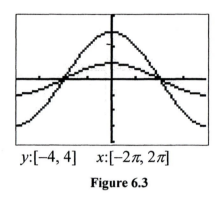

$y:[-4, 4]$ $x:[-2\pi, 2\pi]$

Figure 6.3

The location of the x-intercepts and of the highest or lowest points on the graphs may be determined by using the trace key as before. In order to determine these points to more precision, we may use the zoom function or the built-in function to find maximum or minimum points as in Procedure G12.

The cosine graph of Example 6.2 was stretched in the y-direction by multiplying the function by a constant greater than one. If we multiply by a constant less than one, the graph will be compressed in the y-direction. The graph may also be stretched or compressed in the x-direction or reflected about the x-axis. These properties will be further investigated in the exercise set.

Example 6.3:   Graph $y = 3\sin \pi t$ and determine the first 3 non-negative zeros of this function and find the first maximum value for y when x is positive.

Solution: 1. First we graph this on the trigonometric viewing rectangle. The graph obtained is given in Figure 6.4.
2. In this case having the x scale in terms of π is not the best way. By using the trace function we determine that the graph crosses the x-axis near 1 and near 2. Since we really only need to see one cycle of the graph on the positive x-axis, we change the viewing rectangle. In this case it is appropriate to use decimal numbers. We change our window so that minX = –2.5, maxX = 2.5. The resulting graph is in Figure 6.5.
3. Now, it is easier to determine the non-negative zeros (or x-intercepts) of the function. These are found to be 0.0, 1.0, and 2.0. The first maximum point for positive x is found to be (0.5, 3) and the maximum value for y is 3.

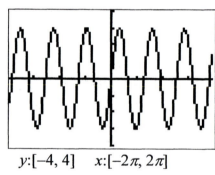

$y:[-4, 4]$ $x:[-2\pi, 2\pi]$

Figure 6.4

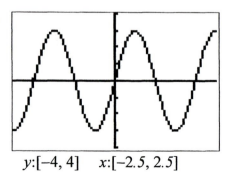

$y:[-4, 4]$ $x:[-2.5, 2.5]$

Figure 6.5

Care is needed in determining the period of a function from the graph on the calculator screen. Because of the way the calculator plots the graph, things may not be exactly as they appear. (See Exercises 9 and 10 in Exercise Set 6.1) **Before graphing trigonometric functions on the graphing calculator it is best to first determine the amplitude and period of the functions if this is possible. This will allow the correct choice of a viewing rectangle.**

Exercise 6.1

For each of the following use the trigonometric viewing rectangle. Make sure calculator is in radians. Find values to two decimal places.

1. Graph the function $y = 3\sin x$ on your calculator then estimate the y values in the following table.

0	$\frac{\pi}{2}$	π	$\frac{3\pi}{2}$	2π

2. Graph the function $y = 4\cos x$ on your calculator, then estimate the y-values in the following tables.

0	$\frac{\pi}{2}$	π	$\frac{3\pi}{2}$	2π

3. Use the calculator to graph $y = \sin x$, $y = 2\sin x$, and $y = 3\sin x$.
   (a) In each case, give the y-coordinates of the local maximum and local minimum points.
   (b) What is the amplitude of each function?
   (c) What is the effect of changing a in the equation $y = a\sin x$?
   (d) Sketch the graph of $y = 5\sin x$ by hand. Check your answer on the calculator.
4. Use the calculator to graph $y = \cos x$, $y = 3\cos x$, and $y = 6\cos x$.
   (a) In each case, give the y-coordinates of the local maximum and local minimum points.
   (b) What is the amplitude of each function?
   (c) What is the effect of changing a in the equation $y = a\cos x$?
   (d) Sketch the graph of $y = 4\cos x$ by hand. Check your answer on the calculator.

5. Use the calculator to graph $y = 2 \sin x$, $y = 3 \cos x$, $y = -2 \sin x$ and $y = -3 \cos x$.
   (a) In each case, give the y-coordinates of the local maximum and local minimum points.
   (b) What is the amplitude of each function?
   (c) What is the effect of changing the sign of a in the equation $y = a \sin x$ or in the equation $y = a \cos x$?
   (d) Sketch the graph of $y = -4 \sin x$ and $y = -6 \cos x$ by hand. Check your answers on the calculator.

6. Use the calculator to graph $y = 2 \sin x$, $y = 2 \sin 2x$, and $y = 2 \sin 4x$.
   (a) In each case, what are the first three x-intercepts (zeros) that have a value greater than or equal to zero? Determine the difference between the first and third. Give in terms of π.
   (b) What is the amplitude and period of each function?
   (c) What is the effect of changing b in the equation $y = a \sin bx$?
   (d) Sketch the graph of $y = 5 \sin 3x$ by hand. Check your answer on the calculator.
   (e) Sketch the graph of $y = -3 \sin 2x$ by hand. Check your answer on the calculator.

7. Use the calculator to graph $y = 4 \cos x$, $y = 4 \cos 2x$, and $y = 2 \cos 3x$.
   (a) In each case, what are the first three x-intercepts (zeros) that have a value greater than or equal to zero? Determine the difference between the first and third. Give in terms of π?
   (b) What is the amplitude and period of each function?
   (c) What is the effect of changing b in the equation $y = a \cos bx$?
   (d) Sketch the graph of $y = 5 \cos 2x$ by hand. Check your answer on the calculator.
   (e) Sketch the graph of $y = -6 \cos 5x$ by hand. Check your answer on the calculator.

8. Determine the amplitude and period of each of the following functions, then sketch one cycle each graph by hand. Indicate on the x-axes the intercepts and on the y-axis the amplitude. You may use your calculator to check your answer.
   (a) $y = 2.5 \sin 3x$ (b) $y = 15.0 \cos 2x$
   (c) $y = 6 \sin \pi x$ (d) $y = -5 \cos (\frac{\pi}{3} x)$
   (e) $y = 110 \cos 60 \pi x$ (f) $y = -0.5 \sin 2.5x$

9. For the function $y = 3 \sin 50x$:
   (a) Graph on the standard trigonometric viewing rectangle. Estimate the period from this graph.
   (b) What is the real period of this function?
   (c) Graph this function so that two cycles appear on the screen for positive x-values. What are the minimum and maximum values of x that describe the viewing rectangle?

10. For the function $y = 2 \cos 60 \pi x$:
    (a) Graph on the standard trigonometric viewing rectangle. Estimate the period from this graph.
    (b) What is the real period of this function?
    (c) Graph this function so that two cycles appear on the screen for positive x-values. What are the minimum and maximum values of x that describe the viewing rectangle?

In Exercises 11-14, graph the function so that two cycles appear on the screen for positive x-values. In each case, give the minimum and maximum values of x and y that describe the viewing rectangle and estimate the x-intercepts.

11. $y = 5 \sin 25x$

12. $y = 0.25 \cos 18x$

13. $y = 0.75 \cos 0.024\pi x$

14. $y = 50 \sin 0.10x$

15. Use the calculator to graph $y = 20 \sin 300x$ in the trigonometric viewing rectangle. Is the graph you obtain correct? Can you explain what happened? What would be a better viewing rectangle in which to view this graph? Find the smallest positive x-intercept to two significant digits.

16. Use the calculator to graph $y = 120 \cos 120\pi x$ in the trigonometric viewing rectangle. Is the graph you obtain correct? Can you explain what happened? What would be a better viewing rectangle in which to view this graph? Give the values of Xmin, Xmax, Ymin, and Ymax that you use. Find the smallest positive x-intercept to two significant digits.

♦ 17. (Washington, Exercises 10.2, #53) The standard electric voltage in a 60-Hz alternating current circuit is given by $V = 120 \sin 120\pi t$ where t is the time in seconds.
   (a) Graph the function and from the graph determine the values for V when $t = 0.008$ and when $t = 0.010$ seconds.
   (b) For what values of t ($0 \le t \le 0.02$) is the voltage zero?
   (c) For what values of t ($0 \le t \le 0.02$) is the voltage 90 volts? (Hint: see Problem 15, Exercise 4.1.)
   (d) Find the first and second positive times when the voltage is at its maximum.

♦ 18. (Washington, Exercises 10.2, #54) To tune the instruments of an orchestra before a concert, an A note is struck on a piano. The piano wire vibrates with a displacement y (in mm) given by $y = 3.45 \cos 880\pi x$, where x is the time in seconds.
   (a) Graph this function and from the graph determine the values of y when $x = 0.0015$ seconds and when $x = 0.0035$ seconds.
   (b) For what values of x ($0 \le x \le 0.005$) is $y = 1.60$? (Hint: see Problem 15, Exercise 4.1.)
   (c) For what values of x does the graph reach a maximum point?.

6.2 Graphing of Sine and Cosine Functions with Displacement
(Washington, Section 10.3)

We now investigate the graphs of functions of the form $y = a \sin (bx + c)$ and $y = a \cos (bx + c)$. The value c is referred to as the **phase angle**. The affect of c is to shift the graph horizontally, either to the right or left. The amount of shift is often referred to as the **displacement** or **phase shift**. Example 6.3 will illustrate the effect of a phase angle.

Example 6.4:   Graph the functions $y = 3 \sin x$ and $y = 3 \sin(x + \frac{\pi}{4})$ on the same graph.

Determine how far and in which direction the second graph is shifted relative to the first graph and express in terms of π.

Solution:

1. Enter the first function for Y_1 and the second function for Y_2. Be sure the calculator is in radian mode and use the trigonometric viewing rectangle. The result is shown in Figure 6.6.

2. To determine how far the second curve is shifted horizontally from the first, determine the first positive x-intercept of the first graph. This occurs at $x = 3.1415927$. Then determine the first positive x-intercept of the second graph. This is at $x = 2.3561945$. The difference between the two intercepts is 0.7853982 which is the same as $\frac{\pi}{4}$.

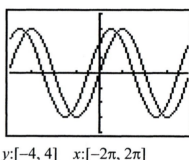

$y: [-4, 4]$ $x: [-2\pi, 2\pi]$

Figure 6.6

The second curve is shifted to the left relative to the first curve.

The addition of the phase angle $\frac{\pi}{4}$ to the function $y = \sin x$, displaces the curve $\frac{\pi}{4}$ units to the left. Another way of stating this is that when we replace x by $(x + \frac{\pi}{4})$ in the function $y = \sin x$, the graph is shifted $\frac{\pi}{4}$ units to the left.

The exercise set includes discovery of the effect of phase angles on various types for functions. Based on what we know from Section 6.1 and what is illustrated in the exercises, we summarize the important quantities about the graphs of sine and cosine curves.

> For equations of the form $y = a \sin(bx + c)$ and $y = a \cos(bx + c)$:
>
> The amplitude is given by $|a|$
>
> The period by $\frac{2\pi}{b}$
>
> The phase shift or displacement by $-\frac{c}{b}$

If the quantity $-\frac{c}{b}$ is positive the displacement is to the right and if $-\frac{c}{b}$ is negative, the displacement is to the left. This last statement will be demonstrated in the exercises.

Exercise 6.2

Determine values to two decimal places.

1. Graph the functions $y = 2 \sin x$, $y = 2 \sin(x + \frac{\pi}{4})$, and $y = 2 \sin(x + \frac{\pi}{2})$.

   (a) What is the amplitude and period of each graph?
   (b) Determine the x-values where each graph crosses the x-axis for $0 < x \le \pi$. How much is the second graph shifted with respect to the first? What about the third with respect to the first? In what direction? Express each difference in terms of π.

   (Example: $1.57 = \frac{\pi}{2}$.)

   (c) What is the effect of c in the equation $y = a \sin(x + c)$?
   (d) Does the amplitude have any effect on the displacement? If so, what?

   (e) Sketch one cycle of the graph of $y = 3 \sin(x + \frac{\pi}{3})$. Check your answer on the calculator.

2. Graph the functions $y = 3 \cos x$, $y = 3 \cos(x + \frac{\pi}{4})$, and $y = -2 \cos(x - \frac{\pi}{4})$.

   (a) What is the amplitude and period of each graph?
   (b) Determine the x-values where the graph crosses the x-axis for $0 < x \le \pi$. How much is the second graph shifted with respect to the first? What about the third with respect to the first? In what direction? Express each difference in terms of π.

   (Example: $1.57 = \frac{\pi}{2}$.)

   (c) What is the effect of c in the equation $y = a \cos(x + c)$?
   (d) Does the amplitude have any effect on the displacement? If so, what?

   (e) Sketch one cycle of the graph of $y = -3 \cos(x - \frac{\pi}{2})$. Check you answer on the calculator.

3. Graph the functions $y = 2 \sin 2x$, $y = 2 \sin(2x - \frac{\pi}{2})$, and $y = 2 \sin(2x - \frac{\pi}{4})$.

   (a) What is the amplitude and period of each graph?
   (b) Determine the x-values where the graph crosses the x-axis for $0 \le x \le \pi$. How much is the second graph shifted with respect to the first? What about the third with respect to the first? In what direction? Express each difference in terms of π.
   (c) Divide the phase angle by the coefficient of x in each case and note how this relates to the displacement. What is the effect of $-\frac{c}{b}$ in the equation $y = a \sin(bx + c)$?

   (d) Sketch one cycle of the graph of $y = 3 \sin(2x + \frac{\pi}{4})$. Check your answer on the calculator.

4. Graph the functions $y = 2.5 \cos \pi x$, $y = 2.5 \cos(\pi x + \frac{\pi}{3})$, and $y = 2.5 \cos(\pi x - \frac{\pi}{6})$.

    (a) What is the amplitude and period of each graph?

    (b) Determine the x-values where the graph crosses the x-axis for $0 < x \le 1$. How much is the second graph shifted with respect to the first? What about the third with respect to the first? In what direction?

    (c) Divide the phase angle by the coefficient of x in each case and note how this relates to the displacement. What is the effect of $-\frac{c}{b}$ in the equation $y = a \sin(bx + c)$?

    (d) Sketch one cycle of the graph of $y = 2 \cos(\pi x + \frac{\pi}{4})$. Check your answer on the calculator.

5. For each of the following equations, determine (1) the amplitude, (2) the period, and (3) the displacement. Then make a sketch of one cycle of each by hand. Check your answer with the calculator.

    (a) $y = 5 \cos 0.5x$ (b) $y = 3 \sin(4\pi x - \pi)$

    (c) $y = 2.5 \cos(x - \frac{\pi}{4})$ (d) $y = -3.5 \sin(2x + 0.5)$

6. For each of the following equations, determine (1) the amplitude, (2) the period, and (3) the displacement. Then make a sketch of one cycle of each by hand. Check your answer with the calculator.

    (a) $y = 12 \sin(0.25x + 0.25)$ (b) $y = -6 \cos(3\pi x - 0.50)$

    (c) $y = -4.5 \cos(6x - 3\pi)$ (d) $y = 24.5 \sin(6\pi x - 3\pi)$

7. Use the graphing calculator to graph each of the following. In each case for $0 \le x \le \pi$, determine (1) the x-intercepts, and (2) the local maximum and local minimum points. Find all values to two decimal places.

    (a) $y = 2 \sin(\pi x)$ (b) $y = 3.5 \cos(2.4x + 1.63)$

8. Use the graphing calculator to graph each of the following. In each case for $0 \le x \le \pi$, determine (1) the x-intercepts, and (2) the local maximum and local minimum points. Find all values to two decimal places.

    (a) $y = 2 \sin(\pi x)$ (b) $y = 4 \sin(2x - \frac{\pi}{2})$

9. Graph one cycle of the function $y = 57.5 \sin(0.50\pi x - 0.33)$. Then for $0 \le x \le 4.5$,

    (a) Determine y when $x = 1.75$ and $x = 2.75$.

    (b) Determine x when $y = 50.0$ and $y = -45.5$.

◆ 10. *(Washington, Exercises 10.3, #34)* The electric current i, in microamperes, in a certain circuit is given by $i = 3.8 \cos 2\pi(t + 0.20)$, where t is the time in seconds. Graph three cycles of this graph then determine:

    (a) The current at $t = 1.00$ seconds and at $t = 2.50$ seconds.

    (b) The times in the first three cycles that the current is 2.50 microamperes.

♦11. *(Washington, Exercises 10.3, # 35)* A certain satellite circles the earth such that its distance *y*, in miles north or south (attitude is not considered) from the equator is given by $y = 4500 \cos(0.025t - 0.25)$, where *t* is the time (in min) after launch. Consider a positive *y*-value as being north of the equator. Determine the period and graph two cycles of this curve then

    (a) Determine how far north or south of the equator the satellite is after one hour, after three hours, and after 6.5 hours.

    (b) Determine all the times during the first two cycles that the satellite is 4000 miles above the equator.

12. For the satellite in Exercise 11, assume each cycle corresponds to one revolution of the satellite about the earth. Due to the revolving of the earth about its axis, the satellite does not appear above the same points of the earth on each revolution. To see how the orbit of the satellite might appear when superimposed on a map of the earth, we need to determine the displacement of the earth relative to each orbit. Multiplying the number of radians the earth turns in one minute $[2\pi/(24\times60)]$ by the period gives the displacement D of the earth each orbit. This gives the equation $y = 4500 \cos(0.025t - (0.25 + nD))$ where n is the number of the orbit. To see three orbits, graph this function on the same set of axes for n = 1, 2, and 3.   for *n* = 0, 1, 2, and 3 (*n* is one less than the number of the orbit).

13. The voltage of an alternating electric current is given by $e = 110 \sin(120\pi t + \dfrac{\pi}{6})$.

    (a) Find the values of *t* for which the voltage *e* is zero for *t* < 1/60.

    (b) What is the first time that the voltage reaches a high point?

    (c) What is the second time the voltage reaches a high point? How much does this differ from the value of part (b)? Does this seem reasonable?

    (d) What is *t* when *e* = 50 for *t* < 1/60?

    (e) What is *e* when *t* = 0.010 sec.?

14. An example of damped simple harmonic motion is given by the equation $y = 5e^{-25t} \cos(50\pi t)$. (Here *e* is the mathematical constant 2.7128...)

    (a) Find the positive values of *t* for which the graph crosses the horizontal axis the first and second time.

    (b) Find the *y* and *t* values for which the graph reaches its high points the first and second times.

♦ 15. *(Washington, Exercises 10.5, #11)* For a point on a string, the displacement *y* is given by $y = A\sin 2\pi(\dfrac{t}{T} - \dfrac{x}{\lambda})$. Each point on the string moves with simple harmonic motion according to this equation. Graph two cycles for the case where *A* = 3.25 cm, *T* = 0.0445 s, λ = 42.5 cm and *x* = 5.45 cm. Then, determine

    (a) The displacement after 0.0220 s and after 0.0500 s.

    (b) The times t in the first two cycles when the displacement is −2.50 cm.

♦ 16. *(Washington, Exercises 10.5, #16)* The acoustical intensity of a sound wave is given by $I = A \cos (2\pi f t - \alpha)$, where f is the frequency of the sound. Graph two cycles of I as a function of t if $A = 0.0245$ W, $f = 245$ Hz, and $\alpha = 0.775$. Then, determine
(a) The acoustical intensity when $t = 0.00100$ and when $t = 0.00600$.
(b) The time t in the first two cycles when the acoustical intensity is a maximum.

17. In each of the following cases, graph y_1 and y_2 on the same set of axes and determine how the graphs of the two functions compare. Which of the functions are equivalent?
(a) $y_1 = 2 \sin x$ and $y_2 = 2 \sin(-x)$
(b) $y_1 = 3 \cos x$ and $y_2 = 3 \cos(-x)$
(c) $y_1 = -4 \cos x$ and $y_2 = 4 \cos (x + \pi)$
(d) $y_1 = 3 \sin x$ and $y_2 = 3 \cos(\frac{\pi}{2} - x)$
(e) $y_1 = 3 \cos (x - \frac{\pi}{2})$ and $y_2 = 5 \cos (x + \frac{3\pi}{2})$

6.3 Other Trigonometric Graphs
(Washington, Sections 10.4 and 10.6)

Investigation of the graphs of tangent, cotangent, secant, and cosecant functions are left to the exercise set. To graph the cotangent, secant, and cosecant, enter the functions into the calculator as the reciprocal of tangent, cosine, and sine functions.

Certain applications involve graphs of functions that are formed by combining simpler trigonometric functions using the operations of addition or subtraction. Functions obtained in this way are called **composite functions**. In some cases it is helpful to see the graphs of separate functions, then the graph of the combination of these two functions.

<u>Procedure G16</u>. **To graph each of two functions and then their sum.**

1. Set the correct values for the viewing rectangle.
2. Enter the first function into the function table for one function (Y_1).
3. Enter the second function into the function table for a second function (Y_2).
4. Move the cursor after the equal sign for a third function (Y_3). Press the Y-VARS key (or VARS, then select Y-VARS) and select the first function name (Y_1) from the menu.
5. Press the key: +
6. Press the Y-VARS key (or VARS, then select Y-VARS) and select the second function name (Y_2) from the menu.
    (The third line could now appear as: $Y_3 = Y_1 + Y_2$)
7. Press GRAPH

When graphing composite trigonometric functions, the viewing rectangle should be set such that at least one complete cycle of the composite function is graphed. To do this, select a viewing rectangle where the range of x-values is the least common multiple of the period of each of the separate functions and range of the y-values is equal to the sum of the amplitudes of the separate functions. The graphing of composite trigonometric functions is illustrated in Example 6.5. When graphing functions involving subtraction, it is best to think of the subtraction as adding the negative of a given function.

Example 6.5: Graph the function $y = 2 \sin x + 0.5 \cos 6x$.

Solution: 1. Select an appropriate scale for the x and y variables. The amplitude of the composite functions would not be more than $2 + 0.5$ or 2.5. The period of the first function is 2π and of the second

function is $\dfrac{\pi}{3}$. For the composite

function we take 2π, which is the least common multiple of these two numbers. In order to see at least one complete cycle, the x-scale needs to be at least 2π in length.

Set Xmin = 0, Xmax = 6.28 (2π),

Xscl = 1.57 ($\dfrac{\pi}{2}$),   Ymin = -3,

Ymax = 3, and Yscl = 1.

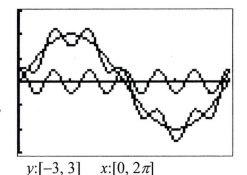

y:[-3, 3] x:[0, 2π]

Figure 6.7

2. The function y is the sum of the two functions $Y_1 = 2 \sin x$ and $Y_2 = 0.5 \cos 6x$. Use Procedure G16 to graph the two functions and their sum. The graph is in Figure 6.7. The first function graphed is $y = 2 \sin x$, the second function is $y = 0.5 \cos 6x$, and the third function is the composite function

$y = 2 \sin x + 0.5 \cos 6x$.

Graphs of some trigonometric functions may be undefined for certain values of x. Graphing of such functions on the graphing calculator may cause lines to appear on the graph that are not really part of the graph. For example, this happens in the graphing of the function $y = \tan x$ as shown in Figure 6.8. This function is undefined and can have no points on the graph at $x = -\dfrac{\pi}{2}, \dfrac{\pi}{2}, \dfrac{3\pi}{2}$, etc. Lines that appear on the graph at these values are not asymptotes and are just lines connecting end points of curves as plotted on the calculator.

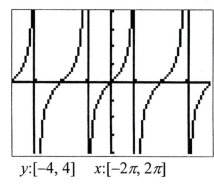

y:[-4, 4] x:[-2π, 2π]

Figure 6.8

Exercise 6.3

In Exercises 1-12 use the trigonometric viewing rectangle.

1. Graph each of the following pairs of functions on your graphing calculator. Estimate to one decimal place (no need to zoom in) the x-intercepts of each graph and the coordinates of any local maximum and/or local minimum points. Give equations of any vertical asymptotes.

   (a) $y_1 = \sin x$ and $y_2 = \csc x$ (b) $y_1 = \tan x$ and $y_2 = \cot x$ (c) $y_1 = \cos x$ and $y_2 = \sec x$

   $\left(\text{ Recall that } \cot x = \dfrac{1}{\tan x}, \sec x = \dfrac{1}{\cos x}, \text{ and } \csc x = \dfrac{1}{\sin x}. \right)$

2. Graph $y = \sec x$, $y = 2 \sec x$ and $y = 3 \sec x$.
   (a) What is the difference between the graphs of these functions?
   (b) What is the effect of a in the equations $y = a \sec x$?
   (c) By hand, sketch a graph of $y = 5 \sec x$. Check your answer on the calculator.

3. Graph $y = \tan x$, $y = 2 \tan x$ and $y = 3 \tan x$.
   (a) What is the difference between the graphs of these functions?
   (b) What is the effect of a in the equations $y = a \tan x$?
   (c) By hand, sketch a graph of $y = 0.5 \tan x$. Check your answer on the calculator.

4. Graph $y = \cot x$, $y = 2 \cot x$, and $y = 4 \cot x$
   (a) What is the difference between the graphs of these functions?
   (b) What is the effect of a in the equations $y = a \cot x$?
   (c) By hand, sketch a graph of $y = 0.5 \cot x$. Check your answer on the calculator.

5. Graph $y = \tan x$, $y = \tan (2x)$ and $y = \tan(4x)$.
   (a) What is the difference between the graphs of these functions?
   (b) What is the effect of b in the equation $y = \sec bx$?
   (c) By hand, sketch a graph of $y = \tan (0.5x)$. Check your answer on the calculator.

6. Graph $y = \csc x$, $y = \csc (2x)$, and $y = \csc (0.5x)$,
   (a) What is the difference between the graphs of these functions?
   (b) What is the effect of b in the equations $y = \csc bx$?
   (c) By hand, sketch a graph of $y = \csc(4x)$. Check your answer on the calculator.

7. Graph $y = 2 \tan x$, $y = 2 \tan \left(x + \dfrac{\pi}{4}\right)$, and $y = 2 \tan \left(x - \dfrac{\pi}{2}\right)$.

   (a) What is the difference between the graphs of these functions?
   (b) What is the effect of c in the equations $y = 2 \tan (x + c)$?

   (c) By hand, sketch a graph of $y = 4 \tan \left(x + \dfrac{\pi}{2}\right)$. Check your answer on the calculator.

8. Graph $y = 3 \cot x$, $y = 3 \cot (x - \frac{\pi}{4})$, and $y = 3 \cot (x + \frac{\pi}{2})$.

    (a) What is the difference between the graphs of these functions?

    (b) What is the effect of c in the equations $y = 2 \cot (x + c)$?

    (c) By hand, sketch a graph of $y = 4 \cot (x + \frac{\pi}{2})$. Check your answer on the calculator.

9. Graph $y = 2 \sec (0.5x)$ and $y = -2 \sec (0.5x)$ on the same graph.

    (a) What is the effect of the negative sign?

    (b) By hand, sketch the graph of $y = -3 \csc x$. Check your answer on the calculator.

10. Graph $y = 2 \tan x$ and $y = -2 \tan x$ on the same graph.

    (a) What is the effect of the negative sign?

    (b) By hand, sketch the graph of $y = -2 \cot x$. Check your answer on the calculator.

♦ 11. *(Washington, Exercises 10.4, #27)* A mechanism with two springs moves according to the equation $b = 4.25 (\sin \frac{\pi}{4}) \csc A$ where A is the angle between the springs and b is the length of one of the springs in centimeters. Graph b as a function of angle A ($0 \le A \le \pi$).

    (a) Determine b when $A = 1.00$ radian and when $A = 2.20$ radians.

    (b) What is the minimum value for b and for what A does it occur?

    (c) What happens to b as A gets closer to π?

♦ 12. *(Washington, Exercises 10.4, #25)* Near Antarctica, a small boat is at a distance s m from the base of an iceberg that has a vertical face 225 m high. From the boat, the angle of elevation θ to the top of the iceberg can be found from the equation $s = 225 \cot \theta$. Graph this function.

    (a) Determine the angle θ when s = 100.0 m and when $s = 250$ m.

    (b) Does it make any sense to try to determine y when the angle $\theta = 2.0$ radians? Explain your answer.

    (c) What happens to the angle θ as the distance s becomes large?

    (d) What is the value of θ when $s = 0$?

In problems 13-22, graph y_1, graph y_2, and then graph the composite graph. $y_1 + y_2$. Graph one complete cycle of the composite graph for values of x greater than or equal to zero. Determine the local maximum and local minimum points of the composite graph and the x-intercepts of the composite graph to two significant digits.

13. $y = 2, y = \cos x$

14. $y = 0.05 x^3, y = \sin x$

15. $y = 0.2 x^2, y = \cos 2x$

16. $y = x + 2, y = 2 \sin 2x$

17. $y = \sin 3x, y = \cos 2x$

18. $y = \sin 3x, y = \cos 4x$

19. $y = 2 \sin 3x, y = 4 \sin 2x$

20. $y = 3 \cos 0.5x, y = 2 \cos x$

21. $y = 2 \sin (2x + \frac{\pi}{4}), y = 2 \cos 3x$

22. $y = 3 \sin (x - \frac{\pi}{6}), y = 2 \cos (x - \frac{\pi}{3})$

23. A higher frequency cosine wave given by $y = 0.4 \cos 25x$ is superimposed on a lower frequency cosine wave $y = 2 \cos 3x$. Graph one complete cycle of the composite graph.
    (a) Determine the point that gives the largest value of y for x between $x = 1$ and $x = 3$.
    (b) Determine all x-intercepts between $x = 0$ and $x = 2$.
24. A sine wave given by $y = 2 \sin 100\pi x$ is superimposed on a lower frequency cosine wave given by $y = 10 \cos 10\pi x$. Graph one complete cycle of the composite graph.
    (a) Determine the point that gives the smallest value for y for x between $x = 0$ and $x = 0.2$.
    (b) Determine all x-intercepts between $x = 0$ and $x = 0.2$.
25. The vertical displacement of a buoy floating in the water is given by $y = 2.50 \cos 0.24t + 1.2 \sin 0.36t$, where y is measured in feet and t is measured in seconds. Graph one complete cycle of this graph.
    (a) What is the t-value of the first point (for $t > 0$) that the curve reaches its maximum value? What is the maximum value?
    (b) Determine the maximum distance from the lowest point on the wave to the highest point on the wave.

◆ 26. *(Washington, Exercises 10.6, #37)* The electric current I (in milliamperes) in a certain circuit is given by $i = 0.25 + 0.55 \sin t + 0.20 \cos 2t$ where t is in milliseconds. Graph one complete cycle of this graph.
    (a) Give all the values of t for which i zero in the first cycle for $t > 0$.
    (b) For what value of t would a maximum point(s) occur?

When certain signals are sent to an oscilloscope special waves may be seen on the oscilloscope screen. These waves may be represented by what are called Fourier series. Exercises 27 and 28 give the first three terms of two Fourier series. In each case, graph the first term of the series, then the sum of first two terms, and then the sum of the first three terms. Guess the next term in the series and graph the sum of the first four terms of the series. On paper sketch the type of wave that the graph approaches as more terms are considered and give the length of one cycle.

27. $y = \sin(\pi x) + \dfrac{1}{2}\sin(2\pi x) + \dfrac{1}{3}\sin(3\pi x)$

28. $y = \sin(\pi x) + \dfrac{1}{3}\sin(3\pi x) + \dfrac{1}{5}\sin(5\pi x)$

6.4 Graphing of Parametric Equations
   (Washington, Section 10.6)

Parametric equations are equations in which two or more variables are expressed in terms of a third variable called the parameter. For the two dimensional case we consider the variables x and y expressed in terms of a third variable t ($x = f(t)$ and $y = g(t)$). The variable t is called the parameter. The letter t is frequently used for the parameter since in many cases it is appropriate to consider time as the third variable. The variable t is the independent variable and x and y are both dependent variables. As t varies, the values for x and y change accordingly. On a graph the points (x, y) are plotted and connected in order of increasing values of t.

Parametric equations are easily graphed on the graphing calculator after first changing to the correct mode. Important applications of graphing of parametric equations are Lissajjous Figures.

Procedure G17. **To change graphing modes for regular functions, parametric, or polar equations.**

1. Set the graphing mode:
   Press the key MODE.
   Make active:

   On the TI-82 or 83 or 84 Plus: On the TI-85 or 86:
   (a) For graphing in **rectangular** coordinates:
       In fourth line: Func   In fourth line: RectC
                                            In fifth line: Func
   (b) For graphing **parametric** equations:
       In fourth line: Par   In fourth line: RectC
                                            In fifth line: Param
   (c) For graphing in **polar** coordinates:
       In fourth line: Pol   In fourth line: PolarC
                                            In fifth line: Pol

   Then, press QUIT or CLEAR.

2. Set the graphing format:
   On the TI-82:   On the TI-85 or 86:
   Press the key WINDOW   Press the key GRAPH
   Highlight FORMAT in the top row.   From the menu, select FORMT (after
   On the TI-83 or 84 Plus: pressing MORE)
   Press the key FORMAT
   Then: Then:
   Make active:
   (a) For graphing in **rectangular** coordinates or for **parametric**
        equations: RectGC
   (b) For graphing in **polar** coordinates: PolarGC
   Then, press QUIT or CLEAR.
   (To make an item active, move the cursor to that item and press ENTER.
   The active item is highlighted.)

After changing to the correct graphing mode, we next enter the parametric equations in the function table in much the same way as we entered regular functions.

Procedure G18. To graph parametric equations.

1. Make sure the calculator is in correct mode for parametric equations. (Procedure G17)

2. To enter the equations:

On the TI-82 or 83 or 84 Plus:
Press the key $Y=$
On the screen will appear:

$X_{1t} =$

$Y_{1t} =$

$X_{2t} =$

$Y_{2t} =$

(etc.)
(enter up to 6 pairs of equations)

On the TI-85 or 86:
Press the key GRAPH.
From the menu, select $E(t) =$
On the screen will appear:
$xt1=$
$yt1=$
(enter up to 99 pairs of equations)

Functions are not turned on (the equal signs are not highlighted) until functions have been entered for both x and y. Functions may be turned on and off as with functions in rectangular coordinates (see Note 3, Procedure G1). Only functions turned on will be graphed.

3. Enter the first equation for x in the first row and the corresponding equation of y in the second row. If there are other sets of functions you wish to graph, enter them for the other x's and y's.
To obtain the variable t

On the TI-82 or 83 or 84 Plus:
Press the key X,T,θ

On the TI-85 or 86:
Select t from the menu.

4. Press or select GRAPH to see the graph or QUIT to exit.

Some adjustment of the viewing rectangle may be necessary. The setting of x and y values are handled as previously mentioned (Procedures G5 and G6). However, with parametric equations we also need to set values for the parameter t.

Procedure G19. To change values for the parameter t.

With the calculator in the mode for parametric equations:
1. Follow Procedure G5 to see the values for the viewing rectangle.
2. Enter new values for Tmin, Tmax, and Tstep and press ENTER or move to the next item by using the cursor keys.
The standard viewing rectangle values for t are:

Tmin = 0, Tmax = 6.28...(2π) , and Tstep = 0.13...($\pi/24$)
(or Tmax = 360 and Tstep = 7.5 if in degree mode).

Tstep determines how often values for *x* and *y* are calculated. The size of Tstep will affect the appearance of the graph. Tmax should be sufficiently large to give a complete graph.

3. Press QUIT to exit.

If the step size that is set for *t* is too large, a series of line segments may result rather than a smooth appearing curve. A step size of approximately 0.1 is usually sufficient. However, be aware that if we zoom in on a portion of the graph, the step size may need to be changed to a smaller value to give a smooth graph. If the graph takes a long time to draw, try increasing the step size for t.

Example 6.6:   Graph the parametric equations $x = 4 \sin 2t$ and $y = 5 \cos (t + \frac{\pi}{4})$ for $0 \le t \le 2\pi$.

If the graph has end points, give the coordinates of the end points and estimate the *x*- and *y*-intercepts to one decimal place.

Solution: 1. Use procedure G17 to change the mode to graph parametric equations and make sure the calculator is set up for using radians.

2. It is generally a good idea to graph parametric equations using the square viewing rectangle. First obtain the standard viewing rectangle, then the square viewing rectangle (see Procedure G6). This will set Tmin = 0 and Tmax = 2π.

3. The function is defined by :

$$x = 4 \sin (2t)$$

$$y = 5 \cos (t + \frac{\pi}{4})$$

Enter these and graph according to Procedure G18.

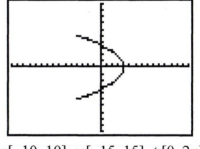

$y{:}[-10, 10]\ x{:}[-15, 15]\ t{:}[0, 2\pi]$

Figure 6.9

4. The graph is drawn in order of increasing values of the variable *t* starting at *t* = 0 and going to *t* = 2π (since this is the maximum of *T*). The graph is in Figure 6.9 and appears to be part of a parabola opening to the left. Using the trace function we determine the end points to be at (−4, −5) and (−4, 5).

The *x*-intercept is 4 and the *y*-intercepts are −3.5 and 3.5.

Notice that when using the trace function the graph is traced out in the order that *t* increases or decreases.

This graph is cyclic. If we increase the maximum value of *t*, we note that we continue to see the same graph.

To obtain a complete graph or to see particular features of a graph it may be necessary to alter the viewing rectangle. If values of t are not given in a problem, select values that give a complete graph. If the graph is cyclic (that is, the same graph is retraced as t increases), select a maximum value of t to give one complete cycle of the graph.

Exercise 6.4

Graph the following sets of parametric equations for the specified values of t. For best results, the calculator should be in radian mode and you should use a square viewing rectangle. (a) Draw a sketch of the complete graph, (b) give the x- and y-intercepts of the graph, and (c) if the graph has endpoints, give the coordinates of the endpoints .
Note: The viewing rectangle may need to be changed to see a complete graph.

1. $x = 3t - 5, y = 6 - t, \; -3 \le t \le 3$

2. $x = 2t + 6, y = 3t + 2, \; -4 \le t \le 4$

3. $x = 3t, y = t^2, \; -5 \le t \le 5$
   What happens to the graph as a is changed in the equations $x = at, \; y = t^2$?

4. $x = t^2, y = t, \; -5 \le t \le 5$
   What happens to the graph as b is changed in the equation $x = t^2, y = bt$?

5. $x = 4t, \; y = 4t^{-1}, \; -8 \le t \le 8$

6. $x = t, y = \sqrt{25 - t^2}, \; -5 \le t \le 5$

In problems 7-12, use $0 \le t \le 2\pi$. The graphs obtained are figures that may appear on an oscilloscope. They are called Lissajous figures and are best graphed using a square viewing rectangle. Draw a sketch of the graph obtained and give the x- and y-intercepts.

7. $x = 3 \sin t, y = 5 \cos t$

8. $x = 5 \sin t, y = 5 \sin t$

9. (a) Graph $x = 3 \sin t, y = 5 \sin t$.
   (b) Graph $x = 3 \sin 2t, y = 5 \sin t$.
   (c) Graph $x = 3 \sin 4t, y = 5 \sin t$.
   What is the effect of changing a in the equations $x = 3 \sin at, y = 5 \sin t$? Is there any difference if a is an even or an odd integer?

10. (a) Graph $x = 3 \sin t, y = 5 \cos t$.
    (b) Graph $x = 3 \sin 2t, y = 5 \cos t$.
    (c) Graph $x = 3 \sin 4t, y = 5 \cos t$.
    What is the effect of changing a in the equations $x = 3 \sin at, y = 5 \cos t$? Is there any difference if a is an even or an odd integer?

11. (a) Graph $x = 5 \sin t$, $y = 5 \cos t$.

   (b) Graph $x = 5 \sin (t + \frac{\pi}{4})$, $y = 5 \cos t$.

   (c) Graph $x = 5 \sin (t + \frac{\pi}{2})$, $y = 5 \cos t$.

   What is the effect of changing c in the equations $x = 5 \sin (t + c)$, $y = 5 \cos t$?
   What happens if $c = \pi$?

12. (a) Graph $x = 5 \sin (2t + \frac{\pi}{2})$, $y = 5 \cos 2t$.

   (b) Graph $x = 5 \sin (2t + \frac{\pi}{2})$, $y = 5 \cos 3t$.

   (c) Graph $x = 5 \sin (2t + \frac{\pi}{2})$, $y = 5 \cos 4t$.

   What is the effect of changing b in the equations $x = 5 \sin (2t + \frac{\pi}{2})$, $y = 5 \cos bt$?

   Does it make a difference if b is even or odd?

◆ 13. *(Washington, Exercises 10.6, #39)* Two signals are seen on an oscilloscope as being at right angles. The equations for the displacements of these signals are $x = 4.5 \cos \pi t$ and $y = 2.5 \sin 3\pi t$. Sketch the graph that appears on the oscilloscope. What are the x- and y-intercepts.

Graph the following parametric equations. Use $-4\pi \leq t \leq 4\pi$. Draw a sketch of the graph obtained.

14. $x = \cot t$, $y = (\sin t)^2$ (Called the Witch of Agnesi)

15. $x = t - \sin t$, $y = 1 - \cos t$ (Called a cycloid)

Chapter 7

Exponent and Logarithm Functions

7.1 Graphing Exponent and Logarithm Functions.
   (Washington, Section 13.2)

Exponent and logarithm functions play important role in mathematics. In this section we study these functions by studying their graphs. The graphing of exponent and logarithm functions is basically handled in the same manner as the graphing of algebraic functions. Care does need to be taken to be sure that we use the appropriate viewing rectangle.

   Exponent functions are functions that have a variable contained in the exponent. A basic exponent function in the real number system is a function of the form $y = b^x$ where b is a positive number not equal to 1.

   Exponent functions and logarithm functions are closely related. From the definition of logarithm it is known that $\log_b x$ is the power to which we must raise the base number b to obtain the value x. (The value of $\log_2 8$ is 3 since $2^3 = 8$.) The basic logarithm function is of the form $y = \log_b x$. In order for $\log_b x$ to be a real number, x must be positive and b needs to be a positive number not equal to 1.

   On the graphing calculator there are two keys for logarithms. One is labeled LOG and the other LN. The LOG key refers to a logarithm of base 10, normally called **common logarithms**, and the LN key refers to logarithms of base e, normally referred to as **natural logarithms**. Pressing the LOG key, entering a number, and then pressing the Enter key then gives a value, called the common logarithm of the number. It is a value such that if 10 is raised to that power, the number entered will result. Thus, $\log x = \log_{10} x$. Pressing the LN key, entering a given number, and then pressing the ENTER key gives a value called the natural logarithm of that number. It is a value such that when a number called e is raised to that power the given number will result. Thus, $\ln x = \log_e x$. The number e is a very important mathematical constant that is approximated by the value 2.71828... (see problem 63, Exercise 1.1). To see the value for e on your calculator, press the key e^x, then press the key 1, and then press ENTER.

   The common logarithm function is the function of the form $y = \log f(x)$ and the natural logarithm function is the function of the form $y = \ln f(x)$, where $f(x)$ is an algebraic function of x. Examples of logarithm functions are $y = \ln x$, $y = \log(2x - 4)$, and $y = \ln(x^2 + 1)$. A logarithmic function may have a different base than 10 or e. For example $y = \log_5(3x)$ is a logarithmic function with a base number 5.

   Exponent functions are function where a constant is raised to a power that is a function of x and are of the form $y = b^{f(x)}$ where f(x) is an algebraic function of x. Examples of exponent functions are $y = 2^x$, $y = e^{5x}$, and $y = 10^{2x+1}$. In this chapter we shall investigate both exponent and logarithmic functions.

Example 7.1:   Graph the exponent functions $y = 2^x$ and $y = 2^{-x}$ and note how the graph of the second function differs from the graph of the first. What are the y-intercepts? Does either graph intersect the x-axis?

Solution: 1. Using the standard viewing rectangle and Procedure G1 enter the functions and graph the two functions on the same graph on the calculator. Use the exponent key and enter 2^x for the first function as Y_1 and 2^{-x} for the second function as Y_2. The resulting graph appears in Figure 7.1.

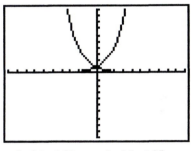

y:[−10, 10] x:[−10, 10]

Figure 7.1

2. The graph of $Y_1 = 2^x$ is always an increasing function. That is, y becomes larger as x becomes larger. Notice that as x becomes larger than 1, the y values increase more rapidly. The graph of $Y_2 = 2^{-x}$ is always a decreasing function since as x becomes larger, y becomes smaller. In both cases the y-values are always positive, the domain is all real numbers, and the y-values get close to zero but never become equal to zero. The y-intercepts are in both cases are 1 and neither graph intersects the x-axis.

Recall that an **asymptote** is a line that a graph gets closer and closer to as the distance from the origin becomes greater. In Example 7.1, the x-axis is an asymptote for both graphs. In order to graph logarithm functions to bases other than e or 10, we use the equation

$$\log_b x = \frac{\log_a x}{\log_a b}.$$ [Equation 1]

Either common logarithms or natural logarithms may be used on the right hand side by letting $a = 10$ or letting $a = e$. This gives two equations either of which may be used when working with logarithms in other bases.

If $a = 10$:   If $a = e$:

$$\log_b x = \frac{\log x}{\log b}$$ and $$\log_b x = \frac{\ln x}{\ln b}$$

Thus, to graph the function $y = \log_5(3x)$ we could either graph $y = \frac{\log(3x)}{\log 5}$ or $y = \frac{\ln(3x)}{\ln 5}$.

♦ <u>Example 7.2</u>:   *(Washington, Section 13.2, Example 9)* Graph the functions $y = \log_3 x$ and

$y = 3^x$ on the same graph and compare the graphs.

<u>Solution</u>:   1. To enter the function $\log_3 x$ use the relation

$$\log_3 x = \frac{\log x}{\log 3}.$$

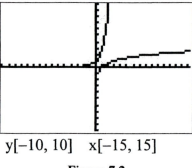

y[−10, 10]   x[−15, 15]

Figure 7.2

For Y_1 enter log x/log 3 for the first function $y = \log_3 x$ and for Y_2, enter 3^x for the second function $y = 3^x$ and graph the functions. In this case it is best to have the same scales on both the x- and y-axes. Use Procedure G6 to first graph on the standard viewing rectangle, then obtain the square-viewing rectangle. The result appears in Figure 7.2.

3. On the graph in Figure 7.2 it is difficult to see what happens as the graph gets close to the x- and y-axes. With the cursor at the origin, we use Procedure G6 to zoom in once on the graph. The resulting graph is given in Figure 7.3.

4. The graph of $\log_3 x$ is graphed first and is an increasing function for all positive x. It has an x-intercept of 1, the domain of this function is all real numbers greater than zero, and the y-axis is an asymptote as x gets close to zero.

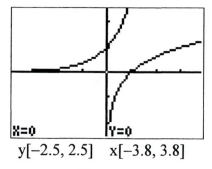

y[−2.5, 2.5]   x[−3.8, 3.8]

Figure 7.3

The graph of $y = 3^x$ is also an increasing function for all x-values, it crosses the y-axis at 1 and the x-axis is an asymptote. By zooming in along the negative axes in each case or tracing along the curves we can verify that the first graph does not cross the y-axis and the second graph does not cross the x-axis.

The two functions in Example 7.2 exhibit an interesting property. Imagine these graphs on a piece of paper on which the line $y = x$ also appears. (The graph of $y = x$ can be added to the graph on your calculator.) If the paper is folded along the line $y = x$ and the x- and y-scales are the same (a square viewing rectangle), the graph of $y = \log_3 x$ will fall on top of the graph of $y = 3^x$. Two functions that have this property are **inverse functions** of each other. This is true of any logarithm function $y = \log_b x$ and its corresponding exponent function $y = b^x$.

The exercise section will illustrate other properties of logarithms.

Exercises 7.1

1. Graph the functions $y = 2^x$, $y = 3^x$, and $y = 4^x$.
   (a) Describe what happens to the graph as the base number of the exponent function is increased.
   (b) For what x-values are the functions increasing?, decreasing?
   (c) What are the asymptotes of these graphs?
   (d) What are the x- and y-intercepts (if any)?
   (e) Determine $f(1)$ in each case.
   (f) What are the domains and ranges of these functions?
   (g) On paper sketch by hand the graph of $y = 5^x$ and determine the equation of the asymptote.   Check your answer on the calculator.

2. Graph the functions $y = 2^x - 1$, $y = 2^x$, and $y = 2^x + 2$.
   (a) Describe what happens to the graph as a constant is added to $y = 2^x$.
   (b) For what x-values are the functions increasing?, decreasing?
   (c) What are the asymptotes of these graphs?
   (d) What are the x- and y-intercepts (if any)?
   (e) What are the domains and ranges of these functions?
   (f) On paper sketch by hand the graph of $y = 2^x + 1$ and determine the equation of the asymptote.   Check your answer on the calculator.

3. Graph the functions $y = \log_2 x$, $y = \log_3 x$, and $y = \log_4 x$.
   (a) Describe what happens to the graph as the base number of the logarithm function is increased.
   (b) For what x-values are the functions increasing?, decreasing?
   (c) What are the asymptotes of these graphs?
   (d) What are the x- and y-intercepts (if any)?
   (e) In each case, determine the value of x that gives a y-value of 1.
   (f) What are the domains and ranges of these functions?
   (g) On paper sketch by hand the graph of $y = \log_6 x$ and determine the equation of the asymptote.   Check your answer on the calculator.

4. Graph the functions $y = -2 + \log x$, $y = \log x$, and $y = 4 + \log x$.
   (a) Describe what happens to the graph as a number is added to the logarithm function.
   (b) For what x-values are the functions increasing?, decreasing?
   (c) What are the asymptotes of these graphs?
   (d) What are the x- and y-intercepts (if any)?
   (e) What are the domains and ranges of these functions?
   (f) On paper sketch by hand the graph of $y = 2 + \log x$ and determine the equation of the asymptote.   Check your answer on the calculator.

5. Graph the functions $y = 4^x$ and $y = 4^{-x}$.
   (a) What is the difference between these two graphs?
   (b) For what x-values are the functions increasing?, decreasing?
   (c) What are the asymptotes of these graphs? Are they the same?
   (d) What are the x- and y-intercepts (if any)? Are they the same?
   (e) What are the domains and ranges of these functions?
   (f) Is one graph the reflection of the other about an axis? If so, which axis?

6. Graph the functions $y = \log x$ and $y = -\log x$.
   (a) What is the difference between these two graphs?
   (b) For what x-values are the functions increasing?, decreasing?
   (c) What are the asymptotes of these graphs? Are they the same?
   (d) What are the x- and y-intercepts (if any)? Are they the same?
   (e) What are the domains and ranges of these functions?
   (f) Is one graph the reflection of the other about an axis? If so, which axis?

7. By hand, sketch graphs of each of the following. Check answers on your calculator.
   (a) $y = 8^x$ (b) $y = 3^x + 2$
   (c) $f(x) = 2^x + 4$ (d) $g(x) = (\frac{1}{2})^x$ (Note: $(\frac{1}{2})^x = 2^{-x}$)

8. By hand, sketch graphs of each of the following. Check answers on your calculator.
   (a) $y = \log_5 x$ (b) $y = 3 + \log x$
   (c) $f(x) = -\log_3(x - 1)$ (d) $g(x) = -4 + \log_5(x + 3)$

For Exercises 9-12, use a square-viewing rectangle.

9. Graph the function $y = \log_2 x$ on the calculator. Guess what the graph of the function $y = 2^x$ looks like and sketch it on a piece of paper. Check your answer on the calculator.

10. Graph $y = \ln x$ and $y = e^x$. How are these functions related?

11. Graph $y = \log(x + 2)$ and $y = 10^x - 2$ on the same graph. How are these functions related?

12. Graph $y = 4 + \log_3 x$ and $y = 3^{x-4}$ on the same graph. How are these functions related?

13. On the Decimal Viewing Rectangle graph the three functions $y = \log(2x)$, $y = \log(10x)$, and $y = \log(20x^2)$ on the same screen. (Some calculators may show a graph for $x < 0$, disregard that part of the graph – we are only considering cases where $x > 0$.) Use the trace function to move the cursor along one of the graphs so that $x = 1$. At $x = 1$, use the up and down cursor keys to jump from curve to curve. Note the y-values at $x = 1$ for each curve. Do you see any relationship between the y-values? Repeat the process at $x = 2$, at $x = 3$, and other values of x and look for a relationship between y-values. What does this indicate about the relationship between the three functions?
   [Hint: Consider the logarithm property $\log_b x + \log_b y = \log_b(xy)$]

14. On the Decimal Viewing Rectangle graph the two functions $y = \log(x^3)$ and $y = \log(x)$ on the same screen. (Some calculators may show a graph for $x < 0$, disregard that part of the graph – we are only considering cases where $x > 0$.) Use the trace function to move the cursor along one of the graphs so that $x = 2$. At $x = 2$, use the up and down cursor keys to jump from curve to curve. Note the y-values at $x = 2$ for each curve. Do you see any relationship between the y-values? Repeat the process at $x = 3$, at $x = 4$, and other values of x and look for a relationship between y-values. What does this indicate about the relationship between the two functions?

   [Hint: Consider the logarithm property $\log_b x^n = n \log_b x$.]

◆ 15. *(Washington, Ch. 13 Review Exercises, #73)* If an amount of P dollars is invested at an annual interest rate r (expressed as a decimal), the value V of the investment after t years is given by $V = P(1 + \dfrac{r}{n})^{nt}$, if interest is compounded n times a year. If $1225 is invested at an annual interest rate of 4.5%, compounded semiannually $(n = 2)$, express V as a function of t and graph for $0 \le t \le 18$ years. Then, from the graph
   (a) Determine the value, V, after 6.5 years and after 9.0 years.
   (b) Determine the time when the investment will have doubled (that is, when $V = \$2450$).

◆ 16. *(Washington, Exercises 13.5, #47)* If interest is compounded continuously (interest compounded daily closely approximates this), with an interest rate of I, a bank account will double in t years according to the function $I = \dfrac{\ln 2}{t}$. Graph this function.

   (a) Determine from the graph the value of I if the time to double is to be 6 years.
   (b) Determine from the graph the time to double if $I = 8.0\%$.

◆ 17. *(Washington, Exercises 13.5, #50)* The velocity v (in m/s) of a rocket increases as fuel is consumed and ejected. Considering fuel as part of the mass of a rocket in flight and if m_0 is its original mass and $m_0 = 9500$, the velocity is given by the function $v = 2500(\ln 9500 - \ln m)$. Graph this function.
   (a) Determine the velocity when $m = 4750$.
   (b) Determine the velocity when $m = 2375$.
   (c) Determine the mass when $v = 2000$ m/s.

18. An original amount of 126 mg of radium radioactively decomposes such that N mg remain after t years. The function relating t and N is
   $t = 2500(\ln 126 - \ln N)$. Graph this function and from the graph
   (a) Determine the time until 100 mg remain and until 63 mg remain.
   (b) Determine how much remains after 600 years and after 15000 years.

19. Considering air resistance and other conditions, the velocity (in m/s) of a certain falling object is given by $v = 95(1 - e^{-0.1t})$, where t is the time of fall in seconds. Graph this function and from the graph
(a) Determine the velocity after 8.5 s and after 15 s.
(b) Estimate what happens as t becomes large. (Hint: let Xmax = 50) Does this graph have an asymptote? What is the limiting velocity (the largest possible value for v)?

7.2 Logarithmic Calculations
(Washington, Sections 13.4 and 13.5)

To find logarithms of numbers on a calculator, we make use of the keys LOG and LN to find logarithms to base 10 and base e. Logarithms using other bases may be found by using the Equation 1 in Section 7.1. Example 7.3 shows how the calculator is used to obtain common or natural logarithm values for any positive number.

The logarithm of a number on the calculator may be displayed with several digits. To achieve approximately the same accuracy the number of decimal places in the logarithm of a number should equal the number of significant digits in the number.

Example 7.3:   Find log 27.2 and ln 6.254

Solution: 1. To find log 27.2, enter exactly as shown. Press the key LOG, enter 27.2, and press ENTER. The result is log 27.2 = 1.435.
(Since 27.2 has 3 significant digits, we round to 3 decimal places.)
2. To find ln 6.254, enter exactly as shown. Press the key LN, enter 6.254, and press ENTER. The result is ln 6.254 = 1.8332
(Since 6.254 has 4 significant digits, we round to 4 decimal places.)

From the definition of logarithm, log 27.2 = 1.435 means that $10^{1.435} = 27.2$ (approximately). This can be verified by using the exponent key on the calculator to find $10^{1.435}$. This will not give exactly 27.2 because of rounding errors. Likewise, ln 6.254 = 1.8332 means that $e^{1.8332} = 6.254$ (approximately).

Example 7.4: Find $\log_6 7.53$.

Solution: 1. Using common logarithms and Equation 1, we have

$$\log_6 7.53 = \frac{\log 7.53}{\log 6}$$

Enter the right side into the calculator as LOG 7.53 ÷ LOG 6
2. Press ENTER. The result shows that $\log_6 7.53 = 1.127$.

Just as important as finding the logarithm of a number is the procedure for finding a number if we know a logarithm. For example, we may wish to find x given that $\log x = 2.3145$. This procedure is often referred to as finding the **antilogarithm**. The fact that $x = b^y$ is equivalent to $\log_b = y$ provides a means for finding the antilogarithm. If N is a known number and if $\log x = N$, then from the definition of logarithm, it must be true that $x = 10^N$. Also, if $\ln x = N$, then $x = e^N$. On the calculator the value for 10^N or e^N are given as secondary functions on the LOG and LN keys. For other bases we raise the base to the appropriate power. If $\log_4 x = N$, then $x = 4^N$.

Example 7.5: If $\ln x = 2.345$, find x.

Solution: Since $\ln x = 2.345$ means that $e^{2.345} = x$, evaluate $e^{2.345}$ by using the e^x key. Since the logarithm has 3 decimal places we round the answer to 3 significant digits and the result is $x = 10.4$.

There are many applications in science and mathematics that require the use of logarithms. One application involves the solving of exponent equations. This makes use of the fact that since the two sides of an equation are equal, the logarithms of the sides are equal. Other applications include equations of the form $d = \log \dfrac{I}{I_0}$ and the calculation of pH in Chemistry.

Example 7.6: Solve the exponent equation $4^{x+1} = 7$

Solution: To solve we take the logarithms of both sides:

$$\log 4^{x+1} = \log 7$$

using the properties of logarithms, this gives: $(x+1) \log 4 = \log 7$

$$\text{or} \qquad x + 1 = \frac{\log 7}{\log 4}$$

$$\text{or} \qquad x = \frac{\log 7}{\log 4} - 1$$

Enter this last expression into the calculator and evaluate to obtain $x = 0.40$.

Example 7.7: In chemistry, the pH of a solution is given by pH $= -\log h$ where h represents the hydrogen ion concentration. Find h if pH $= 3.20$.

Solution: 1. If pH $= 3.20$ then $3.20 = -\log h$, or $\log h = -3.20$.
                   This means that $h = 10^{-3.20}$.

              2. Evaluate $10^{-3.20}$. The result is $h = 6.3 \times 10^{-4}$ or 0.00063.
                   Thus, the hydrogen ion concentration is 0.00063.

Exercises 7.2

In Exercises 1-12, find the required logarithms.

1. log 6.35

2. log 0.001378

3. log 16253

4. $\log 7.5 \times 10^8$

5. ln 1.73

6. ln 256350

7. ln 0.06154

8. $\ln 7.83 \times 10^{15}$

9. $\log_3 17.85$

10. $\log_8 0.0275$

11. $\log_5 1.234 \times 10^6$

12. $\log_{12} 1.024 \times 10^{-7}$

In Exercises 13-22, find the number N, x, or y.

13. $\log N = 5.124$

14. $\log N = -2.3145$

15. $\log x = 0.12483$

16. $\log y = -12.634$

17. $\ln N = 1.53$

18. $\ln N = 7.1542$

19. $\ln x = -12.0$

20. $\ln N = -1.00$

21. $\log_2 x = 2.18$

22. $\log_5 y = -1.72$

Do Exercises 23-38 as indicated:

23. Find x: $\log x - \log 23.4 = 1.563$

24. Find y: $0.065 = \dfrac{\ln y}{8.5}$

25. Find each of the following logarithms:
   (a) log 2.351   (b) log 23.51   (c) log 235.1
   (d) log 2351   (e) $\log 2.351 \times 10^9$
   What do you notice about all these logarithms?
   (The decimal part of a base 10 logarithm is called the mantissa of the logarithm and the whole number part is called the characteristic of the logarithm.)

26. Find each of the following logarithms:
   (a) log 24.2   (b) log 73.5   (c) log 99.9
   (d) log 11.1   (e) log 10.001
   What do you notice about all these logarithms? (See problem 25.)

27. Solve: $2^x = 9$.

28. Solve: $3.5^x = 7.6$

29. Solve: $e^x = 6.5234$

30. Solve: $3^{x+1} = 6.782$

31. In a solution, the hydrogen ion concentration is 0.00031. What is the pH?

32. The pH of a solution is 10.5. What is the hydrogen ion concentration?

33. If $d = \log \dfrac{I}{I_0}$, then find d if $I_0 = 95$ and $I = 163$.

34. If $d = \log \dfrac{I}{I_0}$, then find I if $I_0 = 95$ and $d = 2.4$.

♦ 35. *(Washington, Exercises 13.4, #37)* If the gain G (in decibels) of an electronic device is given by $G = 10 \log \dfrac{P_0}{P_i}$ where P_0 is the output power and P_i is the input power in watts.

   (a) Determine the power gain if the input power is 0.785 W and the output power is 27.25 W.
   (b) Determine the output power if the gain is 4.8 decibels and the input power is 0.850 W.

♦ 36. *(Washington, Ch. 13 Review Exercises, #79)* The approximate formula for the population (in millions) of a certain state since 1970 is given by the equation $P = 2.14 e^{0.00375t}$ where t is the number of years since 1970. Graph this function.
   (a) Find P if t = 30 years.
   (b) Find t when P = 4.28.

37. If an amount of P dollars is invested at an annual interest rate r (expressed as a decimal), the value V of the investment after t years is given by $V = P(1 + \dfrac{r}{n})^{nt}$, if interest is compounded n times a year. If $5000 is invested in an account earning interest compounded 4 times per year, use logarithms to determine how long it will take the value to double the original investment if interest is calculated at a rate of
   (a) 4.0%   (b) 6.0%   (c) 8.0%

38. The effect of inflation on the price, P, of an item may be calculated using the formula $V = P(1 + r)^n$ where r is the annual inflation rate (expressed as a decimal) and V is the price after n years.
   (a) If the ticket at a certain movie house now costs $6.50 and the inflation rate is 3.5%, how long will it take the cost of the movie ticket to become $10.00?
   (b) A child is given a piece of land valued at $25,000 when it is born and later the land is worth $50,000 on the one of the child's birthdays. If the inflation rate is 4.0%, what birthday is the child celebrating?

39. The temperature (in°C) of a gold ingot is given by $T = 23.5 + 98.0(0.290)^{0.20t}$, where t is the time in minutes.
   (a) How long will it take the ingot to cool to 32.5°C?
   (b) Will the temperature ever reach 23.5°C? (Hint: Look at the graph.)

Chapter 8

Systems of Non-linear Equations

8.1 Graphical Solutions to Non-linear Systems
(Washington, Section 14.1)

A non-linear equation is an equation that does not give a straight line when graphed. Such equations may involve either x^2 or y^2 terms or other functions such as trigonometric functions or logarithmic functions. Of special interest are systems involving equations that can be written in the general form $Ax^2 + By^2 = C$ where A, B, and C represent real numbers. However, any system of equations, in which each equation can be solved for y in terms of x, can be solved by graphical means using the graphing calculator.

In order to solve a system of equations by graphical means, it is first necessary to solve each equation for y as a function of x. Equations of the form $Ax^2 + By^2 = C$ are not functions. However, it is possible to solve such an equation for y in terms of x and obtain two functions. Generally, one function will represent the top half of the graph and the other function will represent the bottom half of the graph.

<u>Procedure G20</u>. To graph equations involving y^2.

    1. Solve the equation for y. There will be two functions: $y = f(x)$ and $y = g(x)$. In many cases it will be true that $g(x) = -f(x)$.
    2. Store $f(x)$ under one function name and $g(x)$ for the second function name (see Procedure C10). (Example: $Y_1 = f(x)$, $Y_2 = g(x)$)
      or
      If $g(x) = -f(x)$, we may do the following:
      (a) Store $f(x)$ after a function name (Example: Y_1).
      (b) Move the cursor after the equal sign for a second function name (Example: Y_2).
      (c) Enter the negative sign: (−)
      (d) Obtain and enter the name of the first function. (Example: $Y_2 = -Y_1$) (See Procedure C11.)
    3. Graph the functions.

Note: Often when equations involving y^2 are graphed on a graphing calculator, the top and bottom parts of the graph do not meet as expected. This is due to the way the points are plotted on the calculator. For example, when the circle $x^2 + y^2 = 25$ is graphed on the standard viewing rectangle, the top and bottom halves of the circle may not meet on the x-axis as expected. (See Figure 8.1) The student must recognize that in such cases the graphs do indeed meet. Be aware that what appears on the calculator screen is not always the whole truth.

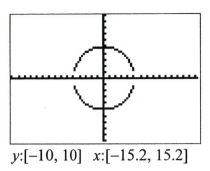

y:[−10, 10] x:[−15.2, 15.2]

Figure 8.1

After the equations of a system of equations are graphed, the points of intersection can be found using Procedure G15.

<u>Example 8.1</u>:   Solve the following system by graphing. Determine answers to two decimal places.

$$y = x^2 - 4$$
$$x^2 + 8y^2 = 64$$

<u>Solution</u>:   1. The first equation can be entered as is appears. However, the second equation must be solved for y in terms of x.

$$x^2 + 8y^2 = 64$$
$$8y^2 = 64 - x^2$$
$$y^2 = 8 - \frac{x^2}{8}$$
$$or \ \ y = \pm\sqrt{8 - \frac{x^2}{8}}$$

2. On the standard viewing rectangle, graph the second equation making use of Procedure G20 and graph the first equation in the form it is given. The graph is shown in Figure 8.2.

3. To find all the solutions to the system of equations it will be necessary to use Procedure G15 to find each of the four points of intersection. The points of intersection are:

(−2.58, 2.68), (2.58, 2.68),
(1.09, −2.80), (−1.09,−2.80).

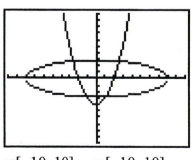

y:[−10, 10] x:[−10, 10]

Figure 8.2

The graphs in Example 8.1 were a parabola and an ellipse. Depending on the coefficients in the equations, a parabola and an ellipse may not intersect at all, in which case there are no solutions, or may intersect at up to four points. This is generally true in systems of quadratic equations (equations that involve x, y, x^2, and y^2 and no higher powers). Each point of intersection gives a pair of values that is a solution.

Sometimes special methods are needed to solve one of the equations for y in terms of x. An example of this is Example 8.2.

Example 8.2:   Solve the following system by graphing. Determine answers to two decimal places.

$$10^y = 5x$$
$$4x^2 - y^2 = 4$$

Solution: 1. We need to first solve the first equation for y in terms of x. To do this we apply the properties of logarithms and see that $10^y = 5x$ is the same as

$$y = \log(5x) \text{ where } x > 0$$

2. The second equation needs to be solved for y:

$$4x^2 - y^2 = 4 \text{ (This is an equation of a hyperbola.)}$$
$$y^2 = 4x^2 - 4$$
$$y = \pm\sqrt{4x^2 - 4} = \pm 2\sqrt{x^2 - 1}$$
$$y = 2\sqrt{x-1} \text{ and } y = -2\sqrt{x-1}$$

3. We enter these equations on the calculator and graph them on the standard viewing rectangle. The graph obtained is shown in Figure 8.3. We then use Procedure G15 to find the point of intersection which is in the first quadrant. There is only one solution which is

$$x = 1.06$$
$$y = 0.73$$

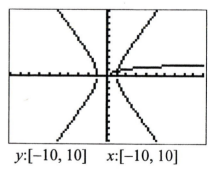

$y:[-10, 10]$ $x:[-10, 10]$

Figure 8.3

In some cases it may be helpful to consider another method of finding the point of intersection of two graphs. For the system $y_1 = f(x)$ and $y_2 = g(x)$, consider the difference $y_1 - y_2 = f(x) - g(x)$. Let $F(x) = f(x) - g(x)$ and graph $y = F(x)$. A zero of the function $F(x)$ will be the x-coordinate to a solution to the system $y_1 = f(x)$ and $y_2 = g(x)$. This is true since at a point of intersection y_1 must equal y_2 and $F(x)$ must be zero. Thus, another way to solve the system $y_1 = f(x)$ and $y_2 = g(x)$, is to graph $y = f(x) - g(x)$ and find the x-intercepts (the zeros of the function). The y-coordinates may be obtained by evaluating either the function $f(x)$ or $g(x)$ at the x-values.

Exercise 8.1 *(Washington, Exercises 14.1)*

Solve each of the following systems of equations by graphing. Draw a sketch of the graphs obtained. Find answers to 3 significant digits. Identify any systems that do not have solutions.

1. $2x + 3y = 5$
   $y = x^2 - 8$

2. $5x - 2y = 7$
   $y = 6 - x^2$

3. $x^2 + y^2 = 25$
   $2x^2 - y = 11$

4. $3x^2 - 5y^2 = 24$
   $2x^2 + 2y^2 = 9$

5. $3x^2 - 12x - y = 6$
   $6x + y = -21$

6. $x^2 - 3x + y = 7$
   $3x^2 = 38 - 2y^2$

7. $0.5x^2 + 0.2y^2 = 48$
   $5x + y = 6$

8. $5x^2 - 12y^2 = 19$
   $x^2 - 4y = 15$

9. $xy = 8$
   $8x^2 - y^2 = 9$

10. $xy = 12$
    $3x + y = 12$

11. $y = \sin x$
    $x^2 - y = 1$

12. $y = \cos 2x$
    $y - x^3 = 2$

13. $y = \log x$
    $x^2 + y^2 = 2$

14. $y - 2^{x+3} = 5$
    $xy = 10$

15. $e^y = 4x$
    $y = x^2 - 2$

16. $5^y = x^2$ (There are 4 solutions)
    $x^2 + y^2 = 8$

17. $y = 3^{x-5}$
    $y = \log_8 x$

18. $y = \sec x$
    $y = e^x$

The method of finding intersection points may also be used to solve a single equation. We let one function be the left side of the equation and another function be the right side of the equation and solve the resulting system of equations. This is illustrated in Exercises 15 and 16.

19. Solve: $x^3 = 5x^2 + 6$.

To do this let $y_1 = x^3$ and $y_2 = 5x^2 + 6$ and find the x-coordinates of the points of intersection.

20. Solve the equation $x^4 = 6 - x^2$ using the method of Exercise 19.

♦ 21. *(Washington, Exercises 14.1, #31)* A helicopter is located 5.358 miles northeast of a radio tower such that it is three times as far north as it is east of the tower. Find the northern and eastern components of the displacement from the tower.

♦ 22. *(Washington, Exercises 14.1, #32)* A 4.125 m insulation strip is placed completely around a rectangular solar panel with an area of 1.158 m^2. What are the dimensions of the panel?

23. An inflatable building has an equation $324.0x^2 + 1225y^2 = 396900$ (the top half). Cables to support the building against the wind are to be tied to one side. These cables can be considered as straight lines that go from the building to points on the positive x-axis and have equations $y = 15.0 - \dfrac{x}{4.25}$ and $y = 18.8 - \dfrac{x}{4.25}$. All measurements are given in feet. Find the points of intersection of the lines with the building. How long do these cables need to be?

24. The outer edge of one gear that is at an extreme point on a movable axis has an equation of $2.000x^2 + 5.120y^2 = 14.0625$ and that of a second gear has the equation $y^2 + x^2 - 11.300x = -25.920$. Measurements are in centimeters. Do the edges of the two gears intersect and if so at what point?

25. Find the solutions in each of the following problems (to 2 decimal places):
(These problems show the difficulty of finding some solutions graphically.)

(a) $4x^2 + y^2 = 8$
     $y = 3 - x^2$

(b) $4x^2 + y^2 = 8$
     $y = 2 - x^2$

(c) $4x^2 + y^2 = 8$
     $y = 2.1 - x^2$

(d) $4x^2 + y^2 = 8$
     $y = -4.82x^2 + 2.82$

Chapter 9

Matrices

9.1 Matrices and Matrix Operations
(Washington, Sections 16.1 and 16.2)

A **matrix** can be defined as an ordered rectangular array of numbers. That is, it is a set of numbers arranged in rows and columns such that the order in which the numbers are arranged is important. The numbers within a matrix are referred to as **elements** of the matrix. One way of classifying matrices is by size. The **size** (or dimension) depends on the number of rows and columns and is given in the form $m \times n$ where m is the number of rows and n is the number of columns. Thus, a 2×3 matrix contains numbers arranged in two rows and three columns. A 4×1 matrix would be a matrix of four rows and one column. A matrix containing only one column is called a **column matrix**. A matrix containing one row and three columns is a 1×3 matrix. A matrix containing only one row is referred to as a **row matrix**.

In order to indicate that a rectangular array of numbers is a matrix, the array is enclosed either with large parentheses or with square brackets. The square brackets will be used here. Upper case letters are often used to name matrices -- such as matrix A or matrix B.

The elements of a matrix are identified according to the location of the element within the matrix by using a double subscript on a lower case letter. The first number in the subscript is the number of the row in which the element appears and the second number in the subscript is the number of the column in which the element appears. For example, a_{23} would represent the element in the second row, third column of matrix A and b_{41} would represent the element in the fourth row, first column of the matrix B. On a calculator or a computer the double subscript is often indicated by using two numbers (often within parentheses) separated by a comma. In this case a_{23} would appear as $A(2,3)$ (or just as 2,3 when working with matrix A) and b_{41} would appear as $B(4,1)$ (or as 4,1 when working with matrix B).

Example 9.1: For each of the following matrices, (a) give the size of each matrix,
(b) identify the position occupied by the number 2 in each matrix and
(c) give a_{32}, b_{31}, c_{12}, and d_{14}.

$$A = \begin{bmatrix} 5 & -4 & 1 \\ 4 & 4 & 2 \\ 1 & -1 & 0 \end{bmatrix}, \quad B = \begin{bmatrix} 2 \\ 4 \\ 3 \end{bmatrix}, \quad C = \begin{bmatrix} 1 & -2 \\ 2 & 3 \\ 0 & -1 \\ 5 & 4 \end{bmatrix}, \quad D = \begin{bmatrix} 1 & 0 & -3 & 2 \end{bmatrix}$$

Solution: (a)   The size of A is 3x3.
                   The size of B is 3x1.
                   The size of C is 4x2.
                   The size of D is 1x4.

            (b)   The number 2 occupies the position given by $a_{23}, b_{11}, c_{21},$ *and* d_{14}. In the notation that may be used on a calculator or computer these are represented as $A(2,3)$, $B(1,1)$, $C(2,1)$, and $D(1,4)$.

            (c) $a_{32} = -1, b_{31} = 3, c_{12} = -2,$ *and* $d_{14} = 2$.

Notice that, in Example 9.1, B is a column matrix and D is a row matrix. A matrix that contains the same number of rows as columns is referred to as a **square matrix**. In Example 9.1, matrix A is a square matrix.

The **identity** matrix is a square matrix with one's as elements in the positions where the number of the row and the number of the column are the same and zeros elsewhere. The identity matrix is denoted by the letter I. The identity matrix has the property such that $AI = A$ and $IA = A$ for any matrix A. When I appears in a matrix expression, it is assumed to have a size that will make the operations of the expression valid. The 2x2 and 3x3 identity matrices are

$$I = \begin{bmatrix} 1 & 0 \\ 0 & 1 \end{bmatrix}, \quad I = \begin{bmatrix} 1 & 0 & 0 \\ 0 & 1 & 0 \\ 0 & 0 & 1 \end{bmatrix}$$

Matrices are important in modern day mathematics for working a variety of problems. One application, the solving of systems of equations, is covered in Section 9.2. Graphing calculators have the ability of handling matrices and can be of great help in working matrix problems.

Procedure M1. To enter or modify a matrix.

1. Press the key MATRX or MATRIX. This gives the matrix menu.
2. Enter EDIT mode.

On the TI-82 or 83 or 84 Plus:
Highlight EDIT in the top row.

On the TI-85 or 86:
Select EDIT from the menu.

3. Select a name for the matrix.

On the TI-82 or 83 or 84 Plus:

Select one of the matrices [A], [B], [C], ... This will be the new matrix or the matrix to be edited.

On the TI-85 or 86:

Enter the name for a new matrix or select the name of a matrix to be edited, then press ENTER. The name may be up to eight characters long (only 5 letters will show in name box).

4. The first line on the screen will contain the word MATRIX followed by the name of the matrix and two numbers separated by ×. These two numbers represent the size of the matrix. First enter the number of rows and press ENTER, then the numbers of columns and press ENTER. To keep the values the same, use the cursor keys to move the cursor or just press ENTER.

5. Enter the elements of the matrix.

On the TI-82 or 83 or 84 Plus:

The notation of the element of the matrix to be entered or changed is highlighted and is shown at the bottom of the screen by the row and column number of the element followed by its value.

As a value is entered, the number at the bottom of the screen will change. To enter this value into the matrix, press ENTER. Values are entered by rows. Continue until all elements are correctly entered.

The cursor keys may be used to move the highlight to particular element(s) to be changed.

On the TI-85 or 86:

On the left of the screen are two numbers separated by a comma. These are row and column numbers of elements of the matrix. Enter one value at a time. After each value is entered, press ENTER to go to the next value. Values are entered by row. (As each value is entered the calculator will jump to the next column.)

The cursor keys may be used to go from one element to the next if only certain elements are to be changed.

On the TI-85, the menu selections ◁Col and Col▷ may be used to change columns.

6. Exit from matrix edit mode by pressing the key: QUIT. (Pressing CLEAR will clear the value for a given element.)

After entering a matrix, it is a good idea to view the matrix and to check to make sure it has been entered correctly.

Procedure M2. **To select the name of a matrix and to view the matrix.**

1. Press the key MATRIX. This gives the matrix menu.
2. Select the name of the matrix.

On the TI-82 or 83 or 84 Plus:
With NAMES highlighted in the top row, select one of the matrices [A], [B], [C], ...

On the TI-85 or 86:
Select NAMES from the matrix menu. From the secondary menu, select the name of the matrix.

3. View the matrix by pressing ENTER. In some cases the matrix may extend off the screen to the right or to the left. To see that part of the matrix not on the screen use the right or left cursor keys to scroll (move) the unseen part onto the screen.

In Example 9.2 we enter two square matrices into the calculator and check to make sure they are entered correctly.

Example 9.2:   Enter the following matrices into the calculator as matrix A and matrix B.

$$A = \begin{bmatrix} 5 & 2 & 1 \\ 2 & 1 & -3 \\ 3 & -2 & 4 \end{bmatrix} \qquad B = \begin{bmatrix} 3 & -3 & 1 \\ -1 & 2 & 3 \\ 0 & -1 & -2 \end{bmatrix}$$

Solution: 1. Use Procedure M1 to enter matrix A into the calculator. Since A is a 3x3 matrix, enter the size as 3 rows and 3 columns.

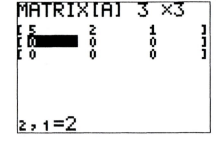

Figure 9.1

2. After entering the size, enter the elements of the matrix. Figure 9.1 shows the calculator screen after entering the first row of matrix A and ready to enter the first element of the second row.
3. Likewise, use Procedure M1 to enter matrix B.
4. View the matrices to check to see if they are entered correctly. Use Procedure M2 to view matrix A, then matrix B. If they have not been entered correctly, use Procedure M1 to make the necessary changes.

There are several operations that may be performed on matrices including the familiar operations of addition and subtraction. In order for matrices to be added or subtracted, the matrices must be of the same size. Division of matrices is not defined, but two kinds of multiplication do exist. One is **matrix multiplication** where two matrices are multiplied together and the other is **scalar multiplication** where a matrix is multiplied by a real number (called a scalar). We only consider the method of performing these operations on the calculator and not the techniques of doing the operations by hand.

Procedure M3. Addition, subtraction, scalar multiplication or multiplication of matrices.
    See Procedure M2 for how to select the names of matrices.
    1. Addition of two matrices:
    Select the name of the first matrix, press the addition sign, select the name of the second matrix, then press ENTER.
    (Example: $[A] + [B]$)
    2. Subtraction of two matrices:
    Select the name of the first matrix, press the subtraction sign, select the name of the second matrix, then press ENTER.
    (Example: $[A] - [B]$)
    3. Multiplication of matrix by a scalar:
    Enter the scalar, select the name of the matrix and press ENTER.
    (Example: 2.5 $[A]$ or 2.5×$[A]$)

4. Multiplication of two matrices:
Select the name of the first matrix, select the name of the second matrix, and press ENTER.
(Example: $[A][B]$ or $[A]\times[B]$)

Many of these operations may be combined into one statement. The result of a matrix calculation is stored under ANS. An error will occur if the operation is not defined for the matrices selected.

Example 9.3 will illustrate the operations of addition, subtraction, and scalar multiplication on the graphing calculator.

Example 9.3: For the matrices A and B given in Example 9.2, find $A + B$, $A - B$, $3A$, and $3A - 2B$.

Solution: 1. Use Procedure M3 to add A and B. Figure 9.2 shows the screen for this sum.

$$A + B = \begin{bmatrix} 8 & -1 & 2 \\ 1 & 3 & 0 \\ 3 & -3 & 2 \end{bmatrix}$$

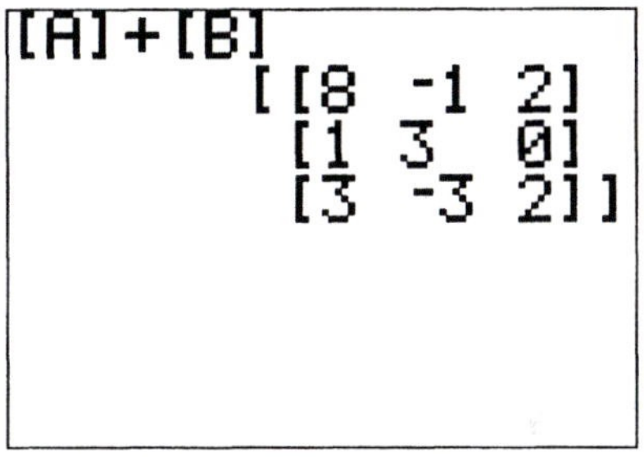

```
[A]+[B]
        [[8  -1  2]
         [1   3  0]
         [3  -3  2]]
```

Figure 9.2

2. Use Procedure M3 to subtract B from A.

$$A - B = \begin{bmatrix} 2 & 5 & 0 \\ 3 & -1 & -6 \\ 3 & -1 & 6 \end{bmatrix}$$

3. Use Procedure M3 to multiply A by the scalar 3.

$$3M = \begin{bmatrix} 15 & 6 & 3 \\ 6 & 3 & -9 \\ 9 & -6 & 12 \end{bmatrix}$$

4. Use Procedure M3 to multiply A by 3 then subtract 2 times B.

$$3M - 2N = \begin{bmatrix} 5 & 12 & 1 \\ 8 & -1 & -15 \\ 9 & -4 & 16 \end{bmatrix}$$

Addition, subtraction, and scalar multiplication of simple matrices are easily done by hand. However, matrix multiplication is not as readily handled by hand. **Matrix multiplication is only defined when the number of columns in the first matrix equals the number of rows in the second matrix**.

Example 9.4:   Use matrices A and B of Example 9.2 to find the matrix product AB and then the product BA.

Solution: 1. Use Procedure M3 to find the product of A and B. Figure 9.3 shows this product.

$$AB = \begin{bmatrix} 13 & -12 & 9 \\ 5 & -1 & 11 \\ 11 & -17 & -11 \end{bmatrix}$$

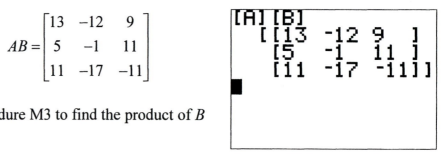

2. Use Procedure M3 to find the product of B and A.

$$BA = \begin{bmatrix} 12 & 1 & 16 \\ 8 & -6 & 5 \\ -8 & 3 & -5 \end{bmatrix}$$

Figure 9.3

In Example 9.4, we see that AB does not equal BA. In general, the commutative law of multiplication in algebra (ab = ba) does not hold for matrix multiplication.

Exercise 9.1

In Exercises 1-22, use the matrices.

$$A = \begin{bmatrix} 1 & 2 & -4 \\ 3 & 2 & -1 \\ 5 & -2 & 7 \end{bmatrix}, \qquad B = \begin{bmatrix} 4 \\ 7 \\ 8 \end{bmatrix}, \qquad C = \begin{bmatrix} 6 & -7 & -1 \\ 3 & 4 & -5 \\ 3 & -5 & 7 \end{bmatrix},$$

$$D = \begin{bmatrix} -3 & 3 & -1 & 7 \end{bmatrix}, \qquad E = \begin{bmatrix} 1 & 2 & 3 & -5 \end{bmatrix}$$

1. Give the size of each matrix A, B, C, and D.
2. Which of the given matrices are square matrices?
3. Identify the position occupied by the number -1 in each matrix in which it appears.

4. Identify the position occupied by the number 7 in each matrix in each matrix in which it appears.

5. List all pairs of these matrices that may be added or subtracted.

6. Give all pairs, in correct order, of these matrices that can be multiplied.

In Exercises 7-22, enter the given matrices into your calculator and use the calculator to perform the indicated operation. If it is not possible to perform the operation, indicate why not.

7. $A + C$

8. $D - E$

9. $D - 2B$

10. $3A$

11. $4A - 2C$

12. $3D - 7E$

13. $-3C - 7A$

14. $4E - C$

15. AB

16. BA (Does this equal AB?)

17. AC

18. CA (Does this equal AC?)

19. $A(CB)$

20. $B + CB$

21. $AB - B$

22. $AC - C$

23. Find AI and IA. Are these two products the same?

24. It can be proven that $x = 2$, $y = -1$ is a solution to the system of equations $\begin{array}{l} 3x - 2y = 8 \\ 4x + 5y = 3 \end{array}$ by showing that the matrix product $\begin{bmatrix} 3 & -2 \\ 4 & 5 \end{bmatrix} \begin{bmatrix} 2 \\ -1 \end{bmatrix}$ is equal to the matrix $\begin{bmatrix} 8 \\ 3 \end{bmatrix}$. Verify that this is true.

◆ 25. *(Washington, Exercises 16.2, #33)* Find $A^2 - I^2$ and $(A + I)(A - I)$ using matrix A given at the beginning of this exercise set. Is it true $A^2 - I^2 = (A + I)(A - I)$?

◆ 26. *(Washington, Exercises 16.2, #36)* In analyzing the motion of a robotics mechanism, the following matrix multiplication is used. Perform the multiplication.

$$\begin{bmatrix} \cos 35.7° & -\sin 35.7° & 0 \\ \sin 35.7° & \cos 35.7° & 0 \\ 0 & 0 & 1 \end{bmatrix} \begin{bmatrix} 2.5 \\ 4.2 \\ 0 \end{bmatrix}$$

9.2 The Inverse Matrix and Solving a System of Equations
(Washington, Section 16.3 and 16.4)

A square matrix A has an **inverse** matrix B only if $AB = I$ and $BA = I$. The notation A^{-1} is used to denote the inverse of matrix A. (Note that the -1 as used here does not mean the reciprocal of A.) Later in this section we will make use of the inverse of a matrix in solving systems of equations. Before we do that, we will look at a few other items.

Other operations on matrices that are of interest are those of finding the determinant of a matrix and of finding the transpose of a matrix. Every square matrix has associated with it a number called the **determinant**. A method of solving systems of equations called Cramer's rule makes use of determinants. We will not consider Cramer's rule, but will consider the use of determinants to determine whether or not a matrix has an inverse. If the determinant of a square matrix is not zero then the matrix has an inverse. Such a matrix is said to be **non-singular**. If the determinant is zero, then the matrix does not have an inverse and the matrix is said to be **singular**.

The **transpose** of a matrix is the matrix formed by interchanging the rows and columns of a matrix, that is a_{12} becomes a_{21}, a_{23} becomes a_{32}, and so on. In general a_{ij} becomes a_{ji} where i is the row number and j the column number of the original matrix. The transpose of matrix A is indicated by the notation A^T. For example:

$$\text{If } A = \begin{bmatrix} 1 & -2 & 3 \\ 2 & 3 & -1 \\ 0 & -1 & -3 \end{bmatrix} \text{ then } A^T = \begin{bmatrix} 1 & 2 & 0 \\ -2 & 3 & -1 \\ 3 & -1 & -3 \end{bmatrix}$$

Procedure M4. To find the determinant, inverse, and transpose of a matrix.

1 To find the determinant:

On the TI-82 or 83 or 84 Plus: On the TI-85 or 86:
Press the key MATRX   Press the key MATRX
Highlight MATH in the top row.   From the matrix menu, select MATH.
From the menu, select det.
Then select the name of the matrix (Procedure M2) and press ENTER.
(Example: det([A]) or det A}

2. To find the inverse:
Select the name of the matrix.
Press the key x^{-1} and press ENTER.

(Example: $[A]^{-1}$ or A^{-1})

3. To find the transpose:
Select the name of the matrix, then
On the TI-82 or 83 or 84 Plus: On the TI-85 or 86:
Press the key MATRX   Press the key MATRX
Highlight MATH in the top row.   From the matrix menu, select MATH.
From the menu, select ^{T} and press ENTER.   (Example: $[A]^T$ or A^T)

Sometimes it is desirable to store one matrix under another name or to store the results of a calculation. Care needs to be taken since whenever something is stored under a variable name, it erases whatever was previously stored under that name.

Procedure M5. To store a matrix.

1. Select the name of a matrix or enter a matrix expression.
2. Press the key STO▷
3. Select the name of the second matrix and press ENTER.
   (Example: $[A] + [B] \rightarrow [C]$ or $A + B \rightarrow C$)
   On the TI-85 or 86, we may also store a matrix by using the equal sign. (Example: $C = A + B$)

   Note: Storing a matrix for a second matrix, erases the previous contents of the second matrix.

Finding the transpose of a matrix is easily done by hand. However, finding the determinant or inverse of a matrix by hand can become rather involved. With the graphing calculator, it is a relatively easy process.

Example 9.5:   Find the determinant of matrix A, then the inverse of matrix A and store this as matrix B. Then show that $AB = I$ and $BA = I$. Also, find the transpose of B. Round numbers to three decimal places.

$$A = \begin{bmatrix} 1 & -2 & 3 \\ 2 & 3 & -1 \\ 0 & -1 & -3 \end{bmatrix}$$

Solution: 1. Enter the given matrix into the calculator as matrix A.
                   (See procedure M1.)

                2. Using Procedure M4, the determinant is: $\det[A] = -28$. Figure 9.4 shows matrix A and its determinant. Since the determinant is not zero, matrix A does have an inverse.

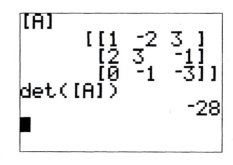

Figure 9.4

3. Since the inverse matrix may be in decimal form and, thus, may be carried out to 10 decimal places, change the mode so that the number of decimal places displayed by the calculator is three. (See Procedure C6)

4. Using Procedure M4 to find the inverse of matrix *A*. Figure 9.5 shows the calculator results. Note here that part of the matrix is of the right of the screen. To see the right part of the matrix, press the right cursor key several times and the number will scroll across the screen.

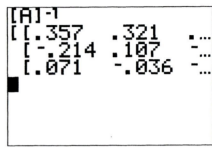

Figure 9.5

We find

$$A^{-1} = \begin{bmatrix} .357 & .321 & .250 \\ -.214 & .107 & -.250 \\ .071 & -.036 & -.250 \end{bmatrix}$$

4. Use Procedure M5 to store the inverse of *A* as matrix *B*.

5. Find the product of matrix *A* and matrix *B* (*A*B*) and the product of matrix *B* and matrix *A*. (*B*A*). In both cases, a matrix similar to the identity matrix appears except that it may contain numbers such as 1.000E–13 (1.0×10^{-13}) or 2.000E–14 (2.0×10^{-14}). These numbers with large negative exponents are essentially zero and may be taken as zero. The matrix obtained in each case is the identity matrix *I* and thus we conclude that *B* is the inverse of *A*.

6. Using Procedure M4 to find the transpose of matrix *B*, we have

$$B^T = \begin{bmatrix} .357 & -.214 & .071 \\ .321 & .107 & -.036 \\ .250 & -.250 & -.250 \end{bmatrix}$$

Since matrix multiplication does not satisfy the commutative property, **to show that *A* is the inverse of *B*, it is necessary to show both that *AB* = *I* and that *BA* = *I*.**

An important use of an inverse matrix is to solve a system of equations. To do this, we first determine the coefficient matrix after writing the system of equations in standard form. **Standard form** for a system of equations means writing the equations so that all the variables are in the same order to the left of the equal sign and the constants to each equation are on the right side of the equal sign.

For the following system of equations

$$2x + 3y = 4 + z$$
$$x + 4z - 8 = 2y$$
$$3x + 2y = 12 + 3z$$

the standard form is

$$2x + 3y - z = 4$$
$$x - 2y + 4z = 8$$
$$3x + 2y - 3z = 12$$

The **coefficient matrix** of a system of equations is the matrix formed by the coefficients of the unknowns when the equations are in standard form. The **constant matrix** is the column matrix formed by the constants to the right of the equal sign when the equations are in standard form. The **unknown matrix** is the column matrix formed by the variables in the system of equations. For the given system, the coefficient matrix, A, the unknown matrix, X, and the constant matrix, C are

$$A = \begin{bmatrix} 2 & 3 & -1 \\ 1 & -2 & 4 \\ 3 & 2 & -3 \end{bmatrix}, \quad X = \begin{bmatrix} x \\ y \\ z \end{bmatrix}, \quad C = \begin{bmatrix} 4 \\ 8 \\ 12 \end{bmatrix}$$

If A and X are multiplied together, the result is the left side of the system of equations

$$AX = \begin{bmatrix} 2 & 3 & -1 \\ 1 & -2 & 4 \\ 3 & 2 & -3 \end{bmatrix} \begin{bmatrix} X \\ Y \\ Z \end{bmatrix} = \begin{bmatrix} 2X + 3Y - Z \\ X - 2Y + 4Z \\ 3X + 2Y - 3Z \end{bmatrix}$$

For a system of three equations this last matrix equals the constant matrix C. In general, any system of equations may be written as a matrix equation in the form

$$A X = C$$

To solve the matrix equation, multiply both sides by the matrix A^{-1}.

$$A^{-1} A X = A^{-1} C$$
$$\text{or} \quad I X = A^{-1} C \qquad [\text{since } A^{-1} A = I]$$
$$\text{or} \quad X = A^{-1} C \qquad [\text{since } I X = X]$$

This indicates that to find the unknown matrix X and, thus, to solve this system of equations, we need to find the inverse of A and multiply it by the constant matrix C. This is called the **solution**

matrix. We can verify that the solution matrix obtained is indeed the solution matrix by multiplying the coefficient matrix by the solution matrix. If we obtain the constant matrix for the system of equations, the solution matrix is correct.

Example 9.6:   Use the inverse of the coefficient matrix to find the solutions to the following system of equations, then verify that the matrix found is the solution matrix. Round values to three decimal places.

$$2x + 3y - z = 4$$
$$x - 2y + 4z = 8$$
$$3x + 2y - 3z = 12$$

Solution: 1. Enter the coefficient matrix into the calculator as the 3x3 matrix A.

2. Enter the constant matrix into the calculator as the 3x1 matrix C.

3. Change the mode of the calculator to display 3 decimal places.

4. Since $X = A^{-1}C$, multiply A^{-1} by C (see Procedure M4). This calculation appears as in Figure 9.6. From the results, the solution to the system of equations, rounded to four decimal places is given by

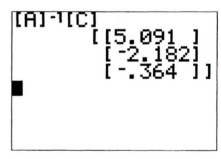

Figure 9.6

$$X = \begin{bmatrix} x \\ y \\ z \end{bmatrix} = \begin{bmatrix} 5.091 \\ -2.182 \\ -.364 \end{bmatrix}$$

By equality of matrices this means that x = 5.091, y = –2.182, and z = – .364.

4. To verify the results, store the product $A^{-1}C$ as matrix B. (See Procedure M5)

5. Next, multiply matrices A and B. We find that

$$AB = \begin{bmatrix} 4.000 \\ 8.000 \\ 12.000 \end{bmatrix}$$

Since this is the same as Matrix C, we conclude that the solution is correct.

The process illustrated in Example 9.6 turns out to be quite powerful. It gives a relatively simple method to solve any system of equations if the system of equations has a unique solution. Since a system of equations may not always have a solution or may have infinite many solutions, it is wise to determine if a solution exists before attempting the matrix calculation. We do this by evaluating the determinant of the coefficient matrix. **If the determinant of the coefficient matrix is zero, then the coefficient matrix is a singular matrix and does not have an inverse and such a system of equations does not have a unique solution.** Such a system may have either no solutions or an infinite number of solutions. When this occurs, other methods need to be employed to determine if there are no solutions or an infinite number of solutions. If there are an infinite number of solutions, the form of these solutions may be obtained from the Gauss-Jordan Method (see Example 9.9).

Example 9.7: Find the determinant of the coefficient matrices for the following system of equations and determine if each system has a unique solution.

(a) $2x + 4y = 3$
      $5x - 3y = 8$

(b) $2x - 3y + z = 2$
      $x + 2y - 2z = 7$
      $3x - 8y + 4z = 9$

Solution: 1. The coefficient matrices are

(a) $A = \begin{bmatrix} 2 & 4 \\ 5 & -3 \end{bmatrix}$ and (b) $B = \begin{bmatrix} 2 & -3 & 1 \\ 1 & 2 & -2 \\ 3 & -8 & 4 \end{bmatrix}$

2. Enter these matrices into the calculator as matrices A and B.

3. Use Procedure M4 to find the determinants of each of these matrices. The determinant of matrix A is −26. The determinant of matrix B is −1.4E−11 or -1.4×10^{-11} (on some calculators this may appear as 0). The value -1.4×10^{-11} is essentially zero and we take the determinant of matrix B to be zero.

4. Since the determinant of matrix A is not zero, system (a) has a unique solution and since the determinant of matrix B is zero, system (b) does not have a unique solution.

Exercise 9.2

In Exercises 1-8, first determine (a) the transpose of the given matrix, (b) the determinant of the given matrix, and (c) if the determinant is not zero, find the inverse of the matrix and prove by multiplication that the inverse is really the inverse. Round all values to 3 decimal places.

1. $\begin{bmatrix} 2 & -5 \\ -3 & 4 \end{bmatrix}$

2. $\begin{bmatrix} -1 & 4 \\ -3 & 12 \end{bmatrix}$

3. $\begin{bmatrix} 3 & 2 & -1 \\ -4 & 0 & 3 \\ 5 & 3 & -2 \end{bmatrix}$

4. $\begin{bmatrix} -6 & 12 & 7 \\ 10 & 5 & -4 \\ -3 & 8 & 11 \end{bmatrix}$

5. $\begin{bmatrix} 1 & -2 & -4 \\ -5 & 3 & 1 \\ 7 & -7 & -9 \end{bmatrix}$

6. $\begin{bmatrix} 10 & 7 & 8 \\ 3 & 2 & -5 \\ -7 & -8 & 15 \end{bmatrix}$

7. $\begin{bmatrix} 0.03 & 0.12 & 0.10 \\ -0.04 & 0.45 & 0.32 \\ 0.11 & 0.95 & 0.04 \end{bmatrix}$

8. $\begin{bmatrix} -0.03 & 0.15 & 0.12 \\ 0.15 & 0.12 & 0.05 \\ -0.24 & 0.33 & 0.31 \end{bmatrix}$

In Exercises 9-20, (a) find the determinant of the coefficient matrix, (b) find the inverse matrix, and (c) using the inverse matrix find the solution to the given system of equations.

9. $3x + 2y = 5$
    $2x - 3y = 7$

10. $5x - 3y = 8$
    $-2x + 4y = 3$

11. $6x - 2y = 5$
    $-9x + 3y = 2$

12. $0.05x + 0.25y = 2.35$
    $0.10x - 0.50y = 7.85$

13. $2x - 3y + 6z = 7$
    $-3x + 2y - 3z = 12$
    $5x + 7y + 3z = 9$

14. $7x - 8y + 6z = -12$
    $4x + 2y - 7z = 9$
    $-5x + 7y + 6z = 15$

15. $3x - 2y = z - 7$
    $2x - 3z = 4 + y$
    $-x = 7 - 2y + 3z$

16. $3x - 4y = 5z - 3$
    $x = -7y - 2z - 7$
    $6y - 6z = 12 - 8x$

17. $0.21x - 0.25y + 0.12z = 265$
    $0.17x + 0.33y - 0.20z = 125$
    $0.11x - 0.45y + 0.16z = 225$

18. $x = 6y$
    $x + y + z = 16$
    $0.04x + 0.07y = 0.25$

19. $2x = 3y$

    $0.10x + 0.15y + 0.25z = 0$

    $5.4x + 8.4y = 1 + 6.2z$

20. $i_1 + i_2 + i_3 = 0$

    $12i_1 - 8.5i_2 = 5.15$

    $8.5i_1 - 15i_3 = -2.14$

Sometimes to solve a system of equations involving x^2 and/or y^2, we can make a substitution for x^2 and/or y^2, solve the system by matrix methods, then by taking the square root of the answer, solve for x and/or y. Care needs to be used in this case to make sure which of the solutions are really answers to the given problem. Use this method to solve exercises 21 and 22. *(Washington, Exercises 16.4, #27-28)*

21. $4x + 7y^2 = 38$

    $12x - 3y^2 = 18$

22. $3x^2 + 2y^2 - 6z = 22$

    $4x^2 - 3y^2 + 3z = -7$

    $5y^2 - 9z = 33$

23. The motion of the rotation of a point through an angle of 37.5° about the origin on a computer screen is given by matrix R. Find R^{-1} that describes the inverse rotation. (Note: In this case we can actually enter the trigonometric function as given, the calculator will then place the decimal equivalent in the matrix.)

$$R = \begin{bmatrix} \cos 37.5° & -\sin 37.5° & 0.0000 \\ \sin 37.5° & \cos 37.5° & 0.0000 \\ 0.0000 & 0.0000 & 1.000 \end{bmatrix}$$

◆ 24. *(Washington, Exercises 16.4, #29)* Two forces A, at an angle of 52.7° to the horizontal, and B, at an angle of 62.5° to the horizontal in the opposite direction, hold up a beam that weighs 325 N. The equations used to find these forces are

$$A \sin 52.7° + B \sin 62.5° = 325$$
$$A \cos 52.7° - B \cos 62.5° = 0$$

Use the methods of this section to solve this system of equations.

Solve Exercises 23 and 24 by using the inverse matrix of the coefficient matrix.

25. Three different alloys are made up of zinc, lead, and copper. Alloy A contains 55.0% zinc, 32.0% lead, and 13.0% copper. Alloy B contains 38.0% zinc, 32.0% lead, and 30.0% copper. Alloy C contains 34.5% zinc and 65.5% lead. It is desired to mix these three alloys to obtain 200.0 grams of a mixture that contains 45.0% zinc, 37.5% lead, and 17.5% copper. To determine how many grams of each alloy are needed, it is necessary to solve the following system of equations where we let x = number of grams of alloy A, y = number of grams of alloy B, and z = number of grams of alloy C.

$$0.550x + 0.380y + 0.345z = 0.450(200)$$
$$0.320x + 0.320y + 0.655z = 0.375(200)$$
$$0.130x + 0.300y \qquad\quad = 0.175(200)$$

♦ 24. *(Washington, Exercises 16.4, #30)* In applying Kirchhoff's laws to an electrical circuit the following equations are found. I_A, I_B, and I_c are three currents, in amperes, in the circuit. Find I_A, I_B, and I_c.

$$I_A + I_B + I_C = 0$$
$$2.35I_A - 4.75I_B = 6.34$$
$$4.75I_B - 1.20I_C = -3.24$$

9.3 Row Operations on Matrices
(Washington, Sections 16.3 and S.2)

Since it not always possible to use the inverse matrix method to solve a system of equations, it is sometimes helpful to use another method called the Gauss-Jordan method. The Gauss-Jordan method may be used to solve any system of linear equations, but it is particularly useful in finding the form of a solution when a system of equations has an infinite number of solutions or in determining if there are no solutions to a system of equations. The Gauss-Jordan method makes use of what are called row operations on matrices.

Row operations may also be used to find the inverse of a matrix. However, we will not use this method to find the inverse, since on a graphing calculator, the inverse matrix may be found directly and quickly.

There are three row operations, which when performed on matrices, yield equivalent matrices (not equal matrices). Many graphing calculators have a provision for performing the matrix row operations.

Matrix Row Operations
1. Any two rows may be interchanged.
2. Any row may be multiplied (or divided) by a non-zero constant.
3. Any row may be multiplied by a non-zero constant and added to a second row, replacing the second row.

Procedure M6: Row operations on a matrix.

1. Obtain the matrix operations menu by first pressing the key MATRX, then:
   On the TI-82, 83, or 84 Plus: On the TI-85 or 86:
   Highlight MATH in the top row.   From the matrix menu,
                                         select OPS
                                         (then press MORE).

2. Select the desired operation from the menu. Names of matrices are selected as in Procedure M1.
   In the following, the left hand column gives the selection, form, and example for the TI-82, 83, or 84 Plus and the right hand column given the selection, form, and example for the TI-85 or TI-86. To obtain some of these operations, it may be necessary to scroll down the given list by using the bottom cursor key.
   (a) To interchange two rows on a matrix, select
       C:rowSwap(RSwap
       Form: Form:
        rowSwap(*name, row1, row2*)   **RSwap**(*name, row1, row2*)
       Example: rowSwap([*A*], 1, 3)   Example: Rswap(*A*, 1, 3)
       (Interchanges rows 1 and 3 of matrix *A*.)
   (b) To multiply a row by a constant, select:
       E:*row(MultR
       Form: Form:
        ***row**(*multiplier, name, row*)   **MultR**(*multiplier, name, row*)
       Example: rowSwap(1/2, [*A*], 3)   Example: multR(1/2, *A*, 3)
       (Multiplies row 3 of matrix *A* by ½.)
       To divide a row by a constant, multiply by the reciprocal.
   (c) To multiply a row by a constant and add to another row, select
       F:*row+(MRAdd
       Form: Form:
        ***row+**(*mult,name,row1,row2*)   **mRAdd**(*mult,name,row1,row2*)
       Example: *row+(−5,[*A*], 1, 3)   Example: mRAdd(−5, *A*, 1, 3)
       (Multiplies row 1 of matrix *A* by −5 and adds the result to row 3 and replaces row 3.)

3. Press ENTER.

Note: The matrix operation *ref* gives the row echelon form directly for a matrix and the operation *rref* gives the reduced row echelon form directly for a matrix. The forms are: ref (*A*) and rref(*A*) where *A* is the name of the matrix to be reduced.

To solve a system of system of equations by the Gauss-Jordan method, the system of equations is first written in standard form. As we have previously seen the matrix form of a system of equations is $AX = C$. For the Gauss-Jordan method, we form an **augmented matrix** of the form $(A|C)$ where A is the coefficient matrix and C is the constant matrix. We then perform matrix row operations on this matrix as demonstrated in Example 9.8. The goal is to get this augmented matrix into the form $(I|Y)$, where I is the identity matrix. If this can be accomplished, then the column Y is the solution matrix.

Example 9.8:   Solve the system of following system of equations by the Gauss-Jordan method.

$$x + 2y = 5$$
$$3x + 8y + z = 5$$
$$2x - z = 12$$

Solution: 1. Since these equations are already in standard form, we write the augmented matrix for this system

$$\begin{bmatrix} 1 & 2 & 0 & | & 5 \\ 3 & 8 & 1 & | & 5 \\ 2 & 0 & -1 & | & 12 \end{bmatrix}$$

2. First, make sure element $a_{11} = 1$. In this case, this is already true. If it were not, we would perform row operation 2 to make it a 1. Next we want to make a_{21} and a_{31} both zero. We do this by performing row operation 3. Use Procedure M6, 2(c) to multiply row 1 by -3 and add to row 2 and to multiply row 1 by -2 and add to row 3. The result is

$$\begin{bmatrix} 1 & 2 & 0 & | & 5 \\ 0 & 2 & 1 & | & -10 \\ 0 & -4 & -1 & | & 2 \end{bmatrix}$$

3. Next, we make $a_{22} = 1$ by multiplying row 2 by $\frac{1}{2}$ using row operation 2. Use Procedure M6, 2(b).

$$\begin{bmatrix} 1 & 2 & 0 & | & 5 \\ 0 & 1 & .5 & | & -5 \\ 0 & -4 & -1 & | & 2 \end{bmatrix}$$

4. Then, we will make a_{12} and a_{32} both zero by using row operation 3. Use Procedure M6. 2(c) to multiply row 2 by -2 and add to row 1 and multiply row 2 by 4 and add to row 3.

$$\begin{bmatrix} 1 & 0 & -1 & | & 15 \\ 0 & 1 & .5 & | & -5 \\ 0 & 0 & 1 & | & -18 \end{bmatrix}$$

5. Next, make sure $a_{33} = 1$. If not, use row operation 2 to make it a one. Make both a_{13} and a_{23} zero by using row operation 3. Use Procedure M6, 2(c)

to multiply row 3 by −.5 and add to row 2 and to multiply row 3 by 1 and add to row 1.

$$\begin{bmatrix} 1 & 0 & 0 & | & -3 \\ 0 & 1 & 0 & | & 4 \\ 0 & 0 & 1 & | & -18 \end{bmatrix}$$

6. We now have achieved the correct form. When we write this last matrix as a system of equations we obtain the solution to the given system of equations

$$x = -3, \ y = 4, \ \text{and} \ z = -18.$$

If, in attempting to obtain the form $(I|Y)$, a row of all zeros is obtained, then the system is a dependent system and has an infinite number of solutions and if a row contains all zeros except for the last value in the row (to the right of the vertical bar, the system is inconsistent and has no solutions.

Example 9.9: Solve the following system of equations using the Gauss-Jordan method.

$$x + 2y - 3z = 5$$
$$2x + y + 2z = 4$$
$$x + 5y - 11z = 11$$

Solution: 1. These equations are already in standard form. Thus, write the augmented matrix for this system.

$$\begin{bmatrix} 1 & 2 & -3 & | & 5 \\ 2 & 1 & 2 & | & 4 \\ 1 & 5 & -11 & | & 11 \end{bmatrix}$$

2. Since element $a_{11} = 1$ we proceed to make a_{21} and a_{31} both zero. We do this by performing row operation 3. Use Procedure M6, 2(c) to multiply row 1 by −2 and add to row 2 and to multiply row 1 by −1 and add to row 3. The result is

$$\begin{bmatrix} 1 & 2 & -3 & | & 5 \\ 0 & -3 & 8 & | & -6 \\ 0 & 3 & -8 & | & 6 \end{bmatrix}$$

3. Next we will make $a_{22} = 1$ by multiplying row 2 by $-\dfrac{1}{3}$ using row operation 2. Use Procedure M6. 2(b). The result, rounded to three decimal places is:

$$\begin{bmatrix} 1 & 2 & -3 & | & 5 \\ 0 & 1 & -2.667 & | & 2 \\ 0 & 3 & -8 & | & 6 \end{bmatrix}$$

4. Next we will make a_{12} and a_{32} both zero by using row operation 3. Use Procedure M6, 2(c) to multiply row 2 by –2 and add to row 1 and multiply row 2 by –3 and add to row 3.

$$\begin{bmatrix} 1 & 0 & 2.333 & | & 1 \\ 0 & 1 & -2.667 & | & 2 \\ 0 & 0 & -2E-13 & | & 0 \end{bmatrix}$$

5. The quantity –2E–13 (this is -2×10^{-13}) in the last row is for most practical purposes, zero. Thus, the last row is all zeros and this system has an infinite number of solutions. The equations given by the first two rows are

$$x + \frac{7}{3}z = 1 \qquad\qquad x = 1 - \frac{7}{3}z$$
$$\text{or}$$
$$y - \frac{8}{3}z = 2 \qquad\qquad y = 2 + \frac{8}{3}z$$

The solutions are the set of numbers (x, y, z) such that if we know any one of the three variables, we determine the other two variables from these equations. For example, to obtain one of the solutions let $z = 3$. Then, $x = -6$ and $y = 10$. Another solution may be obtained by letting $z = 0$. Then, $x = 1$ and $y = 2$.

Exercise 9.3

Solve each of the following systems of equations by setting up the augmented matrix and using row operations. Identify those that have no solutions and those that have an infinite number of solutions.

1. $8x - 5y = 25.5$
   $5x = 32.7 - 2y$

2. $3.52x + 6.24y = 125.45$
   $y = 6.54x - 54.60$

3. $3x + 4y = 64.8$
   $8y = 25.5 - 6x$

4. $0.22x + 0.35y = 2.45$
   $0.64x - 0.65y = 3.47$

5. $x + 2y - z = -1$
   $-5y - 3x + z = -7$
   $2x + 2z - y = 1$

6. $2x - 3y + 5z = 13$
   $-y + z - 3x = 9$
   $3x + 16z - 10y = 48$

7. $6x - 3y + 2z = 29$
   $x + 2y - 3z = 7$
   $3x - 9y + 11z = 8$

8. $z = x - 2y$
   $0.3x + 0.4y - 0.5z = 1.5$
   $1.1x - 5.2y = 0.5z = 0$

9. $x + y + z = 15$
   $2.5x + 3.0y - 1.5z = 7$
   $6.0x + 7.0y - 2z = 29$

10. $25x + 30y - 15z = 125$
    $10x - 15y + 12z = 173$
    $5x - 34z = 221$

11. $3.5x + 2.1y + 2.8z = 55$
    $1.4x - 1.3y - 2.3z = 17$
    $0.7x + 5.7y + 7.4z = 21$

12. $3x + 5y + 4z + w = 7$
    $2x - 3y - 6z + w = 9$
    $x + 4y + 2z - w = 5$
    $4x - 2y - 4z = 6$

♦ 13. *(Washington, Exercises S.2, #31)* Three machines together produce 672 parts each hour. Twice the production of the second machine is 12 parts more than the sum of the other two machines. If the first operates for 3.5 hours and the others operate for 2.5 hours, 1992 parts are produced. Find the production rate of each machine.

♦ 14. *(Washington, Exercises S.2, #32)* A total of $12,550 is invested, part at 6.5%, part at 6.0%, and part at 5.5%, yielding a total annual interest of $773.20. The income from the 6.5% part yields $35.40 more than that for the other two parts combined. How much is invested at each rate?

Chapter 10

Graphing of Inequalities

10.1 Graphing of Inequalities
(Washington, Section 17.5)

In this section we are interested in solving inequalities in two variables by graphing. Inequalities contain two algebraic expressions separated by one of the inequality signs ($>$, $<$, $\leq$, or $\geq$). A **solution to an inequality** in two variables consists of all pairs of numbers (x, y) that satisfy the inequality. The pairs of numbers $(0, 1)$, $(1, 0)$, $(-2, 2)$ and $(1, -1)$ are all solutions of the inequality $3x - 2y \leq 5$ since when the values are substituted for x and y the inequality is a true statement. Inequalities generally have an infinite number of solutions. To visualize the solutions to an inequality, we graph the inequality and shade in the region corresponding to all pairs of numbers that are solutions.

Before graphing an inequality, it is first necessary to solve the inequality for y as we did when graphing equations. Rules for working with inequalities are similar to, but not the same as, those for working with equations. A major difference occurs when both sides of an inequality are multiplied or divided by the same negative value.

Rules for Inequalities

1. The same term may be added or subtracted to both sides of the inequality without changing the direction of the inequality sign.
2. Both sides of an inequality may be multiplied or divided by the same positive value without changing the direction of the inequality sign.
3. If both sides of an inequality are multiplied or divided by the same negative value, then the direction of the inequality sign is reversed.

To graph an inequality on a graphing calculator, first graph the equation corresponding to the inequality and then shade the region that indicates the solution to the inequality by using the shade function of the calculator. When using the shade function it is sometimes desirable to use special variables such as Ymin, Ymax, Y_1, etc. Procedure G21 indicates how to obtain these special values and Procedure G22 indicates how to use the shade function.

Procedure G21. **To obtain special *Y*-variables.**

On the TI-82 or 83 or 84 Plus:
1. To obtain window quantities
   Ymin, Ymax, Xmin, or Xmax:
   press the key: VARS
   From the menu:
   Select: Window..
   Then select desired quantity.

2. To obtain variable names Y_1, Y_2,
   etc.:
   On the TI-82,
   press the key: *Y*-VARS
   On the TI-83 or 84 Plus,
   press the key: VARS,
   highlight *Y*-VARS at the top
   From the menu,
   select: Function
   then, select the function name.

On the TI-85 or 86:
The best way perhaps is to just enter the name from the keyboard. When entering the name from the keyboard, be sure the correct upper or lower case is entered.
or

1. Press the key VARS
2. Select ALL
3. Press F1 to page down till desired variable is on screen.
4. Use cursor keys to select desired variable.
5. Press ENTER

Procedure G22. **To shade a region of a graph.**

1. Enter the desired function for a function name. Make sure all functions not wanted are deleted or turned off.
2. Select appropriate window values and graph the function.
3. Select the Shade command:
   On the TI-82, 83, or 84 Plus:
   Press the key: DRAW
   From the menu, select 7:Shade(
   On the TI-82 the form is
   Shade(*L, U, D, Lt, Rt***)**
   On the TI-83 and 84 the form is
   Shade(*L, U, Lt, Rt, Pat, D***)**

   On the TI-85 or 86:
   From the GRAPH menu,
   select DRAW (after pressing MORE) and Select Shade
   On the TI-85, the form is
   Shade(*L, U, Lt, Rt***)**
   On the TI-86, the form is
   Shade(*L, U, Lt, Rt, Pat, D***)**

   Where:
   L is lower boundary (value or function) of the shaded area.
   U is upper boundary (value or function) of the shaded area.
   D is the density value (a digit from 1 to 8) that determines the spacing of the vertical lines the calculator draws to shade the region. The higher the value for the density, the more widely spaced the shading lines. If the density value is omitted, the shading is solid. (optional)
   Lt is the left most *x*-value (optional).
   Rt is the right most *x*-value (optional).
   On the TI-83, *Pat* is the pattern. (optional)
   (A digit from 1 to 4 – each digit gives a different fill pattern.).

[D, Lt, Rt, Pat are optional - but on the TI-82, if Lt and Rt are included, D must be also included.]

(Example: Shade(−10, Y_1) shades in the area above $y = -10$ and below the graph of Y_1)

4. Enter the appropriate values for L, U, and/or D and left and right x-values. Then, press ENTER.

5. To clear the shading:

On the TI-82, 83, or 84 Plus:	On the TI-85 or 86:
Press the key: DRAW	From the GRAPH menu,
From the menu,	select DRAW (after pressing
Select 1:ClrDraw	MORE) and Select CLDRW
	(After press MORE twice).

Notes: To shade above the function represented by Y_1, use Shade(Y_1, Ymax).

To shade below the function represented by Y_1, use Shade(Ymin, Y_1).

The TI-83 Plus and TI-84 Plus may have an application program called Inequalz that allows the insertion of inequality symbols for the equal signs in the function table. To run this or to turn it off, press the key APPS and select :Inequalz and press ENTER. The inequality symbols may be selected by using the function keys ($F_1, F_2, etc.$)

When graphing inequalities involving ≤ or ≥, the curve is included as part of the solution, but when graphing inequalities involving > or <, the curve itself is not included as part of the solution. On the graphing calculator it is not possible to show the difference between these cases. However, when sketching the graph by hand, use a solid curve for the graph of the equation for inequalities involving ≤ or ≥, and a dashed curve for the graph of the equation for the inequalities involving > or <. After the curve is drawn, shade in the proper area.

Example 10.1: Graph the solution to the inequality

$$x - 2y > 6.$$

Solution: 1. First, solve the inequality for y.

$$x - 2y > 6$$
$$-2y > 6 - x \quad \text{[subtract } x \text{ from both sides]}$$
$$y < -3 + \frac{1}{2}x \quad \text{[divide both sides by −2, change direction of}$$
$$\text{inequality]}$$

2. Graph the equation $y = -3 + \frac{1}{2}x$ by entering $-3 + \frac{1}{2}x$ for Y_1.

Y_1 should be the only function turned on. Select the standard viewing rectangle.

3. Next shade in the area corresponding to the solution. Since the inequality is $y < -3 + \frac{1}{2}x$, we want the area such that

y is less than the line $y = -3 + \frac{1}{2}x$

4. Use Procedure G22 to shade the region below Y_1. For the lower boundary of the shaded area, we select Ymin. For the upper boundary of the shaded area, we want the graph of

$$y = -3 + \frac{1}{2}x \text{ or } Y_1. \text{ (Procedure G21)}.$$

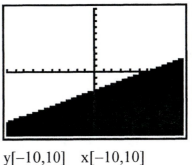

y[−10,10]   x[−10,10]

Figure 10.1

The statement needed is
Shade(Ymin, Y_1). Press ENTER. The graph is in Figure 10.1.

Any pair of numbers x and y that correspond to points in the shaded region will be solutions to the given inequality. As a check some selected points (such as $(3, -5)$ or $(10, 1)$ may be substituted into the inequality to determine if the inequality is true at these points. Remember when sketching this graph on a sheet of paper, to make the line $y = -3 + \frac{1}{2}x$ a dashed line.

In certain applications, it is helpful to find solutions to systems of inequalities. A system of inequalities is a set of two or more inequalities with two or more variables. A **solution to a system of linear inequalities** in two variables x and y consists of all pairs of numbers (x, y) that satisfy all the inequalities in the system. To visualize the solution to a system of inequalities, we graph the solution to each inequality in the system. The area where these solutions overlap is the region containing the solution to the system.

The corners of the region on a graph that represents the solution set of a system of inequalities are called the **vertices** of the region. The vertices may occur at the intersection of two graphs or at the x- or y-intercepts. The y-intercepts, in general, are easily found by letting $x = 0$ and solving for y. The x-intercepts of the function are the same as the roots of the corresponding equation. (The x-intercept of $y = 3x + 2$ is the same as the root of the equation $3x + 2 = 0$.) Procedures G11 and G15 describe the method of finding roots of equations and points of intersection of graphs.

◆ Example 10.2: *(Washington, Section 17.5, Example 5)* Graph the region defined by the inequalities $y \geq -x - 2$ and $y + x^2 < 0$. Find the vertices of the region.

Solution: 1. First, solve each inequality for y.

$$y \geq -x - 2 \text{ and } y < -x^2$$

2. Next enter the corresponding equations in for function names and graph the two curves. Use the standard viewing rectangle. Let

$$Y_1 = -x - 2$$
$$Y_2 = -x^2$$

3. To shade the area between the two graphs, use the shade command in Procedure G22 by letting the lower boundary be Y_1 and the upper boundary Y_2. The command is

Shade(Y_1, Y_2).

The shaded area represents that set of points that satisfies both inequalities at the same time. When sketching this graph on paper it needs to be remembered that the curve $y = -x^2$ should be drawn as a dashed curve and the line $y = -x - 2$ as a solid line.

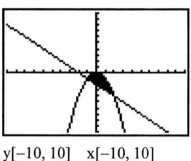

y[–10, 10]   x[–10, 10]

Figure 10.2

4. We use Procedure G15 to find the points of intersection. The vertices are $(-1, -1)$ and $(2, -4)$

Exercise 10.1

In the following exercises, use the graphing calculator to determine the solution set to the given inequality or system of inequalities. Based on your results, hand sketch a graph and shade in an area that represents the solution.

1. $3x - 5y > 15$

2. $4x + 6y \leq 12$

3. $4x + 7y - 18 \geq 0$

4. $-3x + 8y > 0$

5. $y + x^2 < 4$

6. $x^3 - 4x < y$

7. $y \leq \sqrt{x + 3}$
   $y \geq 0$

8. $y \leq \dfrac{1}{x}$
   $y \geq 0, x \geq 1$

9. $y < 2 \log x$
   $y \geq 0$

10. $y < 5 \sin (2\pi x)$
    $x \geq 0, y \geq 0, x \leq 1$

11. $y < |x^3 - 9x|$
    $y > 0, -3 < x < 3$

12. $y < e^{\frac{x^2}{2}}$
    $y > 0, 1 < x < 3$

In Exercises 13-25, give all corner points (vertices) of the solution region and label these vertices on your hand-sketched graph. Round all values to 3 significant digits.

13. $15x + 13y \le 135$
    $x \ge 0, y \ge 0$

14. $23x + 19y < 243$
    $x \ge 0, y \ge 0$

15. $12x + 20y > 47$
    $x \ge 0, y \ge 0$

16. $0.25x - 1.47y \le 23.45$
    $x \ge 0, y \ge 0$

17. $5x + 4y \le 19$
    $y \le 3, y \ge 0, x \ge 0$

18. $3x + 6y \ge 35$
    $y \le x, y \ge 2$

19. $7x + 4y \le 50$
    $x + 5y \le 34$
    $y \ge 5, x \ge 0$

20. $x + y \le 25$
    $3x + 5y \ge 44$
    $x \ge 0, y \ge 0$

21. $x + 5y \le 65$
    $7x + 3y \ge 98$
    $x + 2y \ge 15$
    $x \ge 0, y \ge 0$

22. $2x + y \le 14$
    $5x + 2y \ge 10$
    $-3x + 2y \ge 5$
    $x \ge 0, y \ge 0$

23. $-2x + 3y \le 12$
    $y > x^2$

24. $y \ge 2^x$
    $y - 5x \le 1$

25. $y < \sin x$
    $y \ge \cos x$
    $x \ge 0, x \le 2\pi$

26. $y \le \sqrt{25 - x^2}$
    $y - x^2 + 4 > 0$

In Exercises 27-30 set up the correct inequalities, use the graphing calculator as an aid to determine the region described, hand sketch a graph showing the region, and determine the vertices (corner points) of the region. Show the coordinates of the vertices on the hand-sketched graph.

27. A cereal company manufactures x twelve ounce boxes of cereal and y eighteen ounce boxes every hour. The total number of boxes of cereal made each hour cannot exceed 185. Determine the region that represents the possible values for x and y.

28. A company sells two products, A and B. It makes a profit of $5.50 for each unit of A and $7.25 for each unit of B it sells. The total profit must be at least $425. Determine the region that represents the possible number of units of A and B that need to be sold.

29. ZZZ Manufacturing Company has $47,500 available for manufacturing machine parts. Part of this is to be spent on parts costing $125 each to produce and part is to be spent on parts costing $158 each to produce. At least 100 of the $125 parts must be produced. Determine the region that represents the possible number of $125 parts and $158 parts the company is able to produce under these conditions.

30. A company produces two different types of computers. One type sells for $1250 and the other type sells $1570. It needs to make at least $250,000 off of sales. It also has set a goal of selling at least 100 of the $1250 computers and 75 of the $1570 computers. Determine the region that represents the possible number of computers of each type that must be sold to meet these conditions..

♦ 31. *(Washington, Exercises 17.5, #43)* A telephone company is installing two types of fiber optic cables in an area. It is estimated that no more than 322 m of type *A* cable, and at least 175 m but no more than 475 m of type *B* cable, are needed. Graph the possible lengths of cable that are needed.

♦ 32. *(Washington, Exercises 17.5, #48)* A rectangular computer chip is to be designed so that its perimeter is no more than 25 mm and its width must be at least 3.2 mm and it length at least 7.5 mm. Graph the possible values of the width *w* and length *l*.

Chapter 11

Polar Graphs

11.1 Graphing in Polar Coordinates
(Washington, Sections 21.9 and 21.10)

There are times when it is either easier or better to graph in coordinate systems other than in the familiar rectangular coordinate system. One important system is the polar coordinate system. Polar coordinates were first mentioned in Section 5.1 of this manual and that material should be reviewed. In the polar coordinate system a point is located by using an angle, θ, measured from the positive x-axis and a distance, r, measured from the origin. The positive x-axis is known as the **polar axis** and the origin is known as the **pole**. Points in polar coordinates are expressed in terms of r and θ in the form (r, θ). (For conversion of coordinates see Procedure C15.) In polar coordinates, we normally express r as a function of the variable θ and write $r = f(\theta)$.

Before using the graphing calculator to graph in polar coordinates, it is first necessary to set the calculator to polar graphing mode. After this is done a function may be entered and graphed.

Procedure G23. To graph in polar coordinates

1. Make sure the calculator is in correct mode for polar coordinates. (Procedure G17)
2. To enter polar equations:

   On the TI-82 or 83 or 84 Plus:

   Press the key $Y=$
   On the screen will appear:
   $r1=$
   $r2=$
   $r3=$
   (etc.)
   (up to 6 equations may be entered)
   Functions may be turned on and off as with functions in rectangular coordinates (see Procedure G1, Note 3).

   On the TI-85 or 86:

   Press the key GRAPH.
   From the menu, select $r(\theta) =$
   On the screen will appear: $r1=$
   (up to 99 polar equations may be entered as $r1, r2, r3, ...$)

3. To enter the equation for r.
   For the variable θ

   On the TI-82 or 83 or 84 Plus
   Press the key X,T,θ

   On the TI-85 or 86:
   Select θ from the menu.

4. Select GRAPH to graph the function or QUIT to exit.

Graphs in polar coordinates are plotted in order of increasing values of the angle θ rather than from left to right as in rectangular coordinates. Careful observation while the calculator is drawing the graph or using the trace function will indicate the order in which the points are plotted. As with regular graphs, we may use the zoom function of the calculator to zoom in on special points. Some adjustment of the viewing rectangle may be necessary when graphing in polar coordinates. In addition to the setting of minimum and maximum x- and y-values (Procedures G5 and G6), we now need to set values for θ.

Procedure G24. To change viewing rectangles values for θ.

With the calculator in the mode for polar coordinates:
1. Follow Procedure G5 to see the values for the viewing rectangle.
2. Enter new values for θmin, θmax, and θstep and press ENTER or move to the next item by using the cursor keys.
   The standard viewing rectangle values for θ are

   θmin = 0, θmax = 6.28... (2π), and θstep = 0.13...($\pi/24$)
   (or θmax = 360 and θstep = 7.5 if in degree mode).

   θstep determines how often points are calculated and may affect the appearance of the graph. Leaving θstep at approximately 0.1 is generally sufficient, but θstep may have to be changed if there is a major change in the viewing rectangle. θmax should be sufficiently large to give a complete graph.
3. Press QUIT to exit.

The step size for θ will affect the appearance of the graph. If the step size is too large, a series of line segments may result rather than a smooth appearing curve. If the step size is too small, it may take too long to graph the curve. However, be aware that if we zoom in on a portion of the graph, the step size may have to be decreased to obtain a smooth appearing graph. The smaller the value for θ the longer it takes to graph the curve because more points are being plotted.

Example 11.1: Graph the polar function $r = 8 \cos \theta$, make a sketch of the graph, and give the polar coordinates, rounded to two decimal places, of the points where the graph crosses the x- and y-axes.

Solution:
1. Make sure the calculator is in the correct mode for polar graphing and use the standard viewing rectangle. Use $0 \leq \theta \leq 2\pi$.

2. Graph the function $r = 8 \cos \theta$ using Procedure G23. In the standard viewing rectangle the graph should appear as an ellipse. If we change to a square viewing

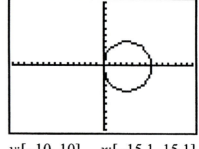

y:[−10, 10] x:[−15.1, 15.1]

Figure 11.1

rectangle we see the graph is really a circle. The graph is in Figure 11.1.

3. With the graph on the screen, use the TRACE function to see points on the graph. As the cursor moves on the curve, the polar coordinates are given at the bottom of the screen. The cursor will move in order of increasing or decreasing value of θ.

The points where the graph intersects the axes (in polar coordinates) are

$(8.00, 0)$ and $(0, \dfrac{\pi}{2})$. Care must be taken in obtaining values near the pole, since values may vary greatly with small movements of the cursor. Values obtained for intersection points should be tried in the equation to be sure they are correct.

Certain forms of functions in polar coordinates, as in rectangular coordinate, give particular types of graphs. We consider some of these standard forms.

Roses

Roses have the standard forms

$$r = a \sin (b\theta) \text{ or } r = a \cos(b\theta)$$

In the exercises, we will see the effect of changing the values of a and b. The value a has an effect on the size of the graph and the value b determines the number of leaves or petals on the rose. Figure 11.2 shows the rose $r = 3\sin 2\theta$. Zooming on the origin will show that the graph passes through the origin.

A **circle** with its center on a coordinate axis and passing through the origin is a special case of a rose when there is only one leaf (that is when $b = 1$).

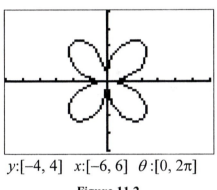

$y{:}[-4, 4]$ $x{:}[-6, 6]$ $\theta{:}[0, 2\pi]$

Figure 11.2

Limaçons

Limaçons are formed by equations of the form

$$r = a + b \sin \theta \text{ or } r = a + b \cos \theta$$

Certain special cases of limaçons result in curves called cardioids. The relative sizes of a and b determine different limaçons. Figure 11.3 shows the limacon $r = 3 + 2\cos\theta$.

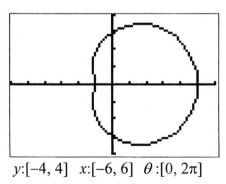

$y{:}[-4, 4]$ $x{:}[-6, 6]$ $\theta{:}[0, 2\pi]$

Figure 11.3

Conics

Various parabolas, ellipses, and hyperbolas are graphed by equations of the form

$$r = \frac{a}{b + c\sin\theta} \quad \text{or} \quad r = \frac{a}{b + c\cos\theta}$$

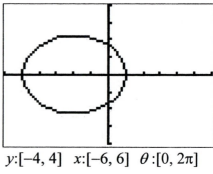

$y:[-4, 4]$ $x:[-6, 6]$ $\theta:[0, 2\pi]$

Figure 11.4

The type of curve and the size will be determined by the relative values of a, b, and c. For such a conic, one focus will always be at the pole. Figure 11.4 shows the ellipse given by the equation $r = \dfrac{5}{3 + 2\cos\theta}$.

Spirals

Certain spirals are graphed by functions of the form

$$r = a\theta^b \quad \text{or} \quad r = a^{b\theta}$$

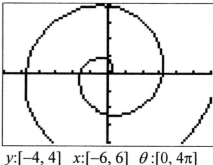

$y:[-4, 4]$ $x:[-6, 6]$ $\theta:[0, 4\pi]$

Figure 11.5

The values of a and b determine the relative size and shape of the spiral. The spiral $r = 0.5\theta$ appears in Figure 11.5.

Exercise 11.1

Use the calculator to graph the following in polar coordinates. Points should all be expressed in polar coordinates. For best results, use a square viewing rectangle and $0 \le \theta \le 2\pi$.

In Exercise 1-10: (a) Determine what type of graph is obtained and draw a sketch of the graph on your paper, (b) Give the maximum distance of a point on the graph from the pole, (c) Give the points, in polar coordinates, (to one decimal place) where the graph crosses the x- and y-axis.

1. $r = 4\cos\theta$

2. $r = -4\sin\theta$

3. $r = -6\cos 2\theta$

4. $r = 8\sin 2\theta$

5. $r = 6 + 4\sin\theta$

6. $r = 5 + 5\cos\theta$

7. $r = \dfrac{1}{5 + \sin\theta}$

8. $r = \dfrac{1}{4 + 4\cos\theta}$

9. $r = 4\theta^2$

10. $r = 1.5(2^\theta)$

Do problem 11-20 as indicated.

11. Graph $r = 3$, $r = 5$, and $r = 7$ in a square viewing rectangle. What kinds of graphs are obtained in each case? Where is the center in each case? What kind of graph is given by the standard form $r = k$, where k is some constant?

12. Graph $r = \dfrac{4}{\cos\theta}$, $r = \dfrac{-3}{\cos\theta}$, $r = \dfrac{4}{\sin\theta}$, and $r = \dfrac{-5}{\sin\theta}$. What kinds of graphs are obtained in each case? What kind of graph is given by the standard forms $r = \dfrac{k}{\cos\theta}$ and $r = \dfrac{k}{\sin\theta}$?

13. Graph $r = \cos\theta$, $r = 2\cos\theta$, and $r = 4\cos\theta$ on the same set of axes. What kinds of graphs are obtained? What happens to a graph of the form $r = A\cos\theta$ as A changes?

14. Graph $r = 5\sin 2\theta$ and $r = 5\cos 2\theta$ on the same axes. Then graph $r = 5\sin 4\theta$ and $r = 5\cos 4\theta$ on the same axes. What is the difference between using $\sin\theta$ and using $\cos\theta$?

15. Graph $r = 8\sin 2\theta$, $r = 8\sin 4\theta$, and $r = 8\sin 6\theta$. Describe the graphs obtained. How does the graph of $r = 8\sin B\theta$ change as B changes? Does this hold for the equation $r = 8\sin 3\theta$? What is the relation of the number of leaves and the number B in the equation $r = A\sin B\theta$?

16. Graph $r = 6\sin 3\theta$, $r = 6\sin 5\theta$, and $r = 6\sin 7\theta$. Describe the graphs obtained. How do these graphs differ from those in problem 15? What is the relation of the number of leaves and the number B in the equation $r = A\cos B\theta$?

17. Graph $r = 5 - 3\sin\theta$. What type of graph is obtained? (Zoom in on the point where $\theta = \dfrac{\pi}{2}$.) Graph $r = 4 - 4\sin\theta$, and $r = 3 - 5\sin\theta$. (Use the square-viewing rectangle.) How do these graphs differ? How do the relative sizes of A and B effect the graphs of the equation $r = A - B\sin\theta$?

18. Graph $r = 6 - 2\cos\theta$, $r = 6 - 6\cos\theta$, and $r = 2 - 6\cos\theta$. How do these graphs differ? How do the relative sizes of A and B effect the graphs of the equation $r = A - B\cos\theta$? How do these graph differ from those in Exercise 17?

19. Graph $r = \dfrac{10}{5 - 3\sin\theta}$, $r = \dfrac{10}{3 - 3\sin\theta}$, and $r = \dfrac{10}{3 - 5\sin\theta}$. What kinds of graphs are obtained? How do these graphs differ? How do the relative sizes of B and C effect the graphs of the equation $r = \dfrac{A}{B - C\sin\theta}$?

20. Graph $r = \dfrac{10}{3 - 2\cos\theta}$, $r = \dfrac{10}{2 - 2\cos\theta}$, and $r = \dfrac{10}{2 - 3\cos\theta}$. What kinds of graphs are obtained? How do these graphs differ? How do the relative sizes of B and C effect the graphs of the equation $r = \dfrac{A}{B - C\cos\theta}$?

21. Graph the two functions $r = 5 \sin \theta$ and $r = 5 \cos \theta$ on the same set of axes. Estimate the points of intersection to two decimal places.

22. Graph the two functions $r = -3 \cos \theta$ and $r = 3 + 5 \sin \theta$ on the same set of axes. Estimate the points of intersection to two decimal places.

23. Graph $r = 0.2\theta^2$. What type of graph is obtained? Give the first 3 points (for $\theta > 0$) that the graph crosses the x- and y-axes.

24. Graph $r = 2^{3\theta}$. What type of graph is obtained? Give the first 3 points (for $\theta > 0$) that the graph crosses the x- and y-axes.

25. The shape of a cam in a piece of machinery is given by the equation $r = 8.7 - 5.2 \cos \theta$. Graph this function and sketch the graph on your paper. Determine the minimum and maximum distance of a point on the edge of the cam from the pole.

26. The polar equation of the path of a weather satellite about the earth is

$$r = \frac{4824}{1 + 0.145 \cos \theta}$$

where r is the distance from the center of the earth and is measured in miles. The path is an ellipse with the center of the earth at one focus. Graph on the calculator and sketch the graph on your paper. Assume the earth is a circle whose radius is 3960 miles. What is the maximum distance the satellite is above the earth's surface and what is the minimum distance?

♦ 27. *(Washington, Exercises 21.10, #41)* An architect designs a patio which is shaped such that it is described as the area within the polar curve $r = 3.25 - 3.25 \sin \theta$. Measurements are in meters. Graph on your calculator and sketch the shape of the patio on your paper. What is the width of the patio along the x-axis? Find the angle between the polar axis where $r = 3.25$ and the line from the pole to the point where $r = 6.00$ in the 4th quadrant.

♦ 28. *(Washington, Exercises 21.10, #44)* A missile is fired at an airplane and is always directed toward the airplane. The missile is traveling at twice the speed of the airplane. An equation that describes the distance r between the missile and the airplane is

$$r = \frac{72 \sin \theta}{(1 - \cos \theta)^2}$$

where θ is the angle between their directions. The plane is assumed to be at the pole at all times. Graph this for $\frac{\pi}{4} \le \theta \le \pi$. Show the graph on your paper. What is the distance of the missile from the plane when $\theta = \frac{\pi}{2}$? When $\theta = \pi$?

Chapter 12

Statistics

12.1 Statistical Graphs and Basic Calculations
(Washington, Sections 22.1, 22.1, and 22.3)

Statistics is the collection, organization, and interpretation of numerical data. This section discusses how data is organized, takes a first look at the interpretation of data, and shows how the graphing calculator may be used to help analyze data.

If the amount of data is relatively small, it is not difficult to look at the data as individual items. However, if the amount of data is large, with many values, each occurring a large number of times, then the data needs to be grouped into what is called a frequency distribution. A **frequency distribution** is a table giving each individual data value and the frequency of each data value. The **frequency** is the number of times a particular data value occurs in that set of data.

To visualize data given in a frequency distribution, it is helpful to see a graph of the data. Three different types of graphs are considered: a histogram, a scatter graph, and a frequency polygon. A **histogram** is a graph in which each data item or group of data items is represented by a rectangle. All the rectangles are the same width and the height of each rectangle represents the frequency of that particular data value. This type of graph is commonly known as a bar graph. In a **scatter graph** points are plotted representing the frequency of each of the data values where the frequency of each item is along a vertical scale. A **frequency polygon** (or **xyLine graph**) is a graph in which the points of a scatter graph are connected by straight-line segments.

Before working with statistical data on the calculator, we first need to organize the data and store it in the calculator.

Procedure S1. **To enter or change single variable statistical data.**

On the Texas Instruments graphing calculators statistical data is stored as lists of numbers.
(See also Procedure C7)
1. Press the key STAT
2. Enter edit mode.

On the TI-82, 83, or 84 Plus: On the TI-85 or 86:
With EDIT highlighted in the top   Select EDIT from the menu.
row, select 1:Edit...

3. Enter the data values and frequencies. All frequencies must be entered as integers.

On the TI-82, 83, or 84 Plus:
Enter the data values under one list name. The list names are L_1, L_2, L_3, L_4, L_5, and L_6. key is pressed. After entering each data value, press ENTER. The data values are entered at the bottom of the screen and don't appear in the list until the ENTER The related frequencies are entered in a similar manner under a second list name. After entering each frequency, press ENTER. Make sure the frequencies are entered in the same order as the data items so that they correspond. Corresponding items should be in the same row of the table.

Note: On the TI-83 and 84 Plus other lists may be placed in the STAT edit table by using the SetUpEditor available on the menu.

On the TI-85:
The calculator will ask for an xlist name and an ylist name. The xlist is for the data values and the ylist is for the frequencies. Enter a name (up to eight characters) for xlist and press ENTER and enter a name for ylist and press ENTER.
Enter each data value as an x-value and the corresponding frequency as a y-value. (y_1 is the frequency for data item x_1, y_2 for x_2, etc.)

On the TI-86:
There are three built in names for lists: xSstat, yStat, or fStat. Data may also be entered in other lists where the user selects the name. For the user to select the name, move the cursor to the name row and move to the right to an unnamed column. Data is entered in a column by moving the cursor to that column (the column name and element number appears at the bottom of the screen), entering a value, and pressing ENTER. Make sure the frequencies are entered in the same order as the data items so that they correspond.

Note: When there are intermediate data values with a frequency of zero, it is best, for graphing purposes, to enter these data values and the zero frequencies.

4. To edit data.

On the TI-82, 83, or 84 Plus:
Move the cursor to the values to be changed, make the change, and press ENTER. As the data item in a list is highlighted, the value of that item appears at the bottom of the screen.

On the TI-85 or 86:
Move the cursor to the values to be changed and make the desired change.

5. To exit, after all data have been entered and the cursor appears on the next value, press QUIT.

6. To view a list, see Procedure C8.

After the data is stored, graphs may be drawn and calculations performed. When working with the calculator, the data values are the x-values and are plotted on the horizontal axis and the frequencies of the data values are the y-values and are plotted on the vertical axis.

Procedure S2. To graph single variable statistical data.

1. Make sure the viewing rectangle is appropriate for the data to be graphed, that the graph display has been cleared, and that all functions in the function table have been turned off.
2. Be sure the data to be graphed is stored in lists in the calculator. (Procedure S1)
3. Define the plot and draw the graph:

On the TI-82 or 83 or 84 Plus:

(a) Press the key STAT PLOT

(b) Select one of the three plots.

The screen will show the plot number selected, followed by several options. The highlighted options are active. Select the plot number desired and press ENTER. There will be several options for the plot number chosen.

(c) Highlight ON to turn that plot on and press ENTER.

(d) Highlight the type of graph. The first symbol following the word Type is for a scatter plot, the second for a xyLine. On the TI-82, the third is for a box plot, and the fourth for a histogram. On the TI-83 or 84 Plus, the third is for a histogram, the fourth and fifth for boxplots, and the sixth for a normal probability plot.

(e) After Xlist enter the list name of the data values. After Ylist (or Freq), enter the name of the data frequencies (On the TI-82, highlight the name of the list of frequencies.). After each entry press ENTER.

On the TI-85 or 86:

(a) Press the key STAT

(b) From the menu, select DRAW

(c) From the secondary menu, to draw a graph, select: HIST for a histogram. SCAT for a scatter graph xyLine for a frequency polygon.

On the TI-86 there are also the options: BOX for a box plot MBOX for a modified box plot.

On the TI-86, before plotting, the plot must be set up as follows: After pressing STAT, from the secondary menu, select PLOT. Then, select PLOT1, PLOT2, or PLOT3. The highlighted options are active.

(a) Highlight ON

(b) From the menu, select the type of plot.

(c) Select the names for the Xlist and Ylist. In this case, the Ylist will be the frequencies.

(f) For Mark, highlight the symbol
     to be used in plotting data.

(g) To exit, press QUIT

(h) To see the graph, press GRAPH.

Be sure to turn the STAT PLOT off
when finished.

(d) Select the symbol to be used.

(e) To exit, press QUIT.

(f) To see the graph, press
     GRAPH key and select
     GRAPH.

Be sure to turn the plot off when
finished.

Note:
  1. For a histogram the value of Xscl will determine width of bars.
  2. One type of graph should be cleared before displaying another.
  3. When graphing statistical data, all functions listed in the function table
      should be turned off.

Procedure S3. To clear the graphics screen.

1. All functions and plots should be deleted or turned off.
2. Clear the screen:

On the TI-82 or 83 or 84 Plus:	On the TI-85 or 86:
Press the key DRAW.	Press the key STAT
With DRAW highlighted,	From the menu, select DRAW.
Select 1:ClrDraw.	From the secondary menu, select
Press ENTER.	CLDRW.

On the TI-83, 84, or 86, a drawing cleared while a plot is turned on will be
redrawn the next time the GRAPH key is pressed.

Example 12.1: Given the following set of numbers, construct a frequency distribution, enter this
distribution into the calculator, and graph the data as a scatter graph, as a
histogram, and as a frequency polygon.

$$5, 7, 3, 6, 5, 9, 5, 6, 7, 3, 4, 8, 6, 7, 5, 8, 3, 4, 6, 6$$

Solution:

1. Notice that the smallest data value is 3
    and the largest value is 9. To construct a
    frequency distribution, make a table of
    two rows. In the first row, list each of the
    integers from 3 through 9. In the second
    row, below the corresponding value in
    the first row, give the frequency (number
    of times) that each value occurs. In this
    case the frequency distribution is

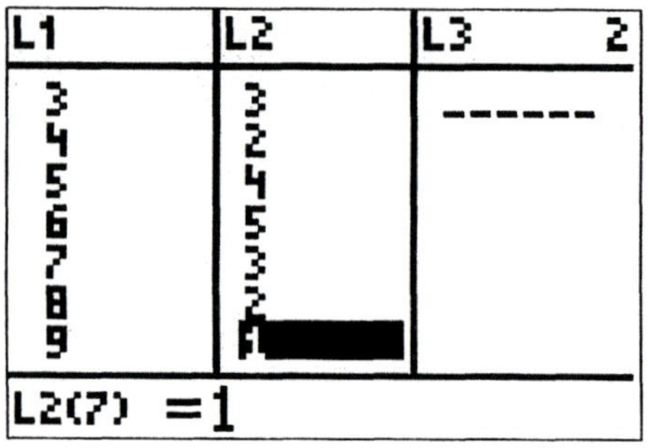

Figure 12.1

x(data): 3 4 5 6 7 8 9
y(freq.): 3 2 4 5 3 2 1

2. Enter this data into the calculator (see Procedure S1).
    The data screen after entering this data is shown in Figure 12.1.
3. Before drawing the graphs, make sure the viewing rectangle is appropriate and
    that previous graphs are cleared from the screen (see Procedure S3). Since the
    x-values vary from 3 to 9, make
    Xmin = 0, Xmax = 11 and
    Xscl = 1 and since the y-values go from 1 to
    5, make Ymin = 0, Ymax = 8, and
    Yscl = 1. These values are taken to allow a
    little extra space around the edges of the
    graph.
4. Use Procedure S2 to graph the data. Clear
    each graph before doing another. The
    scatter graph is in Figure 12.2, the
    histogram in Figure 12.3 and the frequency
    polygon in Figure 12.4.

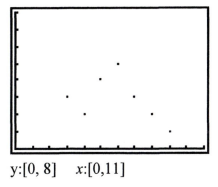

y:[0, 8]   x:[0,11]

Figure 12.2

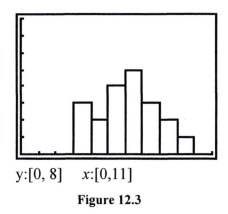

y:[0, 8]   x:[0,11]

Figure 12.3

y:[0, 8]   x:[0,11]

Figure 12.4

    The graphs of Example 12.1 give different representations of the same data. The choice
of which to use will depend on personal preference and the given situation.
    We now proceed with additional analysis of data. Two quantities of particular
importance when working with data are the mean and the standard deviation of the data. The
mean is the sum of all the x-values (data values) divided by the total number of x-values. This is
represented symbolically as

$$\bar{x} = \frac{\sum x_i}{n} \quad \text{or} \quad \bar{x} = \frac{\sum f_i x_i}{\sum f_i}$$

where $\sum x_i$ means the sum of all the x-values (individual values are represented by x_i, for i = 1,
2, 3...), n is the total number of x-values, and f_i represents the frequency of each x-value x_i.

The sum of all the x-values is equal to the sum of the products of each x-value with its frequency, $\sum f_i x_i$. The total number, n, of all x-values is equal to the sum of the frequencies, $\sum f_i$.

The **standard deviation** is an important quantity in statistics and is calculated from one of the following formulas:

$$\sigma = \sqrt{\frac{\sum (x_i - \bar{x})^2}{n}} \quad \text{or} \quad s = \sqrt{\frac{\sum (x_i - \bar{x})^2}{(n-1)}}$$

The formula for σ (the Greek letter sigma) is used to calculate standard deviation when every element of the data under study is used, while the formula for s is used to determine the standard deviation when a sample of data is used as an approximation to the complete set of data. We will call the formula for σ the σ-standard deviation and the formula for s the s-standard deviation. The two values will differ slightly.

Procedure S4. To obtain mean, standard deviation and other statistical information.

Be sure one-variable data has been stored as in Procedure S1.

1. Press the key STAT
2. Obtain the statistical information:

On the TI-82 or 83 or 84 Plus:

(a) Highlight CALC in the top row and from the menu select 1:1-Var Stats

(b) 1-Var Stats will appear on the home screen. After these words first place the list name for the data and if frequencies are given in a second list name, the list name for the frequencies. Example:

$$\text{1-Var Stats } L_1, L_2$$

(c) Press ENTER

or

(a) On the TI-82, with CALC highlighted at the top of the screen, select: 3:Setup Then, under 1-Var Stats, select the list name to be used for the data values and the list name to be used for the frequencies. (If each has a frequency of one, select 1.) and press QUIT

On the TI-85:

(a) From the menu, select CALC.

(b) Make sure the xlist name is correct and press ENTER and the ylist name is correct and press ENTER.

(c) From the secondary menu, select 1-VAR.

On the TI-86:

(a) From the menu, select CALC

(b) From the secondary menu, select OneVa. On the home screen OneVar appears. After these words enter the name of the data list, comma, the name of the frequency list. If the name of the frequency list is omitted the frequencies are taken as 1.

(b) Again press STAT and highlight CALC in the top row.

(c) From the menu, select 1:1-Var Stats, then press ENTER

If both list names are omitted the data list is taken as the list named xStat and the frequency is taken as the list named fStat.

Note: On the TI-86, to get list names, press the key VARS, select STAT (after pressing MORE twice), move the indicator to the variable name desired and press ENTER.

(c) Press ENTER

The following statistical quantities appear:

$\bar{x}$ is the mean of all the x-values (the first list name)

$\sum x$ is the sum of all the x-values

$\sum x^2$ is the sum of all the squares of the x-values

Sx is the s-standard deviation of the x-values

σx is the σ-standard deviation of the x-values

n is the total number of x-values

3. To clear the screen.
   Press CLEAR to clear the screen.

On the TI-85 or 86:
Press QUIT to exit. Then CLEAR to clear screen.

Example 12.2: Find the mean and standard deviation of the data in Example 12.1.

Solution:

1. Check to make sure the data is still stored in the calculator. If it is not, it will have to be reentered. (Procedure S1)

2. Use Procedure S4 to obtain the statistical data. This screen appears in Figure 12.5 and tells us that the mean of the x-values is 5.65, the s-standard deviation is 1.73 and that the σ-standard deviation is 1.68 (values rounded to two decimal places) and that we had a total of 20 data values.

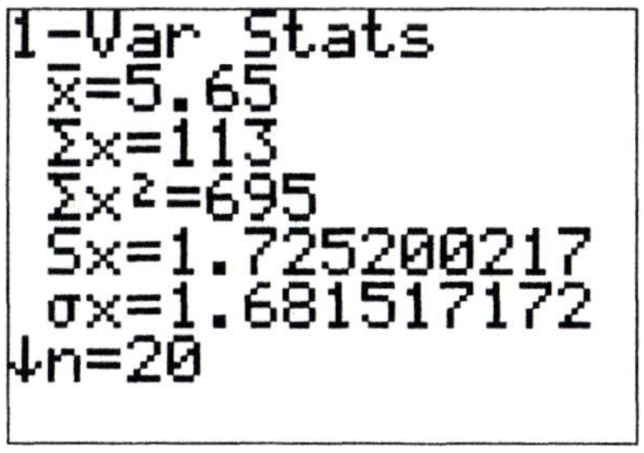

```
1-Var Stats
x̄=5.65
Σx=113
Σx²=695
Sx=1.725200217
σx=1.681517172
↓n=20
```

Figure 12.5

Before entering new data it is always a good idea to clear old statistical data from the memory of the calculator. Otherwise, the two sets of data may become intermixed and the results may be in error.

Procedure S5. To clear statistical data from memory.

1. Press the key: MEM.
2. From the menu, select Delete or MemMgmt/Del….
3. Then select List…
4. Use the cursor keys to move the indicator to the name of the list to be deleted and press ENTER. On the TI-83 or 84 Plus, press the DELETE key will delete the entire list and name.

   On the TI-82, 83, or 84 Plus, data is best deleted by pressing the key STAT, selecting 4:ClrList, entering the names, separated by commas, of the lists to be cleared (by pressing the keys for those names), and then pressing ENTER. This will not clear the list names. (Example: ClrList L_1, L_2)
5. Press QUIT to exit.

Note: **Caution:** After pressing the MEM key, do not select RESET – this will erase everything in the calculator.

Two other important quantities when considering a set of data are the median and the mode of the data. The **median** is that value which lies in the middle when the data is arranged in numerical order. If there are an odd number of values, the median is the middle value. If there are an even number of values, the median is the average of the two middle values. The **mode** is that value that occurs most often. The median and mode are generally not displayed by graphing calculators and must be found by hand.

The quantities mean, median, and mode are called **measures of central tendency**. They are an indication as to where the middle of the data is located. The standard deviation is an indication as to the extent that the data deviates from the mean (on the average). These quantities are used extensively for comparison and analysis of data.

Example 12.3: Find the median and mode for each of the following sets of data.

  (a) 3, 4, 9, 3, 3, 5, 2, 8, 4
  (b) 12, 16, 10, 14, 14, 15, 18, 15, 14, 16
  (c) The data in Example 12.1 .

Solution: 1. First, for parts (a) and (b) rearrange the values in numerical order. For part(c), we will use the frequency distribution of Example 12.1.

  (a) 2, 3, 3, 3, 4, 4, 5, 8, 9
  (b) 10, 12, 14, 14, 14, 15, 15, 16, 16, 18
  (c) (value) x: 3 4 5 6 7 8 9
      (freq.) y: 3 2 4 5 3 2 1

2. (a) For the first set, there are an odd number of values. Therefore the median is the middle value or 4. The values that occurs most often is 3, thus, the mode is 3.

   (b) For the second set, there is an even number of values. Therefore the median is the average of the two middle values (14 and 15) or 14.5. The value, which occurs most often, is 14 and 14 is the mode.

   (c) In the case of the third set, the total number of values (the sum of the frequencies) is 20. Thus, the median will be the average of the 10th and 11th values. To obtain these, add the frequencies starting from one end. Both the 10th and 11th values are 6. Thus, the median is 6. The value, which occurs most often in this case, is also 6, which is the mode.

Exercise 12.1

In Exercises 1-5, 7 and 8, (a) construct a frequency distribution of the data and enter the data into the calculator, then obtain (b) a histogram, (c) a scatter graph, (d) a frequency polygon, (e) the mean, (f) the median, (g) the mode, and (h) the s-standard deviation. Sketch graphs for parts (b), (c), and (d) on your paper.

1. For a set of quiz scores (10 max. score):
   6, 7, 10, 8, 9, 8, 7, 7, 5, 10, 10, 8

2. For a set of quiz scores (10 max. score):
   9, 10, 8, 8, 7, 5, 6, 9, 9, 10, 8

3. For a set of test scores (100 max. score):
   85, 90, 100, 75, 85, 95, 90, 80, 80, 90, 100

4. For a set of test scores (100 max. score):
   65, 75, 85, 100, 85, 75, 85, 95, 90, 90, 95, 85

5. For a set of test scores (100 max. score):
   75, 85, 80, 90, 90, 85, 75, 100, 100, 90, 95, 75

6. Based on your results for problems 3 and 5, which class did the best on the test? On what do you base you answer?

♦ 7. *(Washington, Exercises 22.1, #13)* In testing a computer system, the number of instructions it could perform in 1 ns was measured at different points in a program. The numbers of instructions were as follows:

   19, 21, 24, 25, 22, 21, 23, 24, 18, 18, 19, 22, 24, 22, 19

◆ 8. *(Washington, Exercises 22.1, #27)* The life of a certain type of battery was measured for a sample of batteries with the following results (in number of hours):

34.2, 30.5, 32.3, 35.6, 31.0, 28.7, 29.4, 30.2, 32.7, 25.6, 30.9, 28.7, 36.3

In Exercises 9-12, find (a) the mean and (b) the s-standard deviation.

9.

value	frequency	value	frequency
21.2	2	23.2	6
21.6	4	23.5	5
21.8	5	23.8	5
21.9	8	24.0	3
22.0	7	24.2	2
22.3	8	24.5	1
22.8	11		

10.

value	frequency	value	frequency
50	2	62	8
51	5	64	7
52	8	65	6
53	9	67	3
55	12	69	2
57	12	70	2
59	10	71	1
60	9	72	1

◆ 11. *(Washington, Section 22.3, #18)* The weekly salaries (in dollars) for the workers in a small factory are as follows:

355, 445, 385, 325, 285, 410, 305, 485, 375, 510, 410, 335, 350, 400

◆ 12. *(Washington, Section 22.3, #20)* In testing a braking system, the distance required to stop a car from 70 mi/h was measured in several trials. The results are in the following table:

Stopping distance	155-159	160-164	165-169	170-174	175-179	180-184
Times car stopped	4	17	35	37	23	8

13. A worker measures the diameters of a sample of gaskets and obtains the following frequency distribution table giving the number of times each diameter (in inches) was observed. Find the mean, median, mode, and the s-standard deviation.

diam.	frequency	diam.	frequency	diam.	frequency
3.25"	1	3.29"	14	3.33"	22
3.26"	4	3.30"	20	3.35"	16
3.27"	8	3.31"	23	3.36"	13
3.28"	10	3.32"	25	3.37"	3

14. A lab technician measures the weight (in ounces) of cereal in one-pound boxes and obtains the following frequency distribution table giving the number of times each weight was observed. Find the mean, median, mode, and the s-standard deviation. Based on your results do you think the company is within its rights to sell these boxes as one pound boxes of cereal?

Wt.	frequency	Wt.	frequency
15.5	2	16.1	25
15.6	4	16.2	23
15.7	6	16.3	12
15.8	10	16.4	8
15.9	21	16.5	5
16.0	31	16.6	1

15. A consumer testing company tests a sample of light bulbs for length of life. The following frequency distribution is obtained for the number of bulbs in the sample that lasted the given number of hours. Find the mean, median, mode, and the s-standard deviation. Based on your results, do you think it is proper for the light bulb company to advertise these lamps as having an average life of 1200 hours?

time	frequency	time	frequency
1150	3	1200	125
1155	10	1205	148
1160	35	1210	118
1165	28	1215	85
1170	57	1220	112
1175	65	1225	87
1180	72	1230	67
1185	88	1235	45
1190	92	1240	10
1195	102	1245	5

16. A worker measures the diameters of bolts in a sample taken from a bin containing 10,000 bolts. He measures the diameter (in millimeters) of each bolt in the sample and arrives at the following frequency distribution. Find the mean, median, mode, and the s-standard deviation.

diam.	frequency		diam.	frequency
10.95	1		11.06	24
10.96	4		11.07	15
10.97	2		11.08	16
10.98	12		11.09	11
10.99	24		11.10	8
11.00	36		11.11	5
11.01	43		11.12	3
11.02	48		11.13	0
11.03	52		11.14	2
11.04	49		11.15	1

◆ 17. (Washington, Exercises 22.1, # 31) Toss three coins 100 times and tabulate the number of heads that appear for each toss. Construct a frequency distribution table for your data. Enter the data on your calculator and construct a histogram and a frequency polygon. Determine the mean and standard deviation of your data. To the mean value subtract the value of one standard deviation ($\overline{x} - s$) and then to the mean value add one standard deviation ($\overline{x} + s$). What percentage of your values occurs between the mean minus one standard deviation and the mean plus one standard deviation ($\overline{x} \pm s$)? Compare your data with others in the class. Discuss the differences.

18. Simulate the tossing of two dice and finding their sum by entering and running the following program on your calculator. (Some calculators may have a dice toss program found by first pressing the key APPS, then selecting: Prob Sim.)

```
PROGRAM: DICETOSS
int (6 rand)+1 → X
int (6 rand)+1 → Y
X + Y → S
Disp S
```

[rand is a rand number generator and is found on the calculator by pressing the MATH key and selecting PRB. To get →, press the key STO▷.]

Run this program 50 times by repeatedly pressing ENTER before pressing any other key. Keep track of the data and form a frequency distribution. Enter the data on your calculator and construct a histogram and a frequency polygon. Determine the mean and standard deviation of your data. What percentage of your values occurs between the mean minus one standard deviation and the mean plus one standard deviation ($\overline{x} \pm s$)?

12.2 The Normal Curve
(Washington, Section 22.4)

In Section 12.1, we saw that data is often visualized by frequency polygons or histograms. In many cases, particularly with large amounts of "naturally" occurring data, the frequency polygons or histograms appear bell shaped as in Figure 12.5. The highest point on the graph will occur near the mean for the set of data and, as the distance from the mean becomes greater, the height of the graph will approach zero. Such a distribution of data is called a **normal distribution** and the related graph is called the **normal distribution curve**.

If we consider a normal distribution with a mean of zero and a standard deviation of one, we obtain the **standard normal distribution**. The equation of the standard normal distribution curve is

$$y = \frac{1}{\sqrt{2\pi}} e^{-\frac{x^2}{2}} \quad \text{(Equation 1)}$$

The graph of this function is shown in Figure 12.6. The area under the graph of this function is exactly one square unit. The graph is symmetrical with respect to the y-axis with half of the area being to the right of the mean and half being to the

left of the mean. The y-intercept is $\frac{1}{\sqrt{2\pi}} e^0$ or about 0.40.

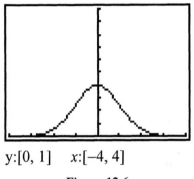

y:[0, 1] x:[–4, 4]

Figure 12.6

For the standard normal distribution curve the values on the horizontal-axis are actually the number of standard deviations from the mean and are called z-**values**. Positive values of z represent the number of standard deviations above the mean and negative values represent the standard deviations below the mean. Thus, a value of zero is at the mean and a value of one represents a location one standard deviation above the mean. For a set of data that has a normal distribution with a mean of 25 and a standard deviation of 3, the location of 1 on the horizontal-axis ($z = 1$) would represent a data value of the mean plus one standard deviation or $25 + 3 = 28$. Likewise, the location of –2.5 ($z = -2.5$) would represent a data value of the mean less two and one-half standard deviations or $25 - 2.5*3 = 17.5$.

The relationship between a data value, x_i, and the corresponding z-value is given by

$$z = \frac{x_i - \mu}{\sigma} \quad \text{(Equation 2)}$$

where we now use σ as the standard deviation and μ is the mean of all the data (for our purposes we will let: $\mu = \bar{x}$ and $\sigma = s$) of the data values.

Of particular interest is the area under the graph of the standard normal distribution curve between two z-values. Since the area under the complete curve (from negative infinity to positive infinity) is one, the area between two z-values represents the likelihood or the probability that data values will fall between these two z-values.

To determine the area between two z-values on the calculator, we will need to write a program for calculating area under the graph. In more advanced mathematics (calculus), the area between the graph, the x-axis, and between two values of x may be found by using something called the definite integral of a function. We will not be concerned at this time with the details of the definite integral but only with the fact that it can be used to find areas. Some graphing calculators have a built in function to evaluate the definite integral and, thus, to find the area under a graph.

To evaluate the area under the graph of the normal distribution function, we first store the standard normal distribution function (Equation 1) for a function name. Then, we enter a short program into the calculator that asks for two z-values and uses the definite integral to evaluate the area between these two z-values. The reader should refer to Section 5 of Chapter 2 in regards to entering programs and using the Input and Disp statements.

Procedure P9.   Finding the area under a graph.

1. The desired function should be stored under a function name in the function table. (Procedure C6)
2. Enter the following program into your calculator to find the area between two values B and C:
   (It is assumed here that the function is stored under the function name Y_5)

> PROGRAM:AREA
> Disp "ENTER FIRST Z"
> Input B
> Disp "ENTER SECOND Z"
> Input C
> fnInt(Y_5,X,B,C)→I
> Disp "AREA IS:"
> Disp I

The form of the definite integral function that is used in this program is
> **fnInt**(*function name, variable, smaller value, larger value*)

(This function may also be used without being in a program.)

To obtain the definite integral function fnInt:

On the TI-82 or 83 or 84 Plus:	On the TI-85 or 86:
Press the key MATH	Press the key CALC
From the menu, select 9:fnInt(	From the menu, select fnInt

(See also Procedure C20.)

Example 12.4:   Use the program AREA to find the area (a) between $z = 0$ and $z = 5$ and (b) between $z = -3$ and $z = 3$ for the standard distribution function (Equation 1). It is important to enter Equation 1 correctly as: $1/\sqrt{(2\pi)} * e(-x^2/2)$.

Solution: 1. Store Equation 1 for a function name (It must be the same function name as that used in the area program) and use Procedure P9 to find the area between the graph, above the x-axis, and between the given z-values.

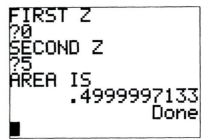

Figure 12.7

2. (a) The area under the standard normal distribution curve between $z = 0$ and $z = 5$ will be given as .4999997.... This indicates that almost all of the area to the right of the mean is contained between $z = 0$ and $z = 5$ (area to the right of the mean is ½ of the total area of one unit or 0.5).
The results of running this program are shown in Figure 12.7. Rounding this number to 4 decimal place accuracy, we obtain 0.5000.

   (b) The area between −3 and 3 is 0.99750... Since the total area under the curve is one unit, most of the area under the curve is between these two values.

It is possible to add a statement to the program so that it will store the standard normal distribution function for a function name. Storing this function will replace any other function that may be stored for that function name. To do this, add a line to the program given in Procedure P9. This is left as an exercise.

Assuming that a set of data has a normal distribution we may use the area under the normal distribution curve to calculate the probability and percent likelihood that certain values of data lie within a given range. If the total number of data items is known, we can use this probability to determine how many of the data items can be expected to lie within the given range.

Example 12.5: Assuming a normal distribution, determine the probability and percent likelihood that a data item lies within 1 standard deviation below the mean and 1.5 standard deviations above the mean.

Solution: 1. We need the area under the standard normal distribution curve between $z = -1$ and $z = 1.5$ as illustrated by the shaded area in Figure 12.8. This area gives the fraction of the total area (one square unit) and thus the probability that the data lies between 1 standard deviations below the mean and 1.5 standard deviations above the mean.

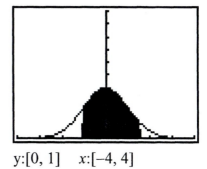

y:[0, 1]   x:[−4, 4]

Figure 12.8

2. Use the calculator and the Procedure P9 to

find the required area (See Example 12.4).

Use −1 for the first z-value and 1.5 for the second z-value and round the answer to 4 decimal places. The area is .7744 or 77.44% of the total area.

Since the area is .7744, the probability that a data item lies in the specified range is .7744. This means that the percent likelihood that a data item lies within this range is 77.44%

Example 12.6: A store receives a shipment of 5000 light bulbs with a mean life span of 1000 hours and standard deviation of 25 hours. Assume the data gives a normal distribution. Determine the percent likelihood that

(a) a bulb will last between 970 hours and 1040 hours.

(b) a bulb will last at least 1050 hours.

Solution:

1. We first use Equation 2 to determine the z-values that correspond to the times. For this data $\mu = 1000$ and $\sigma = 25$. Thus,

$$\text{For } x = 970, z = \frac{970 - 1000}{25} = -1.2$$

$$\text{For } x = 1040, z = \frac{1040 - 1000}{25} = 1.6$$

$$\text{For } x = 1050, z = \frac{1050 - 1000}{25} = 2.0$$

2. Use the Procedure P9 to find areas under the normal curve as in Example 12.6.
   (a) In order to determine the likelihood that a bulb lasts between 970 and 1040 hours we need the area between $z = -1.2$ and $z = 1.6$. We find this area to be 0.8301
   (b) In order to determine if a bulb will last longer than 1050 hours, we need the area to the right of $z = 2.0$. Since the area past $z = 5$ is negligible (to 4 decimal places) find the area between $z = 2$ and $z = 5$.

3. (a) Since the area between $z = -1.2$ and 1.6 is 0.8300, 83.00% of the values lie between $x = 970$ and $x = 1040$. The number of bulbs with a life span between 970 and 1040 hours is 83.00% of 5000 or 4150 bulbs.
   (b) The area above $z = 2$ is 0.0227 and 2.27% of the values lie above $x = 1050$. The number of bulbs with a life span of at least 1050 hours is 2.27% of 5000 or 113.5 bulbs. Since we cannot have a fractional bulb, the number of bulbs that can be expected to last 1050 hours is 113.

Many times it is important to know what values of data fall within a certain percentage range. This is called a **confidence interval**. To determine the 80% confidence interval for a set of data, we determine two data values, equally spaced on either side of the mean, such that 80% of the data lie between these two values.

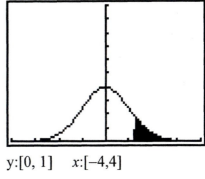

y:[0, 1] x:[−4,4]

Figure 12.9

Example 12.7: Determine the z-value (to two decimal places) such that 10% of the total area is to the right of this z-value as in Figure 12.9.

Solution: We are not able to calculate the z-values directly, thus, we to use a trial and error method with Procedure P9.

1. Since 10% of the area is to be to the right of the z-value and we know that 50% of the area is to the right of the mean, 40% of the area will be between the mean and this z-value.
   Thus, we desire a value of z such that the area is between $z = 0$ and this z-value is 0.4000.

2. Start by finding the area between $z = 0$ and $z = 1$ which is .3413.
   Thus, the z-value we seek must be greater than 1.

3. Next, find the area between $z = 0$ and $z = 2$. This gives an area of .4772. The z-value we seek must lie between 1 and 2.

4. Then, find the area between $z = 0$ and $z = 1.5$. The area is .4332 and the z-value must lie between 1 and 1.5. We continue this process of narrowing the interval until we obtain the required z-value. With the first z value equal to zero, we obtain on successive calculations

   For second $z = 1.25$, area = 0.3943
   For second $z = 1.30$, area = 0.4032
   For second $z = 1.28$, area = 0.3997
   For second $z = 1.29$, area = 0.4015

   Since the area for 1.28 is closer to 0.4000 than that for 1.29, we conclude that, to two decimal places, the required z-value is 1.28. Thus, under the normal distribution curve, 10% of the total area is to the right of $z = 1.28$.

Example 12.8:   For the shipment of light bulbs of Example 12.8, with a mean life span of 1000 hours and a standard deviation of 25 hours, determine the 80% confidence interval.

Solution: To determine the 80% confidence interval, we determine the life span values between which 80% of the data will lie. Since these are equally spaced on either side of the mean, 40% will lie above the mean and 40% below the mean.

1. We first determine the z-value such that 40% of the area under the normal distribution curve will be above the mean. Using the procedure of Example 12.9, we find this value to be 1.28.

2. The z-value of 1.28 indicates 1.28 standard deviations above the mean or $(1.28)(25) = 32$ hours above the mean is the upper limit for the 80% confidence level. Since, the graph is symmetrical, 32 hours below the mean will be the lower limit.

3. Thus, an 80% confidence interval will be those values within 32 hours of the mean. This means that 80% of the light bulbs can be expected to have a life span of between 968 hours $(1000 - 32)$ and 1032 hours $(1000 + 32)$.

Exercise 12.2

1. Use Procedure P9 to find the area under the normal distribution curve between
   (a) $z = -1$ and $z = 1$
   (b) $z = 0$ and $z = 2$
   (c) $z = 3$ and $z = 5$

2. Use Procedure P9 to find the area under the normal distribution curve for
   (a) $z = -2$ and $z = 2$
   (b) $z = -4$ and $z = 0$
   (c) $z = -5$ and $z = 5$

3. Add a line to the program in Procedure P9 to store the standard normal distribution function for a function name in the function table. Then, find the area under the normal distribution curve between $z = -1$ and $z = 1$.

4. Use the program of Exercise 3 to find the area under the normal distribution curve between
   (a) $z = -1.5$ and $z = 2.5$
   (b) $z = -2.0$ and $z = 1.4$
   (c) $z = 1$ and $z = 2.3$

5. For data that has a normal distribution, find the probability and percent likelihood that the data
   (a) Lies between the mean and 2.2 standard deviations above the mean.
   (b) Lies between 1.45 standard deviations below the mean and 2.25 standard deviations above the mean.
   (c) Lies to the left of 1.82 standard deviations below the mean.
   (d) Lies within 1.80 standard deviations of the mean.

6. For data that has a normal distribution, find the probability and percent likelihood that the data
   (a) Lies between the mean and 1.3 standard deviations below the mean.
   (b) Lies between 2.22 standard deviations below the mean and 1.25 standard deviations above the mean.
   (c) Lies to the right of 2.20 standard deviations above the mean.
   (d) Lies more than 1.75 standard deviations from the mean.

In Problems 7-16, assume a normal distribution of data values.

7. A set of data has a mean of 73.5 with a standard deviation of 2.5. Find the percent likelihood that a data value
   (a) Lies between 68.5 and 76.0.
   (b) Is greater than 74.7.
   (c) Is greater than 69.2 and less than 74.2.
   (d) At least 75.2

8. A set of data has a mean of 0.712 with a standard deviation of 0.012. Find the percent likelihood that a data value
   (a) Lies between 0.700 and 0.725.
   (b) Is greater than 0.730.
   (c) Is greater than 0.715 and less than 0.718.
   (d) At least 0.705.

9. A sample of half-inch bolts shows a mean diameter of 0.5030 inches with a standard deviation of 0.0025 inches. Find the probability that
   (a) A bolt will exceed 0.5050 inches.
   (b) A bolt will lie between 0.5000 and 0.5040 inches.
   (c) A bolt will be less than 0.4995 inches.
   (d) A bolt will exceed 0.5030 inches.

10. A sample of 20-ounce cereal boxes has a mean weight of 20.35 ounces with a standard deviation of 0.25 ounces. Find the probability that
    (a) A box of cereal will exceed 20.50 ounces.
    (b) A box of cereal lie between 20.00 and 21.00 ounces.
    (c) A box of cereal will be less than 20.00 ounces.
    (d) A box of cereal will exceed 20.35 ounces.

11. A sample of resistors shows a mean resistance of 1.205 ohms with standard deviation of 0.015 ohms. Out of 2000 resistors, determine how many will
    (a) Lie between 1.195 ohms and 1.210 ohms.
    (b) Have a resistance greater than 1.212 ohms.
    (c) Have a resistance less than 1.184 ohms.
    (d) Have a resistance less than 1.196 ohms or greater than 1.214 ohms.

♦ 12. *(Washington, Exercises 22.4, #17-20)* The lifetimes of a certain type of automobile tire have been found to be distributed normally with a mean lifetime of 62,500 miles with a standard deviation of 6,250 miles.
   (a) How many tires can be expected to last between 60,000 and 62,500 miles?
   (b) How may tires will last between 60,000 and 70,000 miles?
   (c) How may tires will last more than 65,000 miles?
   (d) If the manufacturer replaces all tires that last less than 60,000 miles, how many tires will he replace?

♦ 13. *(Washington, Exercises 22.4, #13-16)* Each battery in a sample of 500 batteries is checked for its voltage. It has been previously established for this type of battery (when newly produced) that the voltages are distributed normally with $\mu = 1.50$ V and $\sigma = 0.04$ V.
   (a) Find how many batteries have voltages between 1.46 V and 1.54 V.
   (b) Find how many batteries have voltages between 1.42 V and 1.53 V.
   (c) What percent of batteries have voltages between 1.40 V and 1.50 V?
   (d) What percent of batteries have voltages above 1.55 V?

14. Use the data of Exercise 12 of Section 12.1 and determine the likelihood of a car stopping
   (a) In less that 162 feet.   (b) In more than 177 feet.
   (c) In more than 182 feet.   (d) Between 162 and 177 feet.

15. In Problem 8 of Exercise 12.1, we found the mean battery life to be 31.2 and the standard deviation to be 2.98.   Use the data of Exercise 12.1, #21 to determine the actual number of data values that lie between the mean minus one standard deviation and the mean plus one standard deviation. What percentage of the total number of data value is this? How does this compare to that predicted by the normal curve?

16. Use the data and mean and standard deviation of Problem 14 in Exercise 12.1 to determine the number of data values that lie between the mean minus one standard deviation and the mean plus one standard deviation. What percentage of the total number of data value is this? How does this compare to that predicted by the normal curve?

17. For a normal distribution, find the z-values such that
   (a) 25% of the total area is to the right of the z-value.
   (b) 5% of the total area is to the left of the z-value.
   (c) 90% of the total area is around the mean (45% above and 45% below)
   (d) 99% of the total area is around the mean.

18. For a normal distribution, find the z-values such that
   (a) 15% of the total area is to the right of the z-value.
   (b) 8% of the total area is to the left of the z-value.
   (c) 95% of the total area is around the mean (45% above and 45% below)
   (d) 98% of the total area is around the mean.

19. For the data in Exercise 9, find
   (a) A 90% confidence interval.
   (b) A 95% confidence interval.
   (c) A 99% confidence interval.

20. For the data in Exercise 10, find
    (a) A 90% confidence interval.
    (b) A 95% confidence interval.
    (c) A 99% confidence interval.
21. For the data in Exercise 11, find
    (a) A 80% confidence interval.
    (b) A 90% confidence interval.
    (c) A 97.5% confidence interval
22. For the data in Exercise 12, find
    (a) A 80% confidence interval.
    (b) A 90% confidence interval.
    (c) A 97.5% confidence interval.
23. Review your data of tossing 100 coins from Exercise 17 of Exercise Set 12.1. How close does your data come to being a standard distribution?

12.3 Regression - Finding the Best Equation
(Washington, Sections 22.6 and 22.7)

Two variables are frequently related to each other in such a way, that as one changes, this causes a change in the other variable. This relationship can be expressed by writing one variable as a function of a second variable. The first variable is considered the **dependent variable** and the second the **independent variable**. For example, from Ohm's law, we know that in a simple direct current circuit containing a 12-ohm resistor the voltage equals 12 times the current in the circuit. If we let E represent the voltage and I represent the current then we arrive at the equation $E = 12I$. Here the E is considered the dependent variable and is expressed as a function of I the independent variable.

Many times the equation, which relates two variables, is not known and must be found from a collection of known data or by collecting data experimentally. A first step in attempting to determine such an equation is to graph the data. This will often indicate the type of equation, such as linear, logarithmic, etc., that will best represent the data.

We will consider five types of equations that are used to represent data: **linear regression** and the **non-linear regressions**: quadratic, logarithmic, exponential, and power functions. Other types of regression equations that will not be considered here, include formulas that involve polynomials of degree higher than 2 and equations that involve trigonometric functions. The type of equation that best represents a given set of data is determined by trying different graphs and determining when a straight line is obtained. These graphs often involve logarithms of one or both variables and may be graphed by plotting the logarithm of the variable or by using semi-logarithm or logarithmic paper.

The type of equation for each type of graph is as follows

Type of Equation	Data gives a straight line when
Linear	y is graphed as function of x
Quadratic	y is graphed as a function of x^2 (or of $(x - k)^2$ for some value k)
Logarithmic	y is graphed as function of $\log x$
Exponential	$\log y$ is graphed as function of x
Power	$\log y$ is graphed as function of $\log x$

From the graph we could, by using a few well-selected points, come up with an equation that "fits" the graph. However, we can find a better equation, by making use of a mathematical method called **least squares regression.**

Many graphing calculators have several mathematical methods built in that may be used to determine an equation that best "fits" a given set of data. Although many calculators are capable of determining other types of equations, we will only consider the five types of regression equations listed here. After the data has been properly entered into the calculator, the calculator gives the constants in the corresponding regression equations. The line obtained by a linear regression is also known as the **least-squares line.**

Type of Regression	Formula
Linear Regression	$y = a + bx$ or $y = ax + b$
Quadratic Regression	$y = ax^2 + bx + c$
Logarithmic Regression	$y = a + b \ln x$
Exponential Regression	$y = ab^x$
Power Regression	$y = ax^b$

In these formulas, the constants (a and b) are calculated for a given set of data where x represents the independent variable and y the dependent variable. For each regression calculation, the calculator will also provide a value for the **correlation coefficient**, r, which is an indication as to how well a given equation "fits" a set of data. The closer r is to 1 or to -1 the better the equation represents the given set of data.

Data of this type is two variable statistical data and is entered into the calculator in much the same way as one variable statistical data.

Procedure S6. To enter or change two variable statistical data.

On the TI-82, 83, 84 Plus, 85, and 86 calculators, data is stored as lists of numbers. (See Procedure C7) Data stored in one list are the x-values and data stored in a second list are the y-values. The two lists must correspond in values.

1. First press the STAT.

2. Enter edit mode to edit data stored in lists.

On the TI-82 or 83 or 84 Plus:
With EDIT highlighted in the top row, select 1:Edit...

On the TI-85 or 86:
Select EDIT from the menu.

3. Enter the x-values as one list and the y-values as a second list:

On the TI-82 or 83 or 84 Plus:
Enter the x-values under a list name. After each value is entered, press ENTER.
Enter the related y-values under a second list name. After each value is entered, press ENTER.
Make sure the y-values are entered in the same order as the x-values so that they correspond by being in the same row.

On the TI-85:
The calculator will ask for an xlist name and an ylist name. The xlist is for the x-values and the ylist is for the y-values. Enter a name (up to eight characters) for xlist and press ENTER and enter a name for ylist and press ENTER .
Enter the first x-value and the corresponding y-value, the second x-value and the second y-value, etc.

Note: On the TI-83 and 84 Plus other lists may be placed in the STAT edit table by using the SetUpEditor available on the menu.

On the TI-86:
There are three built in names for lists: xStat, yStat, or fStat. Data may also be entered in these lists or in other lists where the user selects the name. For the user to name a list, move the cursor up to the name row and move to a column to the right. Data is entered in a column by moving the cursor to that column (the column name and element number appears at the bottom of the screen), entering a value, and pressing ENTER.
Make sure the y-values are entered in the same order as the x-values items so that they correspond.
Normally the x-values would be entered under xStat, the y-values under yStat, and the frequency of each point under fStat.

4. To edit data.

On the TI-82 or 83 or 84 Plus:
Move the cursor to the values to be changed, make the change, and press ENTER. As the data item in a list is highlighted, the value of that item appears at the bottom of the screen.

On the TI-85 or 86:
Move the cursor to the values to be changed and make the desired change.

5. To exit, after all data have been entered and the cursor appears on the next value, press QUIT.

6. To view a list, see Procedure C8.

After the data is stored in the calculator, we are ready to determine regression equation.

Procedure S7. **To determine a regression equation.**

1. Be sure two-variable data has been stored as in Procedure S6.
2. Obtain the proper menu:

On the TI-82 or 83 or 84 Plus:

(a) Press the key STAT

(a) Highlight CALC in the top row and from the menu select the type of regression equation required.

(b) The words indicating the type of regression equation to be selected will appear on the home screen. After these words enter the list name for the *x*-values and secondly, after a comma, place the list name for the *y*-values.

A third list name may be added to indicate the frequencies of each pair of values.
Examples:
$$\text{LinReg } L_1, L_2$$
$$\text{QuadReg } L_1, L_2, L_3$$

(c) Press Enter to see the constants in the equation.

or

(a) On the TI-82, with CALC highlighted at the top of the screen, select: 3:Setup
Then, under 2-Var Stats, select the list name to be used for the *x*-values and the list name to be used for the *y*-values. Then select the list name to be used as a frequency of each point.
To exit, press QUIT

(b) Again press STAT and highlight CALC in the top row.

(c) From the menu, select the regression equation required, then press ENTER

On the TI-85 or 86:

(a) Press the key STAT

(b) From the menu, select CALC.

Then, on the TI-85:

(c) Make sure the xlist name is correct and press ENTER and the ylist name is correct and press ENTER.

(d) Select the type regression equation from the menu.

or

On the TI-86:

(c) From the secondary menu, select the type of regression equation required. On the home screen the words indicating the type of equation selected will appear. After these words enter the name of the *x*-value list, comma, and the name of the *y*-value list. A third list name may be added to indicate the list that contains the frequencies of each point. If the name of the frequency list is omitted the frequencies are taken as 1.
Example:
$$\text{LinReg LISTA, LISTB}$$
If the list names are omitted the *x*-value list is taken as the list named xStat, the *y*-value list is taken as the list named yStat and the frequency is taken as the list named fStat.

Note: On the TI-86, to get list names, press the key VARS, select STAT (after pressing MORE twice), move the indicator to the variable name desired and press ENTER.

(d) Press ENTER

3. The types of regression equations are listed below. (The form of the menu item for the TI-82, TI-83, and TI-84 Plus is on the left and for the TI-85 and TI-86 is on the right.)

(a) For linear regression ($y = a + bx$), select:
LinReg($ax + b$) or   LinR
LinReg($a + bx$)

(b) For logarithmic regression ($y = a + b \ln x$), select
LnReg   LnR

(c) For Exponential Regression ($y = ab^x$), select
ExpReg   ExpR

(d) For Power Regression ($y = ax^b$), select
PwrReg   PwrR

(e) For Quadratic Regression ($y = ax^2 + bx + c$), select
QuadReg   P2Reg

(f) For Cubic Regression ($y = ax^3 + bx^2 + cx + d$), select
CubicReg   P3Reg

(g) For Quartic Regression ($y = ax^4 + bx^3 + dx^2 + ex + f$), select
QuartReg   P4Reg

Note: The TI-83, TI-84 Plus, and TI-86 also have a sinusoidal regression using the formula $y = a \sin(bx + c) + d$ and a logistic regression of the form

$$y = \frac{c}{1 + ae^{-bx}}$$ for the TI-83 and 84 Plus and $$y = \frac{a}{1 + be^{cx}} + d$$ on the TI-86.

The calculator gives values for constants in the equations and in some cases, the correlation coefficient as *corr* or *r*. To see the *r*-value on the TI-83, it is necessary to turn diagnostics on. Do this by pressing the key CATALOG, moving down the list until the words DiagnosticOn appears and press ENTER, then press ENTER again.

◆ Example 12.9:   *(Washington, Section 22.6, Example 5)* In a research project to determine the amount of a drug that remains in the bloodstream after a given dosage, the amounts y (in mg of drug/dL of blood) were recorded as in the following table. Find the linear regression equation that represents the following set of data.

t (hours)	1.0	2.0	4.0	8.0	10.0	12.0
y (mg/dL)	7.6	7.2	6.1	3.8	2.9	2.0

Solution:

1. First use Procedure S5 to clear any previously entered data from the statistics memory.
2. Then, use Procedure S6 to enter this two-variable data into the calculator.
3. Now, use Procedure S7 to find the linear regression equation of the form $y = a + bx$.
4. The screen (see Figure 12.10) shows the values $a = 8.158..$, $b = -0.523..$, and $r = -0.999$. This means that the equation of the line (of the form $y = a + bx$) which best represents this set of data is $y = 8.16 - 0.523x$. The r-value of -0.990 is very close to -1 indicating a good fit to the data.

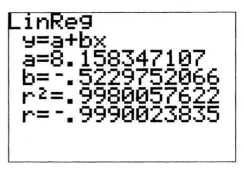

Figure 12.10

The data in Example 12.9 is to calculated to two significant digit accuracy. When writing the coefficients for the equation, we use one additional significant digit to insure an accuracy of two significant digits in calculated values of x or y.

Two variable statistical data may be graphed on the calculator screen by making a scatter graph. This allows us to compare visually the data and the corresponding regression equation.

Procedure S8. To graph data and the related regression equation.

1. Make sure the viewing rectangle (Window values) is set properly to graph the set of data, that the graph display has been cleared (see Procedure S3), and that all functions in the function table have been removed or turned off and that all statistical plot graphs have been turned off.
2. If the data has not already been entered into the calculator, it will need to be entered (Procedure S6) and the regression equation calculated (Procedure S7).

3. Store the regression equation for a function name and graph:

On the TI-82 or 83 or 84 Plus:

(a) Press the key *y=*

(b) Move the cursor after an unused function name.

(c) Press the key VARS

(d) From the menu, select Statistics...

(e) Move the highlight in the top row to EQ

(f) Select RegEQ

(g) Press the key GRAPH

On the TI-85 or 86:

(a) Press the key GRAPH

(b) From the menu, select *y*(x)=

(c) Move the cursor so it comes after an unused function name.

(d) Press the key VARS

On the TI-85 or 86:

(a) Press the key GRAPH

(b) From the menu, select *y*(x)=

(c) Move the cursor so it comes after an unused function name.

(d) Press the key VARS

This graphs the regression equation.

Note: On the TI-83, TI-84 Plus and TI-86, the regression equation may be stored directly for a function name by adding the function name to the lists when the equation is calculated. (Example: LinReg L1, L2, *Y*1)

4. Define the plot and draw a scatter graph:

On the TI-82, 83, or 84 Plus:

(a) Press the key STAT PLOT

(b) Select one of the three plots. (Press ENTER) The screen will highlight the plot number selected, followed by several options. The highlighted options are active. (To make an option active, move the cursor to that option and press ENTER.)

(c) Highlight the option ON.

(d) Highlight the type of graph. The first symbol following the word Type is for a scatter plot.

(e) On the TI-82, for Xlist, highlight the name of the *x*-values. For Ylist, highlight the name of the *y*-values.

On the TI-83 or 84 Plus, enter names for the Xlist and for the Ylist.

(f) For Mark, make active the symbol to be used in plotting data.

On the TI-85:

(a) Press the key STAT

(b) From the menu, select DRAW

(c) From the secondary menu, to draw a graph, select: SCAT for a scatter graph.

On the TI-86, before plotting, the plot must be set up.

Do this as follows:

After pressing STAT, from the secondary menu, select PLOT. Then, select PLOT1, PLOT2, or PLOT3. The highlighted options are active.

(a) Highlight ON

(b) From the menu, select the type of plot.

(c) Select the names for the Xlist and Ylist.

(d) Select the symbol to be used.

(e) To exit, press QUIT.

(g) To exit, press QUIT
(h) To see the graph with the data points, press GRAPH
Be sure to turn the plot off when finished.

(f) To see the graph, press GRAPH key and select GRAPH.
Be sure to turn the plot off when finished.

Example 12.10: Graph the data and the linear regression equation for the data of Example 12.9.

Solution: 1. If the data from Example 12.4 is still in your calculator continue with the next step, otherwise, reenter the data and calculate the regression equation (see Procedure S7).

2. Clear the graphics screen (Procedure S3) and make sure the viewing rectangle is set correctly. For this data, set the following values: Xmin = 0, Xmax = 15, Xscl = 1, Ymin = 0, Ymax = 10, Yscl = 1. This allows for some space around the graph and shows the graph relative to the x- and y-axis.

3. Use Procedure S8 to graph the data and the related linear regression equation. The graph is in Figure 12.11.

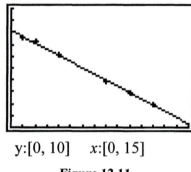

y:[0, 10]   x:[0, 15]

Figure 12.11

Other regression formulas may be used in a similar manner. If it is not known which formula would give the best fit, it may be helpful to do a calculation for each formula and note the corresponding values for the coefficient of regression.

Exercises 12.3

In Exercises 1-10, find the regression equation, sketch a graph of the resulting equation on your paper and show the data points on the graph.

1. Find the linear regression equation which best "fits" the data. Also, give the correlation coefficient. Plot the equation and the data on the same screen. How well does the equation "fit" the data?

$$
\begin{array}{ccccccccc}
x & 25 & 50 & 59 & 80 & 40 & 41 & 49 & 52 \\
y & 11 & 14 & 18 & 24 & 12 & 13 & 15 & 16
\end{array}
$$

2. Find the linear regression equation which best "fits" the data. Also, give the correlation coefficient. Plot the equation and the data on the same screen. How well does the equation "fit" the data?

$$
\begin{array}{lccccccccc}
x & 0 & 2 & 45 & 64 & 79 & 93 & 95 & 127 & 141 \\
y & 0 & 0.3 & 21.5 & 29.8 & 36.9 & 43.4 & 43.9 & 62.1 & 68.4
\end{array}
$$

♦ 3. *(Washington, Exercises 22.6, #9)* The pressure p was measured along an oil pipeline at different distances from a reference point, with results as shown in the following table. Find the linear regression equation which best "fits" the data. Also, give the correlation coefficient. Plot the equation and the data on the same screen.

$$
\begin{array}{lccccc}
x \text{ (ft)} & 0 & 112 & 220 & 320 & 405 \\
p \text{ (lb/in}^2) & 651 & 628 & 603 & 585 & 568
\end{array}
$$

♦ 4. *(Washington, Exercises 22.6, #10)* The heat loss L per hour through various thicknesses of a particular type of insulation was measured as shown in the following table. Find the linear regression equation which best "fits" the data. Also, give the correlation coefficient. Plot the equation and the data on the same screen. How well does the equation "fit" the data?

$$
\begin{array}{lcccccc}
x \text{ (in)} & 2.8 & 4.1 & 4.9 & 6.2 & 7.4 & 7.9 \\
y \text{ (Btu)} & 5910 & 4750 & 3880 & 3130 & 2440 & 1670
\end{array}
$$

5. Find the logarithmic regression equation which best "fits" the data. Also give the correlation coefficient. Plot the equation and the data on the same screen. How well does the equation "fit" the data? Find y, when $x = 10$.

$$
\begin{array}{lccccccc}
x & 2.4 & 3.5 & 7.6 & 8.4 & 9.9 & 11.7 & 14.6 \\
y & 4.90 & 5.40 & 7.10 & 7.35 & 7.48 & 7.80 & 8.48
\end{array}
$$

6. Find the logarithmic regression equation which best "fits" the data. Also give the correlation coefficient. Plot the equation and the data on the same screen. How well does the equation "fit" the data? From the equation, find y, when $x = 55.0$.

$$
\begin{array}{lcccccc}
x & 11.1 & 23.2 & 25.6 & 37.8 & 49.6 & 57.3 \\
y & -1.26 & -3.55 & -3.85 & -5.06 & -5.09 & -6.35
\end{array}
$$

7. Find the exponential regression equation which best "fits" the data. Also give the correlation coefficient. Plot the equation and the data on the same screen. How well does the equation "fit" the data? From the equation, find y, when $x = 4.0$.

$$
\begin{array}{lcccccccc}
x & 0.8 & 1.4 & 1.8 & 2.5 & 2.9 & 3.4 & 3.8 & 4.6 \\
y & 18.5 & 28.7 & 40.1 & 67.7 & 91.3 & 133.0 & 173.8 & 328.2
\end{array}
$$

8. Find the exponential regression equation which best "fits" the data. Also give the correlation coefficient. Plot the equation and the data on the same screen. How well does the equation "fit" the data? From the equation, find y when $x = 1.00$.

x	0.15	0.47	0.56	0.63	0.85	0.92	1.08	1.20
y	3.71	3.65	3.51	3.45	3.36	3.24	3.10	3.02

◆ 9. *(Washington, Exercises 22.7, #7)* The pressure p at which Ammonia vaporizes for temperature T is given in the following table. Find the quadratic regression equation which best "fits" this data. Plot the equation and the data on the same screen. From the equation or graph, find the value of p when $T = 55°F$ and the value of T when $p = 30$ lb/in^2.

T (°F)	−30	−20	−10	0	10	20
p (lb/in^2)	13.90	18.30	23.74	30.42	38.51	48.21

◆ 10. *(Washington, Exercises 22.7, #6)* The increase in length y of a certain metallic rod was measured in relation to particular increases x in temperature. Find the quadratic regression equation that best "fits" this data. Plot the equation and the data on the same screen. From the equation or the graph, find the value of y when x is 125 °C and the value of x when $y = 220.500$.

x (°C)	50.0	100.0	150.0	200.0	250.0
y (cm)	220.180	220.372	220.580	220.795	220.015

◆ 11. *(Washington, Exercises 22.7, #9)* The makers of a special blend of coffee found that the demand for the coffee depended on the price charged. The price P per pound and the monthly sales S are shown in the following table. Find the power regression equation which best "fits" the data. S is the independent variable. Also give the correlation coefficient. Plot the equation and the data on the same screen. How well does the equation "fit" the data? From the equation, find P, when the sales are 500,000.

S (thousands)	244	308	419	484	562
P (dollars)	5.62	4.43	3.18	2.83	2.42

◆ 12. *(Washington, Exercises 22.7, #10)* The resonant frequency of an electric circuit containing a 4-μF capacitor was measured as a function of an inductance in the circuit. The following table gives the data obtained. Find the power regression equation which best "fits" the data. Also give the correlation coefficient. Plot the equation and the data on the same screen. How well does the equation "fit" the data? From the equation, find f, when $L = 5.00$.

L (henrys)	1.1	2.1	4.2	6.2	9.1	12.1
f (hertz)	492	363	251	204	175	156

In Exercises 13-17, enter the given data into the calculator. Try the different regression methods available on the calculator and select the equation with the regression coefficient nearest to −1 or to 1 to represent the data. Then, use the equation to answer the questions in the exercise.

13. The resistance in a resistor is measured at various temperatures. The results are summarized in the following table. Resistance is a function of temperature.

Temperature (°C)	4.2	8.0	12.1	15.8	20.3
Resistance (ohms)	0.880	0.918	0.941	0.980	1.014

(a) What is the resistance at 10.0°C?

(b) What is the resistance at 22.0°C?

(c) At what temperature is the resistance 1.000 ohms?

(d) At what temperature is the resistance 1.020 ohms?

14. The pressure, in atmospheres, in a steel cylinder is measured at various temperatures, in degrees Celsius. The results are summarized in the following table. Temperature is the independent variable.

Temperature (°C)	20.5	29.6	39.7	50.6	62.3	70.5
Pressure (atm)	1.22	1.77	2.37	3.03	3.73	4.22

(a) What is the pressure at 45.0°C?

(b) What is the pressure at 10.0°C?

(c) At what temperature is the pressure 2.00 atm.?

(d) At what temperature is the pressure 1.00 atm.?

15. A student measures the resistance in a certain wire as a function of diameter of the wire. The data is summarized in this table.

Diameter (inches)	0.128	0.104	0.080	0.064	0.048
Resistance (ohms)	0.302	0.466	0.775	1.232	2.190

(a) What is the resistance of 0.036-inch diameter wire?

(b) What is the resistance of 0.160-inch diameter wire?

(c) What diameter wire would have a resistance of 1.500 ohms?

16. In a laboratory experiment a student measures the period (in seconds) of a simple pendulum for various lengths (in centimeters) and arrives at the following data. Find the period as a function of length.

Length (cm) 200.0 220.0 240.0 260.0 280.0 300.0 320.0
Period (sec) 21.2   22.3   23.1   24.3   25.0   26.0   26.8

(a) What would the period be for a pendulum of length 250 cm.?

(b) What would the period be for a pendulum of length 350 cm.?

(c) What length would the pendulum have to be to have a period of 10.0 sec.?

17. To measure the atmospheric pressure at different altitudes, a balloon is released containing an instrument which radios back pressure readings at various altitudes. The resulting data is as follows. Pressure is a function of altitude. (Note: For this exercise, do not round constants used in calculated equation.)

Altitude (feet)   721.0   3115   6335   8224 10032
Pressure (lb./in^2) 14.25   13.08 11.61 10.81 10.14

(a) The height of Mt. Rainier in Washington State is 14,410 feet. What is the pressure on the top of Mt. Rainier?

(b) According to this data, what is the air pressure at sea level (0 feet)?

(c) At what altitude would the pressure be 5.00 lb./in^2?

18. First class postage rates for United States (since 1919) are listed in the following table.

Year (19--) 19 32 58 63 68 71 74 75 78 81 81 85 88 91 95 99 2001 2002
Rate (cents) 2   3   4   5   6   8 10 13 15 18 20 22 25 29 32 33 34   37

(a) By hand, graph postage as a function of year on regular graph paper.

(b) In some cases, when working with data, it is necessary to disregard some of the data. In this case, for finding an equation we will disregard the data prior to 1960. Based on the graphs in parts (a), why might we do this?

(c) Enter the data into your calculator, but use the year 1960 as time (the x variable) zero. Then 1963 will be $x = 3$, 1968 will be $x = 8$, 1971 will be $x = 11$, etc. These correspond to the data points (3, 5), (8, 6), and (11, 8). (For 1981, enter both data points.) Then, based on the coefficient of regression, which of the following types of regression equations given in this section best "fits" the data: linear, logarithmic, exponential, or power,? Give this equation.

(d) Graph the equation on the same graph as the data of part (a). How do the graphs compare?

(e) Use the equation of part (d) to calculate the rate for years 1980, 1995, and the present year. How much different are these than the actual rates?

(f) Calculate the rates that could be expected five years from now and ten years from now.

(g) Determine the year that the rate could be expected to be 50 cents.

19. The public debt of the United States is given in the following table for certain years since 1940.

Year	1940	1950	1960	1970	1975	1980	1985	1990	1995	1999
Debt	50	256	291	381	542	909	1,817	3,206	4,921	5,606

(in **billions** of dollars)

(a) Graph debt as a function of year on regular graph paper.

(b) If you have semi-logarithm graph paper, graph the debt on the logarithmic scale and the year on the linear scale. If you don't have semi-logarithm graph paper, it will be necessary to first use a calculator to find the logs of the debt values. Then, plot the logs of the debts as the dependent variable and the year as the independent variable.

(c) In some cases, when working with data, it is necessary to disregard some of the data. In this case, when we try to find an equation relating this data, we will disregard the data prior to 1965. Based on the graphs in parts (a) and (b), why should we do this?

(d) Enter the data into your calculator, but use the year 1965 as time (the x variable) zero. Then 1970 will be $x = 5$, 1975 will be $x = 10$, 1980 will be x = 15, etc. These will give the data points (5, 381), (10, 542), (15, 909), etc. Then, based on the coefficient of regression, which of the following types of regression equations given in this section best "fits" the data: linear, logarithmic, exponential, or power,? Give this equation.

(e) Graph the equation on the same graph as the data. How do the graphs compare?

(g) Using the equation you found in part (d), calculate the debts for years 1995, 1999, and 2020. For 1995 and 1999, how much different are these than the actual rate?

(h) Estimate from the calculated equation the year when the debt would be 20,000 billion dollars.

Chapter 13

Limits and Derivatives

13.1 Limits
(Washington, Section 23.1)

In this chapter we open the door to an area of mathematics called calculus. Calculus is very important from many standpoints in today's world. To understand calculus, it is first necessary to understand the concept of limit. With limits we are concerned with the value that a function approaches as the independent variable comes close to a given value.

For example, the function $f(x) = x^2$ has a value of $f(4) = 16$ at $x = 4$. We can also say that the limit of $f(x)$ as x approaches 4 is 16. This last statement follows because as we let take values of x that get closer to 4 -- values that are either greater than or less than 4 -- the corresponding values of the function get closer to 16. We write this as $\lim_{x \to 4} x^2 = 16$.

In general, the **limit L of a function** $f(x)$ is the value L that the function approaches as x approaches a number a. We write this as $\lim_{x \to a} f(x) = L$ and read it as: *the limit of f(x) as x approaches a is L*. In finding a limit, we are not concerned with the value of the function exactly at a, or even if the function exists at a, but we are concerned with what value the function approaches as x gets very close to a.

Limits of a function as x approaches a given value may be estimated numerically by evaluating the function at several x-values where each value is closer to the given value. For example, if we wish to find $\lim_{x \to 1} f(x)$, we might first evaluate the function at values greater than 1 such as 1.5, 1.2, 1.1, 1.05, 1.01, 1.001, etc. This would give what is called the **right hand limit.** For the right hand limit we use the notation $\lim_{x \to 1^+} f(x)$. We could also evaluate the function at values less than 1 such as 0.5, 0.8, 0.9, 0.95, 0.99, 0.999, etc. This gives what is called the **left-hand limit** and is indicated by the notation $\lim_{x \to 1^-} f(x)$. **Only when the right hand limit and the left hand limit both exist and are equal does the limit itself exist.**

It is important to note that estimating a limit numerically does not mathematically prove that the limit actually exists or that the limit is the number we estimate. In this material, we shall only consider the numerical estimation of limits and shall not determine limits by other methods.

The estimation of a limit using the graphing calculator requires the evaluation of the function at several values of x. For methods of evaluating functions, refer to Section 1 of Chapter 2 and to Procedure C11 (using a list or creating a table may be helpful). It may also be helpful to write a short program as in Example 2.11. Another method that offers more insight is to evaluate the function while the graph is on the screen by using the trace function or as in method described in Procedure G13.

♦ <u>Example 13.1</u>: *(Washington, Section 23.1, Examples 7 and 8)* Estimate $\lim\limits_{x \to 2} \dfrac{2x^2 - 3x - 2}{x - 2}$ by graphing the function and evaluating both the right hand and left hand limits.

Solution:

1. Graph this function using the standard viewing rectangle. A straight line is obtained and there appears to be a point on the graph for $x = 2$. When we use Procedure G13 to evaluate the function at $x = 2$ no value appears for y. (What happens when we try substituting $x = 2$ directly into the function?)

2. Select the decimal viewing rectangle (Procedure G6) and change the viewing rectangle so that Ymin = 0 and Ymax = 7. The graph is shown in Figure 13.1 has a hole appearing at $x = 2$ (since the function is undefined at $x = 2$).

3. Use the trace function and notice the y values as x approaches 2 from the left side. What number is approached by the y-values? Repeat the process as x approaches 2 from the right hand side. With the cursor near where x = 2, zoom in on the graph and repeat the process. What number is approached by the y-values? From this experiment we conclude that the left-hand limit and the right hand limit are both 5 since that is the value that y approaches. Thus, we say that

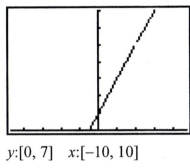

y:[0, 7] x:[−10, 10]

Figure 13.1

$\lim\limits_{x \to 2} \dfrac{2x^2 - 3x - 2}{x - 2} = 5$. We could continue to zoom in to obtain x-values that are closer to 2 to help verify the limit.

4. Next, evaluate the function at several values of x with the graph on the screen. To estimate the right hand limit, evaluate the function at $x = 1.9, 1.95, 1.99, 1.999$, and 1.9999. Then, estimate the left-hand limit by evaluating the function at $x = 2.1, 2.05, 2.01, 2.001, 2.0001$. (See Procedure G13 or the program in Exercise 2.5, Problem 5 for evaluating functions.) Rounding all values to 4 decimal places we obtain the following values:

x	1.9	1.95	1.99	1.999	1.9999
y	4.8	4.9	4.98	4.998	4.9998

x	1.9	1.95	1.99	1.999	1.9999
y	4.8	4.9	4.98	4.998	4.9998

5. Based on this table of values we estimate that

$$\lim_{x \to 2^-} \frac{2x^2 - 3x - 2}{x - 2} = 5.0 \text{ and that } \lim_{x \to 2^+} \frac{2x^2 - 3x - 2}{x - 2} = 5.0.$$

Since both the right hand and left-hand limits are the same we conclude

that $\lim_{x \to 2} \dfrac{2x^2 - 3x - 2}{x - 2} = 5.0.$

Some limits may only be one-sided limits. That is, they do not have both a right and left hand limit. We consider such a case in Example 13.2

Example 13.2:   Estimate $\lim_{x \to 3} \sqrt{x^2 - 9}$ by graphing the function and evaluating the limits.

Solution:

1. Graph this function using the standard viewing rectangle. A graph similar to that in Figure 13.2 is obtained. When we use Procedure G13 to evaluate the function for x values less than 3 we do not get real numbers no matter how close to 3 we try. In this case there is no left hand limit since the function does not exist for $x < 3$.

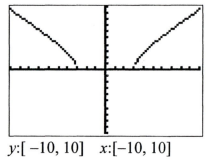

y:[−10, 10] x:[−10, 10]

Figure 13.2

2. If we start at values greater than 3 and use the trace function to see what happens as we move closer to 3, we find the y-values approaching 0. This can be seen better by zooming in on the region at $x = 3$, $y = 0$. Zooming in is particularly helpful in this case since on the calculator screen, the graph does not appear to intersect the x-axis. (We do know that even though it does not appear that way on the graphing calculator, the graph does exist at x = 3 where y = 0.) We thus find that $\lim_{x \to 3^+} \sqrt{x^2 - 9} = 0$ and the right hand limit is 0. Since the left hand

limit does not exist, we also know that $\lim_{x \to 3} \sqrt{x^2 - 9}$ does not exist.

3. Another way to see the right hand limit is to evaluate the function for the x-values 4, 3.5. 3.2, 3.1, 3.05, 3.01… (As x gets closer to 3 from the right side.)

x	4.0	3.5	3.2	3.1	3.05	3.01	3.001
y	2.645	1.803	1.114	0.781	0.550	0.245	0.077

In some cases, the y-values may become very large positive values or very large negative values as x approaches a given value from either the right hand or left-hand side. If the y-values approach a very large positive number we say the limit is $+\infty$ and if they approach a very large negative number we say the limit is $-\infty$. In this case, the limit itself does not exist. (Note: By large negative value we mean a negative number whose absolute value is large.)

In other cases, a function may approach a real value as x becomes a very large positive number (approaches positive infinity, $+\infty$) or as x becomes a very large negative number (approaches negative infinity, $-\infty$). For example, the function $\dfrac{1}{x}$ approaches zero as x approaches ∞ or as x approaches $-\infty$. To see what happens, as x becomes large positive values evaluate the function at numbers such as 100, 1000, 10000, 100000, etc., or as x becomes large negative values, at numbers such as $-100, -1000, -10000, -100000$, etc. If the y-values approach a real number L, we say that a limit exists and is equal to the value L. We write $\lim\limits_{x \to +\infty} f(x) = L$ for a limit as x approaches a large positive number and $\lim\limits_{x \to -\infty} f(x) = L$ for a limit as x approaches a large negative number.

Example 13.3: Estimate $\lim\limits_{x \to +\infty} \dfrac{2x^2 - 1}{x^2 + 2}$ and $\lim\limits_{x \to -\infty} \dfrac{2x^2 - 1}{x^2 + 2}$.

Solution: 1. Graph the function $y = \dfrac{2x^2 - 1}{x^2 + 2}$ on the

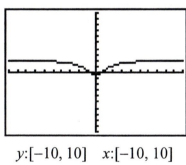

y:$[-10, 10]$ x:$[-10, 10]$

Figure 13.3

standard viewing rectangle. The graph is shown in Figure 13.3. Use the trace function and let x increase in value. The y-values become larger and get closer to the number 2 but never exceeds 2. This is true no matter how large x becomes. If we use the trace function and let x become large negative numbers, we notice that again the y-values become larger and get closer to the number 2 but never exceed 2. We estimate the limit in each case as being 2 and write $\lim\limits_{x \to +\infty} \dfrac{2x^2 - 1}{x^2 + 2} = 2$ and $\lim\limits_{x \to -\infty} \dfrac{2x^2 - 1}{x^2 + 2} = 2$. In some cases it may be necessary to change the window values to see a sufficient part of the graph.

2. The limits may also be estimated numerically by evaluating the function using appropriate values of x to estimate the limit:

x	5	10	100	1000
y	1.814814	1.950980	1.999500	1.999995

x	-5	-10	-100	-1000
y	1.814814	1.950980	1.999500	1.999995

Based on this table we estimate $\lim\limits_{x \to +\infty} \dfrac{2x^2 - 1}{x^2 + 2} = 2$ and $\lim\limits_{x \to -\infty} \dfrac{2x^2 - 1}{x^2 + 2} = 2$

Exercises 13.1

In Exercises 1-12, find the right hand limit, the left hand limit and the limit itself if it exists. Indicate when a limit doesn't exist. Use both graphs and numerical evaluation of the function to estimate the limits. Sketch the graph of the function. When finding limits, indicate the values of x used and the values of the function.

1. $\lim_{x \to 2}(x^2 - 3x)$

2. $\lim_{x \to -3}(4 - x^2)$

3. $\lim_{x \to 2}\dfrac{x^2 - 4}{x - 2}$

4. $\lim_{x \to 9}\dfrac{x^2 - 9x}{x - 9}$

5. $\lim_{x \to \sqrt{3}}\dfrac{x^4 - 6x^2 + 9}{x^2 - 3}$

6. $\lim_{x \to 0}\dfrac{4 - \sqrt{x + 16}}{x}$

7. $\lim_{x \to 0}\dfrac{(2 + x)^2 - 3(2 + x) + 2}{x}$

8. $\lim_{x \to 0}\dfrac{-16(1 + x)^2 + 16}{x}$

9. $\lim_{x \to 1.5}\dfrac{1 - x}{x - 1.5}$

10. $\lim_{x \to 3}\dfrac{x - 2}{x - 3}$

♦ 11. *(Washington, Exercises 23.1, #52)* $\lim_{x \to 0}\dfrac{\sin x}{x}$ [Place calculator in radian mode]

12. $\lim_{x \to 0}4^{\frac{1}{x}}$

In Exercises 13-18 estimate the limit both from the graph of the function and by evaluating the function at several appropriate values of x. Indicate the values of x used and the corresponding values of the function. Estimate limits to three significant digits.

13. $\lim_{x \to \infty}\dfrac{3x + 3}{4x - 1}$

14. $\lim_{x \to \infty}\dfrac{4.7x^2 - 3x + 2}{2x^2 + 4x}$

15. $\lim_{x \to \infty}\dfrac{32 - 12.64x^3}{4x^3 + 24x}$

16. $\lim_{x \to -\infty}\dfrac{125 - 75x^2 + 61x^3}{94x^3 - 72x}$

♦ 17. *(Washington, Exercises 23.1, #51)* $\lim_{x \to \infty}(1 + x)^{\frac{1}{x}}$

♦ 18. *(Washington, Exercises 23.1, #50)* A 5.25 ohm resistor and a variable resistor of resistance R are placed in parallel. The expression for the resulting resistance R_T is given by

$R_T = \dfrac{5.25R}{5.25 + R}$. Estimate the limiting value of the resulting resistance as R becomes large (find $\lim_{R \to \infty} R_T$).

19. The average rate of change of y with respect to x for the function $y = x^2 - 2$ as x changes from x to $x + \Delta x$ is given by the expression $\dfrac{\Delta y}{\Delta x} = \dfrac{(x + \Delta x)^2 - 2 - (x^2 - 2)}{\Delta x}$. To find the instantaneous rate of change at $x = 2$, determine the limit, as Δx approaches 0, of $\dfrac{(2 + \Delta x)^2 - 2 - (4 - 2)}{\Delta x} = \dfrac{(2 + \Delta x)^2 - 4}{\Delta x}$. Do this by graphing the equivalent function $y = \dfrac{(2 + x)^2 - 4}{x}$ (letting $x = \Delta x$) and find the limit as x approaches zero. Estimate the limit both from the graph and by evaluating the expression.

20. If the distance y, in miles, a car travels as a function of time t, in minutes, is given by the expression $y = 30t + 0.5t^2$, then the average rate of change (average speed or velocity) is given by the expression $\dfrac{\Delta y}{\Delta t} = \dfrac{30(t + \Delta t) + 0.5(t + \Delta t)^2 - (30t + 0.5t^2)}{\Delta t}$. To find the instantaneous rate of velocity at t = 5 minutes, substitute 5 for t, simplify, and find the limit as Δt approaches 0. (See Exercise 19.) Estimate the limit from the graph and by evaluating the expression.

13.2 More on Programming

It will be helpful in this chapter and the next to know more about programming of the graphing calculator. In particular we want to consider the creation of loops within a program.

A **loop** within a program is a program structure that causes a set of statements to be repeated a set number of times. There are several different structures for loops. The first that we will consider is the While... loop. (It may be helpful for the student to first review Section 5 of Chapter 2 before proceeding.) For each of the loop procedures, we must first be in programming edit mode (see Procedure P1 or Procedure P2).

Procedure P10. Creating a While... loop (While, End commands).

The construction of the loop is of the form:

```
... (program statements before loop)
While condition
... (program statements within loop if condition true)
End
... (programming statements after loop)
```

To access the While and End statements,
While in programming edit mode (at a step in program):

On the TI-82 or 83 or 84 Plus:	On the TI-85 or 86:
1. To select the While command: (a) Press PRGM (b) With CTL highlighted, select 5:While	1. To select the While command: (a) Select CTL (b) Select While (after pressing MORE)
2. To select the End command: (a) Press PRGM (b) With CTL highlighted, select 7:End	2. To select the End command (a) Select CTL (b) Select End

In this case, if *condition* is a true statement the statements following the word While are performed and if *condition* is not true, the program will then transfer to the statements following the word End. For information on *condition* see Procedure P7.

Example: $1 \rightarrow K$
            While $(K<5)$
            $K + 1 \rightarrow K$
            End
            Disp K

When this program is run, it will cycle through the loop until K is no longer less than 5, then it will display the result. The result of this program is that $K = 5$.

Following Procedure P10 is an example of a very simple program using the While... loop. We next show how the while loop may be used in helping to determine a limit. When using a loop within a program it is often helpful to use a **counter** to count the number of times through the loop. In Example 13.4, K is used as a **counter** and each time through the loop a new value is stored for K that is one more than the previous value. Note that at the beginning of the program K is assigned an initial value. All variables whose initial value does not come from a calculation within a program should be assigned an **initial value** at the beginning of the program.

Example 13.4: Write a program to find the right-hand limit of a function as x approaches a value a. Then use the program to find the right-hand limit of the function

$$f(x) = \frac{x^3 - 4x}{x - 2} \text{ as } x \text{ approaches 2.}$$

Solution: We will use a technique similar to that used in Section 13.1 in that we will calculate values for the function as x gets closer and closer to 2. For this example we will present the program then discuss what it does. Before execution of the program store the function to be evaluated under the function name $Y1$.

PROGRAM: LIMIT

$0 \to K$

$10 \to D$

Prompt A

Prompt I

(*Program continued in next column.*)

(*Program continued from previous column.*)

While $K<5$

$A+I/(D^\wedge K) \to x$

Disp $y1$

$K+1 \to K$

End

In this program, K is a counter set initially to zero and K is increased by 1 each time through the loop. D is a divisor factor set to 10, A is the value which x is approaching in the limit and I is the increment or distance from a where we are starting to calculate values for the function.

As the program proceeds through the While... loop values closer and closer to a are stored for x. The Disp $Y1$ statement calculates the function $Y1$ (at the most recent value stored for x) and displays the value of the function on the screen. Since the condition on the if statement is $K<5$, this calculation will take place 5 times.

Storing the function $f(x) = \dfrac{x^3 - 4x}{x - 2}$ for $Y1$ and entering 2 for A and 1 for I, we find the values displayed (rounded to 5 decimal places) are:

8.61

8.0601

8.00600

8.00060

8.00006

It would appear that this right hand limit is approaching the number 8.

We leave it to the student to determine how to use the same program to find the left-hand limit of a function.

Two other types of loops often used in programming are the Repeat... loop and the For... loop. We next give the procedures for these loops, but leave creation of programs containing these loops as exercises.

Procedure P11. Creating a Repeat... loop (Repeat, End commands).

The construction of the loop is of the form:

... (program statements before loop)
Repeat *condition*
... (statements within loop repeated until condition is true)
End
... (programming statements after loop)

To access Repeat and End statements,
While in programming edit mode (at a step in program):

On the TI-82 or 83 or 84 Plus:	On the TI-85 or 86:

1. To select the Repeat command:
  (a) Press PRGM
  (b) With CTL highlighted,
     select 6:Repeat
	1. To select the Repeat command:
	  (a) Select CTL
	  (b) Select Repeat (after
	     pressing MORE)

2. To select the End command:
  (a) Press PRGM
  (b) With CTL highlighted,
     select 7:End
	2. To select the End command
	  (a) Select CTL
	  (b) Select End

In this case, if *condition* is a false statement the statements following the word Repeat are performed until the condition becomes true. The program will then transfer to the statements following the word End.

Example: $10 \to K$
Repeat ($K<5$)
$K - 1 \to K$
End
Disp K

K is initially set to the value 10, but each time through the loop one is subtracted from K until K is less than 5. At that point the loop ends and the value of K is displayed. The result of this program is that $K = 4$.

Procedure P12. Creating a For... loop (For, End commands).

The construction of the loop is of the form:

    ... (program statements before loop)
    For(*var*, *beg*, *end*, *inc*)
    ... (statements within loop repeated if *end* not exceeded)
    End
    ... (programming statements after loop)

To access For and End statements,
While in programming edit mode (at a step in program):

On the TI-82 or 83 or 84 Plus:	On the TI-85 or 86:

1. To select the For command:
  (a) Press PRGM
  (b) With CTL highlighted,
     select 4:For
	1. To select the For command:
	  (a) Select CTL
	  (b) Select For

2. To select the End command:
  (a) Press PRGM
  (b) With CTL highlighted,
     select 7:End
	2. To select the End command
	  (a) Select CTL
	  (b) Select End

Where   *var* is a variable name   *inc* is an increment value and is optional
          beg is a beginning value   (Increment is 1 if this is omitted.)
          end is an ending value

In this case, the first time through the loop, *var* has the beginning value, and the next time through the loop *var* has the value of *beg* + *inc*. Each time through the loop the value for *var* is increased by *inc*. The loop stops when the value for *var* exceeds the value for *end* and the program will then transfer to the statements following the word End.

Example: $0 \rightarrow K$
For(I,1,5)
$K + I \rightarrow K$
End
Disp K

The first time through the loop, I has a value of 1 that is added to K giving 1. The second time I has a value of 2, which is added to the previous K giving 3. Etc. The program loops 5 times (as I goes from 1 to 5). The result of this program is that $K = 15$.

Another loop that may need to be used on some calculators (ones that don't allow the While..., Repeat..., or For... loops) is the If...Goto... loop.

Procedure P13. Creating an If...Goto... loop (Lbl, If, Goto commands).

The construction of the loop is of the form:

... (program statements before loop)
Lbl *label*
... (program statements within loop)
If *condition*
Goto *label*
... (programming statements after loop)

To access Lbl, If, and Goto statements,
While in programming edit mode (at a step in program):

On the TI-82 or 83 or 84 Plus:
1. Select the label (Lbl) command:
   (a) Press PRGM
   (b) With CTL highlighted, select 9:Lbl
2. Select the If command:
   (a) Press PRGM
   (b) With CTL highlighted, select 1:If
3. Select the Goto command:
   (a) Press PRGM
   (b) With CTL highlighted, select 0:Goto
Note: A label may be a letter or a numerical digit.

On the TI-85 or 86:
1. Select the label (Lbl) command:
   (a) Select CTL
   (b) Select Lbl (after pressing MORE)
2. Select If command
   (a) Select CTL
   (b) Select If
3. Select the Goto command
   (a) Select CTL
   (b) Select Goto (after pressing MORE)
Note: A label may be a name of up to 8 characters, beginning with a letter.

The form of the label command is Lbl *label* (where *label* is a character)
The form of the If command is If *condition*
The form of the Goto command is Goto *label*
The statements within the loop are executed as long as condition is true. When condition ceases to be true, the program will transfer to statements following the Goto statement.

Example: $0 \rightarrow K$
Lbl *G*
$K + 1 \rightarrow K$
If $(K<5)$
Goto *G*
Disp *K*

The program loop through the Goto *G* statement back up to the Lbl *G* statement until *K* is no longer less than 5. It then displays the value for *K*. The result of this program is that $K = 5$.

Exercises 13.2

1. Write a program using a While... loop to input *n* grades (*g*) and find their average. Note: To do this you will need to use an accumulator. An accumulator (*S*) is used in a manner similar to a counter, except that each time the newly entered value for *g* is added to the previously stored value for *S*. A statement like $S + G \rightarrow S$ will be needed in the program.   Test the program with the grades: 75, 84, 92, 88, 99, 77.

2. Write a program containing a While... loop to input the weights (in pounds) of a sample of n boxes of bolts and display the sum. Test the program with the weights: 124.6, 125.7, 123.9, 124.8, 126.8, 125.3, 127.1

3. Write the program of Example 13.3 using a Repeat... loop rather than a While... loop.

4. Write a program to find the limit of $f(x) = e^{1-2x}$ as *x* becomes large. Increase *x* by 5 each time. Execute the loop until successive values of $f(x)$ are within 0.0001 units.

5. Write the program of Example 13.3 using a For... loop rather than a While... loop.

6. Write the program for Exercise 1 of this exercise set using a For... loop.

7. Use a program to find $\lim\limits_{x \to 1} \dfrac{x-1}{x^2 - 3x + 2}$.

8. Use a program find $\lim\limits_{x \to 0} 5^{\frac{1}{x}}$.

9. Write a program using a While... loop to find the limit of the quantity $\dfrac{f(x+h) - f(x)}{h}$ as *h* approaches 0 using the function $f(x) = x^3$ and the following input values for *x*: 2, 5, −3, 0

10. Write a program using a While... loop to find the limit of the quantity $\dfrac{f(x+h)-f(x-h)}{2h}$ as h approaches 0 using the function $f(x) = x^3$ and the following input values for x: 2, 5, –3, 0. Compare with Exercise 9.

The following program draws secant lines on a graph of a function from a point where $x = A$ to a point where $x = B = A + H$. The program illustrates how the secant lines approach the tangent line as H approaches 0.

PROGRAM:SECTLINE
ClDrw   [ClrDraw on TI-82 and 83]
Prompt A
$A \rightarrow x$
$y1 \rightarrow U$
Prompt H
For(K,1,4)
$A + H \rightarrow B$
(*Program continued in next column*)

(*Program continued from previous column*)
$B \rightarrow x$
$y1 \rightarrow W$
Line(A,U,B,W)
$H/1.2 \rightarrow H$
Pause
End
TanLn($y1$,A)   [Tangent on TI82 and 83]

Notes: 1. Before executing this program, the desired function must be stored for the function name $y1$ and all other functions must be turned off.
2. After each secant line is drawn, PAUSE requires the ENTER key to be pressed to continue.
3. The commands ClDrw (or ClrDraw), Line, and TanLn (or Tangent) appear under the DRAW menu (see Procedures G25 and G30).
4. To view the secant lines properly, adjust the viewing rectangle.

Use the SECTLINE program to graph the secant lines and tangent line for functions in Exercises 11-14 at the given values of x.

11. $y = x^2 - 4$ at $x = 1$ [Hint: use $H = 4$, Xmin = –1, Xmax = 5, yMin = –10, yMax = 20]
12. $y = x^3 - 3$ at $x = 1$ [Hint: use $H = 2$, Xmin = –1, Xmax = 5, yMin = –10, yMax = 25]
13. Try Exercise 11 with $H = -4$. (Some adjustment of viewing rectangle is necessary.)
14. $y = \sin x$ at $x = \pi$ with $H = \dfrac{\pi}{2}$, then with $H = -\dfrac{\pi}{2}$.

13.3 The Derivative
(Washington, Sections 23.3 and 23.4)

Consider the ratio $\dfrac{\Delta y}{\Delta x}$ for the function $y = f(x)$ where $\Delta y = f(x + \Delta x) - f(x)$. As Δx approach zero,

we find that this ratio $\dfrac{\Delta y}{\Delta x}$ often approaches a fixed value (see Exercises 19 and 20 of Exercise

Set 13.1). The limit (as $\Delta x \to 0$) of this ratio is a very important quantity in mathematics and is called the derivative of the function. More formally we define the **derivative** of a function $f(x)$ as

$$\lim_{\Delta x \to 0} \frac{\Delta y}{\Delta x} \quad \text{or} \quad \lim_{\Delta x \to 0} \frac{f(x + \Delta x) - f(x)}{\Delta x} \qquad \text{[Equation 1]}$$

Another equivalent definition of the derivative is

$$\lim_{\Delta x \to 0} \frac{f(x + \Delta x) - f(x - \Delta x)}{2\Delta x} \qquad \text{[Equation 2]}$$

To approximate the derivative of the function at $x = a$, either Equation 1 or Equation 2 may be evaluated when $x = a$ for values of Δx as Δx approaches zero. The value obtained is called the numerical derivative. In Problem 9 of Exercise 13.2 we use Equation 1 (with $h = \Delta x$) to calculate the derivative of x^3.

Some graphing calculators determine the value of the numerical derivative directly by a formula similar to Equation 2. Since different notations are used to represent the numerical derivative, we will use nDeriv.

Procedure C19. **To determine the numerical derivative from a function.**

On the TI-82 or 83 or 84 Plus:
1. Press the key MATH
2. Select 8:nDeriv(
   The form is
   nDeriv(*y*, *variable*, *value*)

On the TI-85 or 86:
1. Press the key CALC
2. From the menu, select nDer
   The form is
   nDer(*y*, *variable*, *value*)

where *y* is the function itself or the name of the function, *variable* is the independent variable used in the function, and *value* is the number at which we wish to determine the derivative.

Example: nDeriv(Y_1,*x*,2)

The precision to which the derivative is calculated may be changed by indicating the precision as a fourth value inside the parentheses.

Example: nDer($y1$,*x*,2)

The precision to which the derivative is calculated may be changed.

If the fourth value is not given, the precision is 0.001. (This value determines the Δx.)

Example:

$$\text{nDeriv}(Y_1, x, 2, 0.0001)$$

On the TI-85, press the key TOLER and on the TI-86, press the key MEM and select TOL, Then change the value for δ (this determines the Δx).

Note: The TI-85 and 86, also have the function der1 which determines the numerical first derivative as exactly as is possible and der2 which determines the numerical second derivative as exactly as possible.

<u>Example 13.5</u>: Find the numerical derivative of $y = x^3 - 2x^2 + 4$ at $x = 1$ and at $x = \dfrac{4}{3}$. Round answers to two decimal places.

<u>Solution</u>: Use Procedure C19 to determine the numerical derivative of the function with x as the variable. Evaluate first with 1 substituted for *value,* then with $\dfrac{4}{3}$ substituted for *value.* The derivative at $x = 1$ is $-.999999$ or -1.00. The value at $x = \dfrac{4}{3}$ appears similar to 1E$-$6 (or 0.000001) which rounds off to 0.00.

The value of the derivative of a function at a given x-value indicates the instantaneous rate of change of y with respect to x. This instantaneous rate of change may represent the slope of a tangent line on a graph, the instantaneous velocity of a moving object, or the acceleration of a moving object.

In Example 13.4, the derivative of $y = x^3 - 2x^2 + 4$ at $x = 1$ is -1. This means that the instantaneous rate of change of y with respect to x at $x = 1$ is -1 and implies that, at $x = 1$, the y changes -1 unit for a one unit change in x. This also indicates that, at $x = 1$, the slope of the tangent line is -1 and that the slope of the tangent line at $x = \dfrac{4}{3}$ is 0. (What does this tell us about the tangent line at $x = \dfrac{4}{3}$?)

The following short program may be helpful in determining values for the numerical derivative of a function at several values of x:

On the TI-82 or 83:

```
PROGRAM:DERIV
Disp "ENTER X VALUE"
Input A
nDeriv(Y₁,X,A)→Y
Disp Y
```

On the TI-85 or 86:

```
PROGRAM :DERIV
Disp "Enter x-value"
Input A
NDer(y1, x, A)→Y
Disp Y
```

When using this program, the function of interest must be stored for the name Y_1. The derivative may be determined for several values of x by repeatedly executing the program by pressing ENTER before pressing any other key. The numerical derivative function is obtained as in Procedure C19 while in programming mode and the arrow $\rightarrow$ is obtained by pressing the STO▷ key.

Example 13.6: A baseball is thrown vertically at 90 ft/s. The height s, in feet, of the ball at time t, in seconds, is given by the equation $s = 90t - 16t^2$. Find height of the ball and the instantaneous velocity of the ball when $t = 1.0, 2.0, 2.5, 3.0,$ and 4.0 seconds. What is the significance of the positive and negative values of the derivative (compare to s-values)?

Solution:
1. In this case, the instantaneous velocity is the instantaneous rate of change of s with respect to t. This instantaneous rate of change may be found by determining the value of the numerical derivative at the given values of t.

2. Evaluating the numerical derivative of the function $s = 90t - 16t^2$, we obtain

time, t	1.0	2.0	2.5	3.0	4.0
height, s	74	116	125	126	104
numerical derivative	58	26	10	−6.0	−38

3. From these values we conclude that when the derivative is positive, the height is increasing and when the derivative is negative, the height is decreasing. As the ball approaches its highest point, the value of the derivative approaches zero.

In Example 13.6, we see that when the derivative is positive the dependent variable is increasing and when the derivative is negative the dependent variable is decreasing. This is true of all functions and provides a method of determining when a function is increasing in value or decreasing in value.

Exercise 13.3

In Exercises 1-8, find the numerical derivative of each function for $x = -2.0, -1.0, 0.0, 1.0,$ and 2.0. Compare the value of the function and its derivative at each x-value and determine if there is a relationship between the given function and its derivative.

1. $y = 2x$ 2. $y = 5$

3. $y = x^2$ 4. $y = x^3$

5. $y = x^4$ 6. $y = x^5$

7. $y = x^{-1}$

8. $y = x^{-2}$

9. Modify program DERIV given in this section to use a While... loop. A new x-value should be input and its derivative determined each time through the loop. Have the program terminate when an x-value larger than 50 is entered.

10. Do Exercise 9, but use a Repeat... loop.

In Exercises 11-16, find the numerical derivative at the given values of x and compare to the graph of the function at the given x-values. Determine the values of x for which y is increasing in value and the values of x for which y is decreasing in value. Can you reach any conclusions about when the value of the derivative is zero?

11. $y = x^2 - 2x$ at $x = -1.0, 0.5, 1.0, 1.5,$ and 2.0.

12. $y = 2 - x - x^2$ at $x = -2.0, -0.5, 0.0, 1.0,$ and 2.0.

13. $y = 12x - x^3$ at $x = -4.0, -2.0, 0.0, 2.0,$ and 3.0.

14. $y = 2x^3 + 6x^2$ at $x = -3.0, -2.0, -1.0, 0.0,$ and 2.0.

15. $y = \dfrac{1}{1+x^2}$ at $x = -2.0, -0.5, 0.0, 0.5,$ and 2.0.

16. $y = \dfrac{1}{(1-x)^2}$ at $x = -1.0, 0.0, 1.0,$ and 2.0.

In Exercises 17-20, find several values of the numerical derivative of the function between $x = -5.0$ and 5.0 and compare to the graph of the function. Locate values of x at which the function has horizontal tangent lines.

17. $y = 2x^2 - 10x$

18. $y = 9x - x^2$

19. $y = 5x^3 - 9$

20. $y = 6x - x^3$

In Exercises 21-24, find the value of the distance s, in feet, and the instantaneous velocity for the given values of time t, in seconds. Use the numerical derivative to find the instantaneous velocity. Note the relationship between the numerical derivative and whether the values of s are increasing or decreasing.

21. $s = 5t - 10$ for $t = 0.0, 1.5, 3.0, 5.5,$ and 8.0

22. $s = 24 + 7t$ for $t = 0.0, 2.5, 5.0, 7.5,$ and 10.0

23. $s = 76t - 16t^2$ for $t = 1.00, 2.25, 2.375, 2.40,$ and 2.50.

24. $s = 135 + 125t - 16t^2$ for $t = 2.50, 3.50, 4.00, 4.50,$ and 5.00.

In Exercises 25-28, use the numerical derivative to find the instantaneous rate of change.

25. The cost of producing x parts in a machine shop is given by $C = 4500 - 100x + x^2$. Find the instantaneous rate of change of cost relative to the number of parts produced when $x = 35$, 45, 50, 55, and 65. What is the significance of positive and negative rates of change?

26. A ball bearing is in the shape of a sphere whose volume is given by $V = \frac{4}{3}\pi r^3$, where r is the radius in centimeters. As it is heated the sphere expands. Find the rate of change of volume with respect to the radius when the radius is 6.010 cm, 6.020 cm, and 6.110 cm.

♦ 27. *(Washington, Exercises 23.4, #31)* The total power P, in watts, transmitted by an AM radio station is given by $P = 500 + 250x^2$ where x is the modulation index. Find the instantaneous rate of change of P with respect to x for $x = 0.88$, $x = 0.90$, $x = 0.92$, and $x = 0.94$.

♦ 28. *(Washington, Exercises 23.4, #28)* A load L, in Newton's, is distributed along a beam 10 meters long such that $L = 5x - 0.5x^2$ where x is the distance from one end of the beam. Find the rate of change of L with respect to x when $x = 2.0$ m, 4.5 m, 5.0 m, and 8.0 m.

13.4 Applications of the Derivative
(Washington, Sections 24.1, 24.2, and 24.5)

In this section we study the use of the graphing calculator in applications related to the graph of a function. We will use the derivative in finding the equation of a tangent line and will use graphical means to investigate the relationship between the derivative and the function. We are already aware that the derivative of a function evaluated at a point gives the slope of the tangent line at that point. From analytic geometry, we know that we can find the equation of any line if we know the slope, m, of the line and a point, (x_1, y_1), on the line by using the standard form

$$y - y_1 = m(x - x_1). \qquad \text{[Equation 3]}$$

To find the equation of a tangent line, use the following procedure:

1. Find the numerical derivative for a given x-value (Procedure C19). This gives the slope of the tangent line.
2. Find the y-coordinate for the given x-value (Procedure G13).
3. Substitute the values into Equation 3 and simplify.

After finding the equation of a tangent line it is interesting to see the tangent line on the same graph as the function. We could just graph the line that we obtained from calculations or we can use the function that is built into many calculators to graph the tangent line.

Procedure G25. Drawing a tangent line at a point.

1. Store the function for a function name.
2. Draw the tangent line using the following:

On the TI-82 or 83 or 84 Plus:

(a) While on the home screen, press the key DRAW.

(b) Select: 5:Tangent(
   The form is
      Tangent(y, *value*)
   Where y is the function or function name and *value* is the x-value.
   Example: Tangent(Y_1,2)

(c) Press ENTER

or

The tangent line may be drawn by:

(a) With the graph on the screen, use trace function to place the cursor at the correct point, then press DRAW.

(b) Select 5:Tangent(and press ENTER to return to graph.

(c) Press ENTER a second time. The tangent line will appear on the graph and the slope of the line will be given at the bottom of the screen.

On the TI-85 or 86:

(a) With the graph on the screen select DRAW from the GRAPH menu (after pressing MORE)

(b) From the secondary menu, select TanLn (after pressing MORE twice). The form is
      TanLn(y, *value*).
   where y is the function or function name and *value* is the x-value.
   Example: TanLn(y1,3)

or

The tangent line may be drawn by:

(a) With the graph and the GRAPH menu on the screen, select MATH (after pressing MORE).

(b) From the secondary menu, select TANLN (after pressing MORE twice).

(c) Move the cursor to the desired point and press ENTER. The tangent line will be drawn on the graph and the slope of the tangent line given at the bottom of the screen.

All tangent lines will appear on the screen until cleared. To clear tangent lines from the screen use the same process as in Procedure G22, Step 5.

Example 13.7: Find the equation of the tangent line to the graph of $y = 9x - 3x^2$ at $x = 2.0$ and use the equation of the tangent line to find the value for y at the point on the tangent line where $x = 2.2$.

Solution:

1. Store the function $9x - 3x^2$ for a function name and graph the function.
2. Determine the coordinates of the point and the slope for $x = 2.0$. The y-coordinate of the point is 6.0 and the slope (m) is −3.0. Use Procedure G25 to graph the tangent line. The graphs of the function and tangent line are in Figure 13.4.

3. From the slope-point form for the equation of a line we have

$$y - 6.0 = -3.0\,(x - 2.0)$$
$$\text{or } y - 6.0 = -3.0x + 6.0$$
$$\text{or } y = -3.0x + 12.0$$

4. At $x = 2.2$: $y = -3.0\,(2.2) + 3.0 = -3.3$.

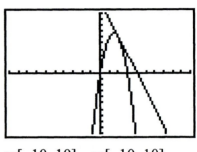

y:[−10, 10] x:[−10, 10]

Figure 13.4

The derivative of a function and the slope of the tangent line may also help us analyze the graph of a function. We next investigate how the slope of the tangent line is related to the graph of a function.

Procedure G26. **Determining the value of the derivative at points on a graph.**

On the TI-82 or 83 or 84 Plus:
1. With a graph on the screen, press the key CALC.
2. Select 6:dy/dx.
3. Move the cursor to the desired point and press ENTER. The value of the numerical derivative appears on the bottom of the screen.

On the TI-85 or 86:
1. With the graph and the graph menu on the screen, press the key MATH (after pressing MORE).
2. From the secondary menu select dy/dx.
3. Move the cursor to the desired point and press ENTER. The value of the numerical derivative appears at the bottom of the screen.

Example 13.8: Graph the function $y = x^2 - 2x$ on the standard viewing rectangle, then change to the decimal viewing rectangle. Find several values of the derivative for x less than one and also for x greater than one. What is true about the derivative for x less than one? For x greater than one? Where is the function decreasing and where is it increasing? What is true about the point where $x = 1$?

Solution: 1. Graph the function on the standard viewing rectangle and then change to the decimal viewing rectangle. Evaluate the derivative at $x = -1.0$, $x = 0.0$, $x = 2.0$, and $x = 3.0$ (Procedure C19). We obtain the following values:

x	−1.0	0.0	2.0	3.0
dy/dx	−4.0	−2.0	2.0	4.0

2. Notice that for x-values less than one, the derivative is negative and the graph is decreasing and for x-values greater than one, the derivative is positive and the graph is increasing. At $x = 1$, the derivative is zero and at $x = 1$, there is a local minimum point.

It is instructive to graph the derivative of a function on the same graph as the function. We can do this even if we do not know the derivative of the function. First enter the function under one name (Y_1) and then enter the numerical derivative under a second function name (Y_2). Enter the variable x in the numerical derivative for *value* ($Y_2 = \text{nDeriv}(y,x,x)$).

Example 13.9: Graph the function $y = x^2 - 6x + 5$ and the numerical derivative on the same graph. What is true about the function when the graph of the derivative is negative (lies below the x-axis)? What is true about the function when the graph of the derivative is positive (lies above the x-axis)? What is true about the function at the x-intercept of the graph of the derivative?

Solution: 1. Enter and graph the function and the numerical derivative of the function on the standard viewing rectangle. To graph the numerical derivative (assuming that the function has been entered under the name Y_1), enter $\text{nDeriv}(Y_1,x,x)$ under a second function name. The graph is in Figure 13.5.

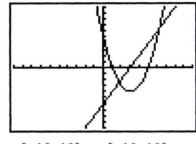

$y:[-10, 10]$ $x:[-10, 10]$

Figure 13.5

2. Notice that when the graph of the derivative is negative, the function itself is decreasing. When the graph of the derivative is positive, the function itself is increasing. At the point where the graph of the derivative has an x-intercept, the derivative must be zero. This corresponds to the local minimum point of the function where there is a horizontal tangent line.

Exercise 13.4

In Exercises 1-8, find the equation of the tangent line at the given x-value. Graph the function and tangent line on your calculator and make a sketch of the graph and tangent line on your paper. Show values for Xmin, Xmax, Ymin, and Ymax at the ends of the axes of your drawing.

1. $y = x^2 - 7$ at $x = 1.5$

2. $y = 5x - 2x^2$ at $x = 1.1$

3. $y = \dfrac{5}{2 + x^2}$ at $x = 0.75$

4. $y = \dfrac{1}{x^2}$ at $x = -1.25$

5. $y = \sqrt{81 - x^2}$ at $x = 5.00$

6. $y = 25x - 2x^3$ at $x = 2.55$

7. $y = x^3 + 5x^2 - 6x - 4$ at $x = 4.75$

8. $y = 125x - 16x^4$ at $x = 0.532$

In Exercises 9-12, graph the given function and use Procedure G26 to determine the derivative at points on the graph.

9. For $y = 8x - x^2$, find the derivative at $x = 1.5, 2.5, 4.0, 4.5$, and 5.5. What is the sign of the derivative for $x < 4.0$? For $x > 4.0$? Is the function increasing or decreasing in each of these regions?

10. For $y = x^3 - 27x$, find the derivative at $x = -5.0, -4.0, -4.1, -2.5, -1.5, 1.5, 2.0, 2.5, 3.5$, and 5.0. What is the sign of the derivative for $x < -3.0$? For $x > -3.0$ and $x < 3.0$? For $x > 3.0$? Is the function increasing or decreasing in each of these regions?

11. For $y = 12x^4 - 6x^2$, find the derivative at $x = -1.0, -0.6, -0.4, -0.1, 0.1, 0.4, 0.6$, and 1.0. For what x-values is the sign of the derivative positive? Negative? For what x-values is the function increasing or decreasing?

12. For $y = \dfrac{25}{x^2}$, find the derivative at $x = -2.5, -1.5, -0.5, 0.5, 1.5$, and 2.5. For what x-values is the sign of the derivative positive? Negative? For what x-values is the function increasing or decreasing?

In Exercises 13-18, graph the function and its derivative on the same axes as in Example 13.9. In each case, answer these questions relative to the given function:
(a) What is true when the graph of the derivative is positive (above the x-axis)?
(b) What is true when the graph of the derivative is negative (below the x-axis)?
(c) What is true at the x-intercepts of the graph of the derivative?

13. $y = 8 - x^2$ What type of graph is obtained for the derivative? Explain.

14. $y = x^3 - 15x^2$ What type of graph is obtained for the derivative? Explain.

15. $y = x^4 - 6x^2$

16. $y = \dfrac{6}{x^2 + 4}$

17. $y = \sqrt{25 - x^2}$ What happens to the graph of the derivative at $x = 5$ and -5? Explain.

18. $y = \dfrac{10}{(x-3)^2}$ What happens to the graph of the derivative at $x = 3$? Explain.

In Exercises 19 and 20 graph the function and its first derivative on the same graph. Use the x-intercepts to determine where any local maximum or minimum points occur.

19. $y = 2x - \dfrac{1}{x^2}$

20. $y = \dfrac{x}{4 + x^2}$

◆ 21. *(Washington, Exercises 24.5, #39)* A batter hits a baseball that follows a path given by $y = x - 0.0025x^2$. Graph the derivative of this function. For what x-value is the derivative zero? What is true at this point? For what x-values is the derivative positive? What is happening to the baseball for these values of x?

◆ 22. *(Washington, Exercises 24.5, #40)* The angle, θ, in degrees, of a robot arm with the horizontal as a function of the time t, in seconds, is given by $\theta = 10 + 12t^2 - 2t^3$ for $0 \le t \le 6$ s. Graph the derivative of this function. For what t-value is the derivative zero? What is true at this point? For what t-values is the derivative negative? What is happening to the angle for these t-values?

In Exercises 23-26, graph the second derivative of the function on the same graph as the function. The second derivative of a function may be graphed by graphing the numerical derivative of the numerical derivative. If Y_1 is the name of the function and $Y_2 = \text{nDeriv}(Y_1, x, x)$ represents the first derivative, then the second derivative by be represented by $Y_3 = \text{nDeriv}(Y_2, x, x)$. In each case determine:

(a) For what values of x is the second derivative negative and for what values of x is the second derivative positive?

(b) What is true about the graph of the function for these values of x?

(c) What is true where the second derivative is zero?

Note: The drawing of these graphs may be slow due to the number of calculations involved.

23. $y = \dfrac{x^3}{4}$

24. $y = 2x^2 - 5x$

25. $y = x^4 - 16x^2$

26. $y = 25x - x^3$

Newton's method is a method of finding the roots of an equation of the form $f(x) = 0$ using the formula

$$x_2 = x_1 - \frac{f(x_1)}{f'(x_1)}$$

where x_1 is the first approximation (or guess) of the root, $f(x)$ is the function from the equation, and x_2 is the second approximation. The value x_2 then is used to replace x_1 and a third approximation calculated. The process is repeated until a root of a desired accuracy is obtained.

A program that will calculate repeated approximations using Newton's method is:

```
PROGRAM:NEWT
0→K
Prompt x
While (K<5)
1+K→K
x − Y₁ / nDeriv(Y₁, x, x) → x
Disp x
End
```

Before running this program, store the function for the function name Y_1. The initial guess is entered as the initial value for x.   This program will go through the loop 5 times. The number of times through the loop may be changed by changing the While statement.

In Exercises 27-30, use the program NEWT to find the real roots of the equations.

27. $x^2 - 14 = 0$

28. $14 - x^3 = 0$

29. $x^4 - 5x^2 - 8 = 0$

30. $x^3 - 7x^2 + 8 = 0$

Chapter 14

Integrals

14.1 The Indefinite Integral
(Washington, Section 25.2)

A concept just as important as finding the derivative of a function is that of finding the function if we know the derivative. This process is called finding the **antiderivative**. Since the derivative of the function $3x^2$ is $6x$, we say that the antiderivative of $6x$ is $3x^2$. But $3x^2 + 1$, $3x^2 + 2$, $3x^2 - 5$, etc. are also antiderivatives of $6x$. In reality there are an infinite number of antiderivatives of $6x$, all differing by a constant. Thus, when writing the antiderivative of $6x$ we write $3x^2 + C$ where C is a constant. (When we take the derivative of $3x^2 + C$, the derivative of C is zero.) In some cases, enough information is given to allow us to determine C.

A notation used to indicate that we are to find the antiderivative of a function $f(x)$ is $\int f(x)\,dx$ and is called an **indefinite integral.** If the derivative of $F(x)$ equals $f(x)$ $[F'(x) = f(x)]$, then the indefinite integral of $f(x)$ equals the function $F(x)$. We write this as

$$\int f(x)\,dx = F(x) + C$$

The function $f(x)$ is called the **integrand** and C is called the **constant of integration.** The process of finding the indefinite integral is called **integration.** Since the antiderivative of $6x$ is $3x^2 + C$, we write that the integral of $6x$ equals $3x^2 + C$:

$$\int 6x\,dx = 3x^2 + C.$$

Our interest is to investigate the graph of the antiderivatives of functions and to find the particular graph that satisfies certain given conditions.

Example 14.1: Determine the indefinite integral of $y = 2x - 3$ and graph the resulting function for $C = -2, 0, 2, 4,$ and 6. Then, determine the equation of the function that passes through the point $(3, 2)$?

<u>Solution:</u>   1. First determine the indefinite integral (or antiderivative) of $2x - 3$ which gives

$$\int (2x - 3)dx = x^2 - 3x + C$$

2. Next graph $y = x^2 - 3x + C$ for $C = -2$, $C = 0$, $C = 2$, $C = 4$, and $C = 6$. It may be helpful here to replace C by the list $\{-2, 0, 2, 4, 6\}$. (See Procedure C11) The resulting graphs are given in Figure 14.1.

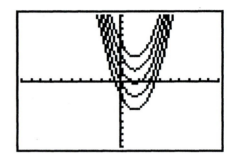

y:[–10, 10] x:[–10, 10]

Figure 14.1

3. The five different graphs lie above each other with the lowest graph given by the smallest value of C. Evaluate the function at $x = 3$ (see Procedure G13) and use the up and down arrow keys to jump from graph to graph until the value 2 is obtained for y. Note which graph this point is on.   In this case, the point $(3, 2)$ is on the third graph where $C = 2$. Thus, the antiderivative that meets the given condition is $y = x^2 - 3x + 2$.

In Example 14.1, it was relative easy to find the right curve. In other cases it may be necessary to zoom in on the graph in the area of the point or to change the viewing rectangle. If one of the graphs do not fall exactly on the given point, we estimate a value of C and check by graphing the function for this value of C.

A set of curves obtained by graphing a function for different values of a constant contained in the function is called a **family of curves**. In Example 14.1, the indefinite integral of a function gives a family of curves. In many problems, we have certain conditions (such as a known point on the graph) that allow us to find a particular curve.   The graphing calculator can be used to determine the constant of integration if the indefinite integral is known.

<u>**Procedure G27.**</u> **To find the constant of integration for an indefinite integral.**

This procedure assumes that the general antiderivative is known.
1. Store a value of 0 for C. $(0 \rightarrow C)$
2. Store the antiderivative function with C attached for a function name and graph.
3. Given the information that $y = y_0$ when $x = x_0$ and with the graph of the antiderivative (for $C = 0$) on the screen, evaluate the antiderivative function at $x = x_0$ (Procedure G13).   With the cursor on the graph of the antiderivative, read the corresponding y-value, y_a. The constant of integration is the difference between the given y-value y_0 and the value from the graph y_a. That is $C = y_0 - y_a$.
4. To check your answer, store this value for C, graph the antiderivative function, and evaluate at $x = x_0$.

♦ Example 14.2: *(Washington, Section 25.2, Example 8)* Find y in terms of x, given that $\dfrac{dy}{dx} = 3x - 1$ and the curve passes through the point $(1, 4)$.

Solution:
1. First determine the general antiderivative by finding the indefinite integral of $3x - 1$.

$$\int (3x - 1)\, dx = \frac{3}{2}x^2 - x + C$$

2. Follow Procedure G27: store 0 for C, enter $\dfrac{3}{2}x^2 - x + C$ for a function name and graph.

3. Evaluate the function at $x = 1$ (Procedure G13). The value obtained from the graph for y_a is 0.5. We desire that the curve pass through the point $(1, 4)$, so the given y-value, y_0, is 4. Thus the constant is $C = 4.0 - 0.5 = 3.5$. The answer is

$$\int (3x - 1)\, dx = \frac{3}{2}x^2 - x + 3.5$$

If the antiderivative of the function is not known, the graphing calculator provides a method of obtaining a graph of one of the antiderivatives of the function. We may then determine a constant to approximate the graph of a particular antiderivative.

Procedure G28. To graph a particular antiderivative of a function when the general antiderivative is unknown.

1. Store a value of 0 for C.
2. Store the function we wish to find the antiderivative of for a function name.
3. Place the cursor following another function name and enter fnInt(

To obtain the function fnInt:

On the TI-85 or 86:	On the TI-82 or 83:
Press the key CALC	Press the key MATH
From the menu, select fnInt	From the menu, select 9:fnInt(

The form of the integral function fnInt is

fnInt(function name, variable, smaller value, larger value)

In this case, store fnInt(fn,X,0,X) + C for the second function where fn is the function or name of the function that we have previously stored and X is the variable x.

(Example: To graph an antiderivative of $f(x) = 3x^2$, let $Y_1 = 3x^2$ and let $Y_2 = $ fnInt(Y_1,X,0,X) + C.)

4. Graph the function Y_2. The graph represents the antiderivative function plus a constant equal to the value of the antiderivative at $x = 0$. (Other members of the family of antiderivative functions may be graphed by storing different values for C).

5. To find the constant, given that $y = y_0$ when $x = x_0$, with the graph of the antiderivative on the screen, evaluate it at $x = x_0$ (Procedure G13) and read the y-value y_a. Subtract this y-value from the given y-value y_0 to obtain a new value for C. $(C = y_0 - y_a)$

6. To check your answer, store the value obtained in Step 5 for C, graph the resulting function, and evaluate at $x = x_0$.

Note: The process of graphing the antiderivative of a function is slow because of the number of calculations involved.

Example 14.3: Graph the indefinite integral $\int \dfrac{1}{x^2 + 2}\, dx$ subject to the condition that $y = 4.00$ when $x = 2.00$ Then, determine the value of the indefinite integral when $x = 0.500$ and when $x = 2.60$.

Solution:

1. Follow Procedure G28 to graph the antiderivative -- be sure to store 0 for C and turn off the function itself so it does not graph.

2. Determine the y_a-value when $x = 2$. The graph is given in Figure 14.2. We find this to be .67551. Subtracting this from the given $y = 4$ we conclude that C is approximately $4 - .67551$ or about 3.32.

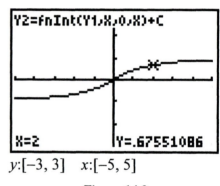

y:[−3, 3]   x:[−5, 5]

Figure 14.2

3. If we now store 3.32 for C, we should obtain a graph for the antiderivative that has a value of 4 when $x = 2$.

4. From this graph for the antiderivative we can determine the value of the antidervative at $x = 0.5$ and 2.6 by evaluating the graph at these x-values. We find that at $x = 0.5$ the antiderivative has a value of 3.56 and at $x = 2.6$ a value of 4.08.

Exercise 14.1

In Exercises 1-10, determine the indefinite integral of the given function and graph the resulting function for the given values of C. Use the graphs to find the particular equation of that function that passes through the point P.

1. $f(x) = 3x - 7$ $C = -4, -2, 0, 2, 4, 6$ $P(2, -4)$

2. $f(x) = 5 - 4x$ $C = -4, -2, 0, 2, 4, 6$ $P(3, -5)$

3. $f(x) = x^2 + 2x$ $C = -5, -3, -1, 1, 3, 5$ $P(3, 15)$

4. $f(x) = 2x - 3x^2$ $C = -5, -3, -1, 1, 3, 5$ $P(2, 1)$

5. $f(x) = x^3$ $C = -5, -3, -1, 1, 3, 5$ $P(-2, 7)$

6. $f(x) = 3x^2 - 4x^3$ $C = -5, -3, -1, 1, 3, 5$ $P(1, 5)$

7. $f(x) = 4(x^2 - 2)^3(2x)$ $C = -4, -2, 0, 2, 4, 6$ $P(2, 14)$

8. $f(x) = 6(1 - x^3)^5(3x^2)$ $C = -4, -2, 0, 2, 4, 6$ $P(-1, 68)$

In Exercises 9-14, for each indefinite integral, use the methods of this section to find the graph of the function that passes through the given point P. Then, use the graph to evaluate the antiderivative at the given value of x. In Exercises 9-12, find the integral exactly using the value for C that you found, then evaluate at the given x-value to check your answer.

9. $\int (3x^2 + 2x)dx$ $P(2, 7)$ Evaluate at $x = 4$.

10. $\int (4 - 6x^2)dx$ $P(1, 5)$ Evaluate at $x = 2.5$.

11. $\int 8(3x + 5)^7(3)dx$ $P(-2, 15)$ Evaluate at $x = -1.2$.

12. $\int (3 - 2x^2)^6(4x)dx$ $P(1, \dfrac{36}{7})$ Evaluate at $x = 0.55$.

13. $\int \dfrac{x}{x^2 - 3}dx$ $P(2.5, 7.3)$ Evaluate at $x = 5.2$.

14. $\int x\sin x^2\,dx$ $P(\dfrac{\pi}{4}, 6.7)$ Evaluate at $x = \dfrac{\pi}{2}$.

♦ 15. *(Washington, Exercises 25.2, #39)* The time rate of change of electric current in a circuit is given by $di/dt = 4t - 0.6t^2$. Use graphical means to find the expression for the current as a function of time if $i = 2.5$ A when $t = 1.0$ s. Then determine the positive time when the current is 2.2 A.

♦ 16. *(Washington, Exercises 25.2, #40)* The rate of change of frequency f of an electronic oscillator with respect to the inductance L is $\dfrac{df}{dL} = 80(4 + L)^{-\frac{3}{2}}$. Use graphical means to find the expression for the frequency as a function of inductance if $f = 75$ Hz when $L = 0$ H.

Exercise 17 is related to the material in Section 14.2 and should be considered and discussed prior to doing the material in Section 14.2.

17. Assume that as a contestant on a TV game show you are placed on an island in the middle of shark infested waters in the Pacific Ocean. The island is of irregular shape and you have in your possession a supply of food and water, a 100-meter tape measure, some wooden stakes, and plenty of string. To get off the island and back to civilization before your food runs out, you must determine the area of the island to within 100 square meters. Explain how you would accomplish this task. How you could obtain the area to a higher degree of precision?

14.2 Area Under a Curve
(Washington, Section 25.3)

We next consider the area between the graph of a function $y = f(x)$, the x-axis, and the two lines, $x = a$ and $x = b$. This area is very closely related to integration and is an important concept in the understanding of integrals. One way of approximating such an area is by the use of rectangles as shown in the Figure 14.3. The interval from $x = a$ to $x = b$ along the x-axis is subdivided into smaller intervals called subintervals. If we divide the large interval into N equal subintervals, then each subinterval has a length of $(b - a)/N$ and is the base of a rectangle. The notation Δx is used to denote the length of each of these subintervals. When using the calculator, we will use H to denote Δx. Thus

$$\frac{b-a}{N} = \Delta x = H$$

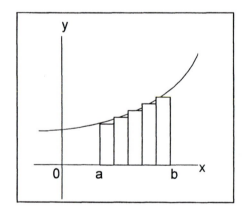

The height of each rectangle is determined by selecting an x-value in each subinterval and evaluating the function $f(x)$ at this value. The height depends on the value of x and on the function used to calculate the height. The area of each rectangle is found by multiplying the base of each rectangle by its height. The sum of the areas of all the rectangles with the subintervals as bases is an approximation to the total area. For convenience, we will determine the heights of the rectangles by evaluating the function at either the left end or the right end of each subinterval.

Figure 14.3

Example 14.4: Use 3 rectangles to estimate the area under the graph of $y = x^2 + 2$, above the x-axis, and between the lines $x = 1$ and $x = 4$. Do calculations (by hand) by first using the left end of each subinterval and then by using the right hand of each subinterval.

Solution: 1. The graph of $y = x^2 + 2$, with the left end of each subinterval used to determine the height of the rectangle, is shown in Figure 14.4. To do these calculations we divide the interval from $x = 1$ to $x = 4$ into 3 equal subintervals, each of one unit length

$$\frac{4-1}{3} = 1 = H$$

2. The left endpoints of each subinterval are 1, 2, and 3. Evaluating the function at the left endpoints gives $f(1) = 3$, $f(2) = 6$, and $f(3) = 11$ for the heights of the rectangles. The area of each rectangle is found by multiplying its width 1 by its height. The area of the first rectangle is $3 \times 1 = 3$, the next is $6 \times 1 = 6$, and the third is $11 \times 1 = 11$. The sum is

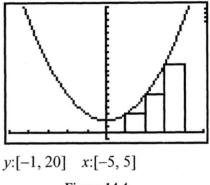

y:[−1, 20] x:[−5, 5]

Figure 14.4

$$3 \times 1 + 6 \times 1 + 11 \times 1 = 3 + 6 + 11$$
$$= 20 \text{ square units.}$$

3. For the right end points we evaluate the function at 2, 3, and 4. This gives $f(2) = 6$, $f(3) = 11$, and $f(4) = 18$. The area of the first rectangle is $6 \times 1 = 6$, the second $11 \times 1 = 11$, and the third $18 \times 1 = 18$. The sum is

$$6 \times 1 + 11 \times 1 + 18 \times 1 = 6 + 11 + 18$$
$$= 35 \text{ square units.}$$

For this function the rectangles using the left hand endpoints all lie under the graph and give a lower bound for the areas -- a value that is less than the exact area. Thus, the exact area of this region must be greater than 20 square units. With the right hand endpoints, the area under the curve lies within the total area of all the rectangles and the calculated area must be an upper bound -- an area that is greater than the exact area. The exact area of the region is less than 35 square units.

In Example 14.4, we found the lower and upper bound on the area. The estimate of the area, may be improved by taking a larger number of rectangles (a smaller base of each rectangle). However, this would require many more calculations. Such calculations are better performed using a calculator or computer.

We now consider a program for the graphing calculator that will both draw the rectangles and find the area of the rectangles. In this program we will need a few additional programming procedures. Information on programming may be found in Sections 2.5 and 13.2 of this manual.

Procedure P14. To display a graph from within a program.

While in programming edit mode (at a step in program):

On the TI-82 or 83:

1. Press the key PRGM
2. Select I/O
3. Select 4:DispGraph

On the TI-85 or 86:

1. Select I/O
2. Select DispG

Procedure P15. Drawing a line on the screen from within a program
(Line command).

The form of the line command is **LINE**(*a, b, c, d*)

While in programming edit mode (at a step in program):

On the TI-82 or 83:

1. Press the key DRAW
2. With DRAW highlighted, select 2:LINE

On the TI-85 or 86:

1. Press the key GRAPH
2. Select DRAW (after pressing MORE)
3. Select LINE

where a and b are the coordinates of one point and *c* and *d* are the coordinates of a second point. This command draws a line from (*a*, *b*) to (*c*, *d*).

Example: LINE(0,2,x, Y_1) draws a line from the point (0,2) to the point (*x*, *f(x)*).

Procedure P16. To temporarily halt execution of a program.
(Pause command)

While in programming edit mode (at a step in program):

On the TI-82 or 83:

1. Press the key PRGM
2. With CTL highlighted, select 8:PAUSE

On the TI-85 or 86:

1. Select CTL
2. Select Pause (after pressing MORE twice)

The Pause command when used in a program halts execution of the steps of a program until the ENTER key is pressed.

The following program asks for values of *A*, *B*, and *N* to be entered and then draws *N* rectangles using the left end points of each subinterval. After the graph is drawn, press ENTER to see the value for the area of the *N* rectangles. The arrow ($\rightarrow$) refers to the STO▷ key. For information on the While... statement see Procedure P10. Before executing the program the function, must be stored for Y_1.

PROGRAM: ARECTG1
ClrDraw
Prompt A
Prompt B
Prompt N
$0 \to S$
$(B-A)/N \to H$
DispGraph
$A \to x$
While $(x<B)$
LINE$(x, 0, x, Y_1)$
(*program continued in next column*)

(*program continued from previous column*)
LINE$(x + H, 0, x + H, Y_1)$
LINE$(x, Y_1, x + H, Y_1)$
$S + Y_1*H \to S$
$x + H \to x$
End
DispGraph
Pause
Disp "AREA IS"
Disp S

<u>Example 14.5</u>: Use the program ARECTG1 to estimate the area under the graph of $y = x^2 + 2$, above the x-axis, and between $x = 1$ and $x = 4$ using left end points of the subintervals. Do the calculation using 3, 6, and 12 rectangles.

<u>Solution:</u>

1. Make sure the function $x^2 + 2$ is entered for Y_1 and all other functions are deleted or turned off and be sure the program ARECTG1 is correctly entered into your calculator. Execute the program (See Procedure P2). After the graph is drawn, press ENTER to see the calculation for the area.

2. Enter 1 for A, 4 for B, and 3 for N. The resulting graph is the same as in Figure 14.3 and the area is 20 square units.

3. Execute the program a second time using $A = 1$, $B = 4$, and $N = 6$. This time the area of the rectangles is 23.375 square units.

4. Execute the program a third time using $A = 1$, $B = 4$, and $N = 12$. This time the area of the rectangles is 25.15625 (or 25.2) square units. (The exact area is 27 square units.)

Exercise 14.2

In Exercises 1-8, use the program ARECTG1 to graph the given function and determine the specified areas for the given values of N.

1. The area bounded by the function $y = 2x + 2$, the x-axis, $x = 1$ and $x = 5$. Determine the area of this region by using the calculator with $N = 2, 4$, and 8, and also by hand by finding the area of the resulting trapezoid.

2. The area bounded by the function $y = 5 - x$, the x-axis, $x = -2$, and $x = 4$. Determine the area of this region by using the calculator with $N = 2, 4$, and 8 and also by hand by finding the area of the resulting trapezoid.

3. The area bounded by the function $y = 9 - x^2$, the x-axis, $x = 0$, and $x = 3$ for $N = 2, 4, 8,$ and 16. Are the values obtained upper bounds or lower bounds for the exact area? Make a guess as to the exact area (you may wish to use larger values for N).

4. The area bounded by the function $y = x^2 + 5$, the x-axis, $x = -3$, and $x = 3$ for $N = 2, 4, 8,$ and 16. Are the values obtained upper bounds or lower bounds for the exact area? Make a guess as to the exact area (you may wish to use larger values for N).

5. The area bounded by the function $y = x^3$, the x-axis, $x = 0$, and $x = 3$ for $N = 2, 4, 8,$ and 16. Are the values obtained upper bounds or lower bounds for the exact area? Make a guess as to the exact area (you may wish to use larger values for N).

6. The area bounded by the function $y = 27 - x^3$, the x-axis, $x = -1$, and $x = 3$ for $N = 2, 4, 8,$ and 16. Are the values obtained upper bounds or lower bounds for the exact area? Make a guess as to the exact area (you may wish to use larger values for N).

7. The area bounded by the function $y = \dfrac{1}{x^2}$, the x-axis, $x = 1$, and $x = 8$ for $N = 2, 4, 8,$ and 16. Are the values obtained upper bounds or lower bounds for the exact area? Make a guess as to the exact area (you may wish to use larger values for N).

8. The area bounded by the function $y = \sqrt{x + 2}$, the x-axis, $x = -2$, and $x = 2$ for $N = 2, 4, 8,$ and 16. Are the values obtained upper bounds or lower bounds for the exact area? Make a guess as to the exact area (you may wish to use larger values for N).

9. Modify the program ARECTG1 so that the right end points of each subinterval will be used to calculate the area of the rectangles. Call the resulting program ARECTG2.

10. Modify the program ARECTG1 so that the mid-point of each subinterval is used to calculate the area of the rectangles. Call the resulting program ARECTGM.

11. Write a program ARECT1 that will calculate the area using rectangles and left hand end points, but which does not show the rectangles or graphs.

12. Write a program to calculate the area of rectangles using the left end points of each subinterval and the right end points of each subinterval and print out both values. This program should not show the rectangles or graphs.

13. Use the program ARECTG2 to do Problem 1. Using these results and the results of Problem 1, give upper and lower bounds for the exact area.

14. Use the program ARECTGM to do Problem 2.

15. Use the program ARECTG2 to do Problem 3. Using these results and the results of Problem 3, give upper and lower bounds for the exact area.

16. Use the program ARECTGM to do Problem 4.

17. Use the program ARECTG2 to do Problem 5. Using these results and the results of Problem 5, give upper and lower bounds for the exact area.

18. Use the program ARECTGM to do Problem 6.

19. Use the program ARECTG2 to do Problem 7. Using these results and the results of Problem 7, give upper and lower bounds for the exact area.

20. Use the program ARECTGM to do Problem 8.

21. Use the program ARECTG1 to estimate the area bounded by $y = 3x^2$, the x-axis, $x = 1$ and $x = 3$. Use N values of 5, 10, 50, and 100.
    (a) Are the areas for these N-values approaching a limit? If so, what do you think it is?
    (b) Find the antiderivative of $3x^2$. Subtract the value of the antiderivative at $x = 1$ from the value of the antiderivative at $x = 3$. How does this number compare to your answer in part (a)?

22. Use the program ARECTG1 to estimate the area bounded by $y = 4 - x^2$, the x-axis, $x = -2$ and $x = 2$. Use N values of 5, 10, 50, and 100.
    (a) Are the areas for these N-values approaching a limit? If so, what do you think it is?
    (b) Find the antiderivative of $4 - x^2$. Subtract the value of the antiderivative at $x = -2$ from the value of the antiderivative at $x = 2$. How does this number compare to your answer in part (a)?

23. Use the program ARECTG1 to estimate the area bounded by $y = x^3$, the x-axis, $x = 0$ and $x = 4$. Use N values of 5, 10, 50, and 100.
    (a) Are the areas for these N-values approaching a limit? If so, what do you think it is?
    (b) Find the antiderivative of x^3. Subtract the value of the antiderivative at $x = 0$ from the value of the antiderivative at $x = 4$. How does this number compare to your answer in part (a)?

24. Use the program ARECTG1 to estimate the area bounded by $y = x^2 + 5$, the x-axis, $x = -1$ and $x = 2$. Use N values of 5, 10, 50, and 100.
    (a) Are the areas for these N-values approaching a limit? If so, what do you think it is?
    (b) Find the antiderivative of $x^2 + 5$. Subtract the value of the antiderivative at $x = -1$ from the value of the antiderivative at $x = 2$. How does this number compare to your answer in part (a)?

25. Use the program ARECTG1 to estimate the area bounded by $y = x^2 - 16$, the x-axis, $x = 0$ and $x = 3$ for $N = 2, 4$, and 8. Does the calculated area of the rectangles come out positive or negative? Can you explain why?

26. Use the program ARECTG1 to estimate the area bounded by $y = x^2 - 4x - 5$, the x-axis, $x = 0$ and $x = 4$ for $N = 2, 4$, and 8. Does the calculated area of the rectangles come out positive or negative? Can you explain why?

27. (a) Use the program ARECTG1 to estimate the area bounded by $y = 8x - x^2$, the x-axis, $x = 0$ and $x = 8$ for $N = 10$ and 50.
    (b) Repeat for the interval $x = 0$ to $x = 12$. Explain the results obtained.
    (c) Repeat for the interval $x = 0$ to $x = 8$ and for the interval $x = 8$ to $x = 12$. How do these two values relate to the answer in part (b)?

28. (a) Use the program ARECTG1 to estimate the area bounded by $y = x^2 - 5x - 6$, the x-axis, $x = 0$ and $x = 6$ for $N = 10$ and 50.
    (b) Repeat for the interval $x = 0$ to $x = 10$. Explain the results obtained.
    (c) Repeat for the interval $x = 0$ to $x = 6$ and for the interval $x = 6$ to $x = 10$. How do these two values relate to the answer in part (b)?

29. Use the program ARECTG1 to estimate the area bounded by $y = \sin x$, the x-axis, $x = 0$ and $x = \dfrac{\pi}{2}$ for $N = 5$, 10, 50 and 100. (Calculator should be in radian mode.)

30. (a) Use the program ARECTG1 to estimate the area bounded by $y = \dfrac{1}{x}$, the x-axis, $x = 1$ and $x = 2$ for $N = 5$, 10, 50, and 100. Then find ln 2 on your calculator using the LN key. How do the results compare?

    (b) Repeat part (a) for the interval $x = 1$ to $x = 3$ and compare the results to ln 3.

    (c) What can you conclude about the area under the graph of $y = \dfrac{1}{x}$ from $x = 1$ to $x = K$ and the value of ln K?

14.3 The Definite Integral and Trapezoidal and Simpson's Rules
(Washington, Sections 25.4, 25.5, and 25.6)

From Section 14.2, we concluded that we could estimate the area under a curve, above the x-axis and between two values of x by considering the areas of a large number of rectangles in this region. The accuracy to which we can calculate the area is limited only by the number of rectangles selected and the corresponding calculations required. If we could take an infinite number of rectangles, we would obtain such an area exactly. This area is also related to the antiderivative (or integral) of the given function (see Exercises 21-24 of Exercise Set 14.2). We sum this up in the following statement: If $F(x)$ is the antiderivative of $f(x)$, then

$\displaystyle\int_a^b f(x)dx$ represents the area of the region bounded by the graph of $y = f(x)$, the x-axis, $x = a$ and

$x = b$. An integral of the form $\displaystyle\int_a^b f(x)dx$ is called a **definite integral**. The value a is referred to

as the **lower limit** of integration and b is referred to as the **upper limit** of integration.

This definite integral may be evaluated by finding the antiderivative of $f(x)$ and subtracting the value of the antiderivative at $x = a$ from the value of the antiderivative at $x = b$. This last statement is summarized as

$$\int_a^b f(x)dx = F(x)\Big|_a^b = F(b) - F(a) \qquad \text{where } F'(x) = f(x) \quad \text{[Equation 1]}$$

The notation $F(x)\Big|_a^b$ means we first evaluate $F(x)$ at $x = b$, then subtract the value of the function at $x = a$.

If we can find antiderivative of $f(x)$ exactly then we can determine the value of the definite integral exactly. However, it is often difficult or impossible to determine the antiderivative of a function directly. Knowing that the definite integral represents an area allows us to approximate the value of a definite integral by numerical means. We have already used one such method -- that of estimating areas by using rectangles. The method of using rectangles is not good because it takes a large number of calculations to achieve a high precision. In this section we consider two other methods of approximating this area.

Even though we are considering the definite integral as representing the area under a graph, applications involving the definite integral often have no direct relationship to an area. For example, definite integrals may be used to find volumes, amount of work or force, and centers of gravity.

One method of estimating the value of a definite integral is to think of it as an area and divide the interval from $x = a$ to $x = b$ up into equal subintervals each of length Δx as we did with rectangles. This time we use trapezoids and come up with a formula for estimating the area. We will not develop this formula here, but only state it as the **trapezoidal rule**:

$$A_T = \frac{H}{2}(y_0 + 2y_1 + 2y_2 + ... + 2y_{n-1} + y_n) \qquad \text{[Equation 2]}$$

where $H = \Delta x =$ the length of a subinterval $= \dfrac{b-a}{2}$ and $y_0 = f(a)$, $y_1 = f(a + H)$,

$y_2 = f(a + 2H)$, etc. The last y-value $y_n = f(b)$. The y-values represent the distance from the x-axis to the graph of the function at the ends of the each subinterval.

The following program estimates the value of a definite integral using the trapezoidal rule. Remember to store the function being evaluated for a function name (Y_1 in this case) before executing the program.

```
PROGRAM: TRAPRULE
0 → S
Prompt A
Prompt B
Prompt N
(B–A)/N → H
A → x
Y₁ → S
For (K,2,N–1)
(program continued in next column)
```

(*program continued from previous column*)
```
A + K*H → x
S + 2Y₁ → S
End
B → x
Y₁ + S → S
S*H/2 → S
Disp "Value is"
Disp S
```

If the exact value of the definite integral is known, we may compare it with the answer found using the trapezoidal rule and may calculate the percentage error. The percentage error is found using the formula

$$\text{Percent Error} = \left| \frac{\text{Exact value} - \text{Calculated value}}{\text{Exact value}} \times 100 \right| \qquad \text{[Equation 4]}$$

Example 14.6: Use the program TRAPRULE to estimate $\int_1^4 (x^2 + 2)dx$. Use $N = 6$, 12, and 100. Compare answers to those in Example 14.3. Determine the percentage error in each case.

Solution: 1. Make sure that the function $x^2 + 2$ is entered for Y_1 and all other functions are deleted or turned off and be sure the program TRAPRULE is correctly entered into your calculator. Execute the program.
2. Enter 1 for A, 4 for B, and 6 for N. The result is 27.125. The value obtained in Example 14.3 for $N = 6$ was 23.375. This value is much closer to the exact value of 27.0 and gives a percentage error of

$$\left| \frac{27.0 - 27.125}{27.0} \times 100 \right| = 0.463\%.$$

3. Enter 1 for A, 4 for B, and 12 for N. This time the result is 27.03125, which gives a percentage error of 0.116%. In Example 14.3 the value for $N = 12$ was 25.25625.
4. Enter 1 for A, 4 for B, and 100 for N. The result is 27.00045, which is within 0.001 units of the exact value for a percentage error of 0.0017%!

In Example 14.6, we see that the trapezoidal rule is much more accurate than using rectangles. A much smaller value of N needs to be taken to obtain a reasonable degree of accuracy.

An even more accurate method of estimating a definite integral is Simpson's rule. Simpson's rule approximates the area under a curve by approximating pieces of the curve using parabolas and considering the area under the parabolas. Again we will not develop the formula here, but just state **Simpson's rule**:

$$A_s = \frac{H}{3}(y_0 + 4y_1 + 2y_2 + 4y_3 + 2y_4 + \ldots + 4y_{n-1} + y_n) \qquad \text{[Equation 5]}$$

Again $H = \Delta x = \dfrac{b-a}{N}$ and $y_0 = f(a)$, $y_1 = f(a + H)$, $y_2 = f(a + 2H)$, etc. When using Simpson's rule, N must be an even number. The program for Simpson's rule is left as an exercise.

Exercise 14.3

In Exercises 1-8, use the program TRAPRULE to estimate the definite integrals. Compare your answers to those found to the related problems in Exercise 14.2.
Use Equation 1 and the antiderivative to find the exact value and then use Equation 4 to find the percentage error in your values.

1. $\displaystyle\int_{1}^{5}(2x+2)dx$ for $N = 4$ and 8

2. $\displaystyle\int_{-2}^{4}(5-x)dx$ for $N = 4$ and 8

3. $\displaystyle\int_{0}^{3}(9-x^2)dx$ for $N = 4$ and 16

4. $\displaystyle\int_{-3}^{3}(x^2+5)dx$ for $N = 4$ and 16

5. $\displaystyle\int_{0}^{3}x^3\,dx$ for $N = 4$ and 16

6. $\displaystyle\int_{-1}^{3}(27-x^3)dx$ for $N = 4$ and 16

7. $\displaystyle\int_{1}^{8}\frac{1}{x^2}dx$ for $N = 8$ and 16   (Exact value is 0.875)

8. $\displaystyle\int_{-2}^{2}\sqrt{x+2}\,dx$ for $N = 8$ and 16   (Exact value is $\dfrac{16}{3}$.)

9. Evaluate $\displaystyle\int_{1}^{2}(4x^3+2)dx$ using the program TRAPRULE for $N = 2, 4, 8,$ and 16. Determine

    the exact value and the percentage error in each case. When we double the value of N by what factor does this decrease the percentage error? Does this agree with your results in Exercises 1-8?

10. Evaluate $\displaystyle\int_{0}^{1}(4x-x^4)dx$ using the program TRAPRULE for $N = 2, 4, 8,$ and 16. Determine

    the exact value and the percentage error in each case. When we double the value of N by what factor does this decrease the percentage error? Does this agree with your results in Exercises 1-8?

11. Use the program TRAPRULE to find each of the integrals $\int_{-2}^{2}(x^3 - 4x)dx$, $\int_{-2}^{0}(x^3 - 4x)dx$,

and $\int_{0}^{2}(x^3 - 4x)dx$ for $N = 20$. Explain the results. Write an equation involving these three integrals.

12. Use the program TRAPRULE to find each of the integrals $\int_{-3}^{3}(9x - x^3)dx$, $\int_{-3}^{0}(9x - x^3)dx$,

and $\int_{0}^{3}(9x - x^3)dx$ for $N = 20$. Explain the results. Write an equation involving these three integrals.

13. Use the program TRAPRULE to find the integral $\int_{1}^{3}6x^3\,dx$ where $A = 1$ and $B = 3$ and the

integral $\int_{3}^{1}6x^3\,dx$ where $A = 3$ and $B = 1$. In both cases use $N = 20$. What is the relationship of these two integrals. Write an equation involving these two integrals.

14. Use the program TRAPRULE to find the integral $\int_{1}^{4}\frac{1}{x^2}dx$ where $A = 1$ and $B = 4$ and the

integral $\int_{4}^{1}\frac{1}{x^2}dx$ where $A = 4$ and $B = 1$. In both cases use $N = 20$. What is the relationship of these two integrals. Write an equation involving these two integrals.

15. Write a program called SIMPRULE for calculating the definite integral using Simpson's rule. It should be similar to the program TRAPRULE but take into account the differences in the formula.

16. Use the program SIMPRULE to evaluate $\int_{1}^{4}\frac{1}{x}dx$ for $N = 10$. Compare the results to the

exact answer ln 4. What is the percentage error?

17.-24. Do each of the Exercises 1-8 using the program SIMPRULE instead of TRAPRULE.

Do Exercises 25-28 using both TRAPRULE and SIMPRULE programs.

♦ 25. *(Washington, Exercises 25.4, #33)* The work W (in ft-lb) in winding up an 85-ft cable is

given by $W = \int_0^{85}(1000 - 5x)dx$. Find W. Use $N = 10$.

♦ 26. *(Washington, Exercises 25.4, #36)* The total force (in Newtons) on the circular end of a

water tank is given by $F = 19600\int_0^6 y\sqrt{36 - y^2}\, dy$. Find F. Use $N = 10$.

♦ 27. *(Washington, Exercises 25.5, #15)* A force F that a distributed electric charge has on a point

charge is $F = k\int_0^2 \dfrac{dx}{(4 + x^2)^{\frac{3}{2}}}$ where x is the distance along the distributed charge and k is a

constant. Find F in terms of k. Use $N = 10$.

♦ 28. *(Washington, Exercises 25.6, #16)* The average value of the electric current i_{av}, in amperes

in a circuit for the first 5.5 s is given by $i_{av} = \dfrac{1}{5.5}\int_0^{5.5}(4t - t^2)^{0.2}\, dt$. Find i_{av} using $N = 10$.

14.4 More on Integration and Areas
(Washington, Section 26.2)

Many graphing calculators have a built in function that numerically calculates the area under a
graph and also the definite integral. In this section, we make use of this built in function along
with some other features of the calculator to determine areas.

<u>**Procedure G29.**</u> **Function to determine the area under a graph.**

With the graph of a function on the screen:

On the TI-82 or 83:
1. Press the key CALC
2. Select 7: $\int f(x)dx$
3. Move the cursor to the first x-value (Lower Limit) and press ENTER.
4. Move the cursor to the second x-value (Upper Limit) and press ENTER.

On the TI-85 or 86:
1. Select MATH (after pressing MORE)
2. Select $\int f(x)$
3. Move the cursor to first x-value (Lower Limit) and press ENTER.
4. Move the cursor to the second x-value (Upper Limit) and press ENTER.

The area between the curve, the x-axis, and the two x-values will be shaded and will be given at the bottom of the screen.

Procedure C20. Function to calculate definite integral.

The form of the function if **fnInt**(*y, variable, a, b*)

On the TI-82 or 83:

1. Press the key MATH
2. Select 9:fnInt(

On the TI-85 or 86:

1. Press the key CALC
2. Select fnint

where y is a function or name of a function, variable is the variable used in the function, and a and b are the lower and upper limits of integration.

(Example: fnInt(x^3 , x, 1,3) is used to evaluate $\int_1^3 x^3 dx$.)

Example 14.7: Use a graphing calculator to evaluate $\int_1^4 (x^2 + 2)dx$.

Solution:

1. Enter $x^2 + 2$ for Y_1 and graph this function. In this case, it better to use the decimal viewing rectangle but with Ymax = 20.
2. Use Procedure G27 to determine the area between $x = 1$ and $x = 4$.
    or
    Use Procedure C20 to determine the definite integral between $x = 1$ and $x = 4$.

With either procedure, we obtain the exact answer of 27.0. Had we not used the decimal viewing rectangle, it would have been difficult to enter the x-values exactly.

To find the area between two curves, we not only need to know the two curves, but to also know the points of intersection of the curves. We should apply one of the following to determine the area between two curves.

Between $x = a$ and $x = b$ with interval along x-axis:

$$\int_a^b (top\ function of\ x - bottom\ function of\ x)dx$$

Between $y = c$ and $y = d$ with interval along y-axis:

$$\int_c^d (right\ function of\ y - left\ function of\ y)dy$$

Graphing procedures could not be used on this last form unless we change the variable y to x and think of the integrand as a function of x. Example 14.6 will demonstrate a method to find the area between two curves.

Example 14.6:   Find the area between the graphs of $y = x^2$ and $y = 4 - x^2$.

Solution: 1. Enter the function x^2 for Y_1 and $4 - x^2$ for Y_2. Graph so that the whole area appears on the screen. On the standard viewing rectangle, the graph is as in Figure 14.5.

2. Use Procedure G15 to determine the left most point of intersection and enter $x \rightarrow A$ to store the x-value for the variable A.

3. Use Procedure G15 to determine the right most point of intersection and enter $x \rightarrow B$ to store this x-value for the variable B.

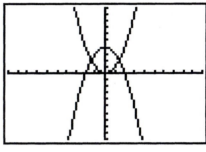

$y:[-10, 10]$ $x:[-10, 10]$

Figure 14.5

4. Use Procedure C20 to find the definite integral between the two curves and between $x = A$ and $x = B$. Use fnInt($Y_2 - Y_1, X, A, B$). This gives the area as 7.5425 rounded to five significant digits.

Exercise 14.4

In Exercises 1-8, use the built in function to determine the following definite integrals.

1. $\displaystyle\int_{1.5}^{2.7} x^2\, dx$

2. $\displaystyle\int_{-1.8}^{1.8} (x^3 - 2)\, dx$

3. $\displaystyle\int_{0.5}^{1.5} (x - \frac{1}{x})\, dx$

4. $\displaystyle\int_{-0.25}^{1.75} (x^4 - 3x^2 + 2)\, dx$

5. $\displaystyle\int_{1.15}^{1.38} 2^{x^3}\, dx$

6. $\displaystyle\int_{-0.5}^{0.5} \frac{1}{1 + x^2}\, dx$

7. $\displaystyle\int_{-5}^{5} \sqrt{25 - x^2}\, dx$

8. $\displaystyle\int_{-\sqrt{2}}^{\sqrt{2}} (x^3 + 5x^2 - 2x - 10)\, dx$

In Exercises 9-20, use the methods of this section to find the areas bounded by the indicated curves.

9. $y = 3x^2$, $y = 0$, and $x = 2.5$

10. $y = 4 - x^2$, $x = 0$, and $y = 0$

11. $y = x^2 - 4$ and $y = x + 2$

12. $y = 2x^2 + 1$ and $y = x + 7$

13. $y = x^2$, $y = 4$, and $x = 0$ $(x > 0)$

14. $y = x^3$, $y = 27$, and $x = 0$

15. $y^2 - 5y = x$ and $y = 2x$

16. $y = 3x - x^2$, $x = 0$ and $y = 1.8$

17. $y = 2\sqrt{x-2}$, $y = 0.5x$, and $y = 0$

18. $y = x^4$, $y = 5 - x$, and $y = 0$

19. $y = x^3$ and $y = 27 - 27x + 9x^2 - x^3$

20. $y = x^2 + 20$, $y = x^2 - 10x + 45$, and $y = 4x^2 - 20x + 40$ for $y < 26.25$

21. If a graph is made of the end of a building where one corner is at the origin, then the top is in the shape of the parabola given by $y = 20 + x - 0.025x^2$, the sides are the lines $x = 0.0$ and $x = 40.0$ and the bottom is along the x-axis. Find the area of the end of the building.

22. The boundary of a piece of land on the curve of a road when sketched on a coordinate system has as its boundaries the circle of radius 6 meters, the line $3x + 4y = 120$ and the x- and y-axes where x and y are measured in meters. Find the area of this piece of land.

♦ 23. (Washington, Exercises 26.2, #31) Since the vertical displacement s, velocity v, and time t, of a moving object are related by $s = \int v\,dt$, it is possible to represent the change in displacement as an area. A rocket is launched such that its vertical velocity v (in km/s) as a function of time in seconds is $v = 1.25 - 0.015\sqrt{2t+1}$. Find the change in vertical displacement from $t = 9.0$ s to 98.0 s.

♦ 24. (Washington, Exercises 26.2, #34) Using CAD (computer-assisted design), an architect programs a computer to sketch the shape of a swimming pool designed between the curves $y = \dfrac{825x}{(x^2+10)^2}$, $y = -\dfrac{825x}{(x^2+10)^2}$, $x = 0$ and $x = 6.5$. All dimensions are in meters. Find the area of the surface of the pool. If the pool is uniformly two meters deep, how many cubic meters of water are needed to fill the pool?

Appendix A

Procedures for the TI-82, TI-83, TI-84, TI-85, and TI-86 Graphing Calculators

A.1 Basic Calculator Procedures

Procedure B1. **To turn calculator on and off.**

1. To turn the calculator on, press the key: ON (in the lower left corner).
2. To turn the calculator off, first press the key: 2nd then the key: ON

Procedure B2. **To clear all memory in the calculator.**

WARNING: The following steps will erase ALL calculator memory.

1. Turn the calculator on.
2. Press the key: MEM
(Do this by pressing the key: 2nd then the key: + or 3)

On the TI-82 or 83 or 84 Plus:	On the TI-85 or 86:
From the menu, select *Reset* to erase everything and Quit to return to calculations screen.	Press the key F3 for RESET. Then, press F1 for ALL. Press F4 to erase everything and F5 to return to home screen.

Note: To check the amount of memory available, after pressing MEM key:

On the TI-82 or 83,	On the TI-86 or 85:
Press the key 1 to check memory. On the 83 Plus or 84 Plus, Press the key 2 to check memory.	Press the key F1 for RAM.

The amount of memory available is given in terms of units of memory called Bytes. The TI-83 has 27,000 bytes of RAM memory available, the TI-83 Plus has 24,000 bytes of RAM memory and 160,000 bytes of Flash ROM memory, and the TI-84 Plus has 24,000 bytes of RAM memory and 480,000 bytes of Flash ROM memory.

Procedure B3. **Adjusting the contrast of the screen.**

To adjust the contrast of the screen you will need to use the key: 2nd and the arrow keys that appear on the upper right of the keyboard.

1. To make lighter: Press the key 2nd, then press the bottom arrow key.
2. To make darker: Press the key 2nd, then press the top arrow key.

The contrast change takes place in a series of increments. Thus, the above steps may have to be repeated to obtain the contrast you desire. Note that the key 2nd must be pressed each time before pressing the arrow key.

Procedure B4. To Clear the Calculator Screen

      1. To clear the calculator screen, press the key: CLEAR twice.

      2. In some cases, pressing the CLEAR key will not clear the screen. To clear the screen, it is necessary to exit the given screen by pressing the key: QUIT

A.2 General Calculator Procedures

Procedure C1. To perform secondary operations or obtain alpha characters.

      1. To perform secondary operations:
         (a) Press the key: 2nd
         (b) Then, press the key for the desired action
            (For example: To find the square root of a number, first press the key 2nd followed by the key: x^2. (The square root symbol $\sqrt{}$ appears above and to the left of the x^2 key.) Then, enter the number and press ENTER.)

      2. To obtain alpha characters:
         (a) Press the key: ALPHA
         (b) Then, press the key for the desired character.
            (For example: To display the letter A on the screen, press the key: ALPHA then press the key that has the letter A above and to the right of the key.)
         On the TI-85 and 86, pressing the key: 2nd before pressing the ALPHA key will give lower case letters.
      Note: The alpha key may be locked down on the TI-82, 83, and 84 Plus by first pressing the key 2nd then pressing the key: ALPHA. Press the ALPHA key twice to lock down the alpha key on the TI-85 and 86.

Procedure C2. Selection of menu items

A menu on the TI-82, 83, and 84 Plus calculators appears as a list of numbers, each followed by a colon and a menu item. A menu item may be selected by:

1. Pressing the number of the item desired

or

2. Moving the highlight to the item desired by pressing the up and down cursor keys, then pressing the key ENTER.

A menu on the TI-85 and 86 calculators appears as a list on the bottom of the screen. Make a selection by pressing the function key just below the item desired. The function keys are the keys labeled F1, F2, F3, F4, and F5. In some cases, selection of a menu item will produce a secondary menu. Items are selected from the secondary menu by pressing the appropriate function key.

(Example: To find the cube root of 3.375, press the key MATH, leave MATH highlighted in the top row and select $4:\sqrt[3]{}$. Then enter the number 3.375 and press ENTER. The result is 1.5.)

In some cases pressing the same key may access more than one menu. This is indicated by more than one title appearing at the top of the screen with one of the words highlighted. The right and left arrow keys are used to move the highlight from one title to another. As this is done different menus appear on the screen. First select the correct menu, then the desired item.
(Example: To find the largest integer which is less than or equal to the number 2.15, first press the key MATH, move the highlight in the top row to NUM, select 5:int((or on the TI-82, select 4:Int). The screen appears as in Figure 1.2. Now enter the number 2.15, and press ENTER. The answer should be 2.)
To exit from a menu without selecting an option, press QUIT.

If the symbol $\rightarrow$ appears on the right of a menu, there are additional items in the menu. To see these additional items, press the key: MORE
To exit from a menu to the previous menu or to the home screen, press the key: EXIT. To exit to the home screen, press QUIT

(Example: To find the largest integer less than or equal to 2.15, press the key: MATH, then press the key F1 for NUM, and from the secondary menu, select **int** by pressing the key F4. Then enter the number 2.15 and press ENTER. The result is 2.)

Procedure C3. To perform arithmetic operations.

1. To add, subtract, multiply or divide press the appropriate key: $+,-,\times,$ or $\div$.
   On the graphing calculator screen, the multiplication symbol will appear as * and the division symbol will appear as /.
   If parentheses are used to denote multiplication it is not necessary to use the multiplication symbol. (The calculation $2(3 + 5)$ is the same as $2\times(3 + 5)$.)

2. To find exponents use the key $\wedge$
   (Example: 2^4 is entered as 2^4)
   For the special case of square we may use the key x^2.

3. After entering the expression press the ENTER key to perform the calculation.

Notes: (a) On the graphing calculator, there is a difference between a negative
number and the operation of subtraction. The key (-) is used for
negative numbers and the key – is used for subtraction.
(b) Enter numbers in scientific notation by using the key EE.
(Example: To enter 2.15×10^5, enter 2.15, press EE, enter 5.)

Procedure C4. To correct an error or change a character within an expression.

1. (a) To make changes in the current expression:
Use the cursor keys to move the cursor to the position of the character to
be changed.
(b) To make changes in the last expression entered (after the ENTER key has
been pressed) obtain a copy of the expression by pressing the key
ENTRY. Then, use the cursor keys to move the cursor to the character
to be changed.
2. Characters may now be replaced, inserted, or deleted at the position of the
cursor.
(a) To replace a character at the cursor position, just press the new character.
(b) To insert a character or characters in the position the cursor occupies,
Press the key: INS.
Then, press keys for the desired character(s).
(c) To delete a character in the position the cursor occupies, press the key:
DEL

Procedure C5. Selection of special mathematical functions.

1. To raise a number to a power:
Enter the number
Press the key ^
Enter the power
Press ENTER.

2. To find the root of a number (other than square root):

On the TI-82, 83, or 84 Plus:	On the TI-85 or 86:
Enter the root index	Enter the root index
Press the key MATH	Press the key MATH
With Math highlighted,	Select MISC
Select $\sqrt[x]{\ }$	Press MORE,
Enter the number	From the secondary menu,
Press ENTER	Select $\sqrt[x]{\ }$
	Enter the number
	Press ENTER.

3. To find the factorial of a positive integer:

On the TI-82, 83, or 84 Plus:	On the TI-85 or 86:
Enter the integer	Enter the integer
Press the key MATH	Press the key MATH
Highlight PRB	Select PROB
Select !	From the secondary menu,
Press ENTER	select !
	Press ENTER.

4. To find the largest integer less than or equal to a given value (This is called the greatest integer function and is indicated by square brackets [ ]):

On the TI-82, 83, or 84 Plus: On the TI-85 or 86:
Press the key MATH   Press the key MATH
Highlight NUM   Select NUM
Select Int   From the secondary menu,
Enter the value   select int
Press ENTER   Enter the value
                                       Press ENTER.

5. For the number π, press the key for π on the keyboard.

6. To find the absolute value of a given value:

On the TI-82:   On the TI-85 or 86:
Press the key ABS   Press the key MATH
on the keyboard.   From the menu, select NUM
Enter the value   Select abs
Press ENTER   Enter the value
                                       Press ENTER.

On the TI-83 or 84 Plus:
Press the key MATH
Highlight NUM
Select abs(
Enter the value
Press ENTER

Procedure C6. Changing the mode of your calculator.

Press the key: MODE You will see several lines. The items that are highlighted in each line set a particular part of the mode. The first two lines set the way that numbers are displayed on the screen. In the first line, Normal (or Norm) is used for normal mode, Sci for scientific notation, and Eng for engineering notation.

In the second line, Float causes numbers to be displayed in floating point form with a variable number of decimal places. If one of the digits 0 to 9 is highlighted, all calculated numbers would be displayed with that number of decimal places.

(Example: All numbers calculated using the settings in Figure 1.4 will be displayed in scientific notation rounded to 2 decimal places.)

The third line is used for trigonometric calculations and indicates if the angular measurements are in degrees or radians. Other lines on this screen will be considered later.

To change mode:

1. Use the cursor keys to move the cursor to the item desired.
2. Press ENTER. The new item should now be highlighted.
3. To return to the home screen, press the CLEAR or QUIT key.

(Example: To change the mode so that numbers are displayed in scientific notation with two decimal places, highlight Sci in the first line and press ENTER, then highlight 2 in the second line and press ENTER. Then, return to the home screen.)

Procedure C7. **Storing a set of numbers as a list.**

On the TI-82, 83, or 84 Plus:
Up to six lists (of up to 99 values each) may be stored using the names L_1, L_2, L_3, L_4, L_5, and L_6
or
On the TI-83 and 84 additional lists may be stored under a user given name up to 5 characters long and starting with a letter.

On the TI-85 or 86:
Lists (of any length) may be stored using a list name of up to eight characters. The first character of the name a must be a letter.

Method I:
1. Press the key {
2. Enter the numbers to be in the list separated by commas.
3. Press the key }
4. Press the key STO▷
5. Press the key for the list name (or enter the list name) and press ENTER.

Method I:
1. Press the key LIST.
2. Select {
3. Enter the numbers to be in the list separated by commas.
4. Select }
5. Press the key STO▷
6. Enter the list name and press ENTER.

Method II:
1. Press the key STAT
2. From the menu, select 1:EDIT
3. Move the highlight to the list and position where numbers are to be entered, and enter the number. The entered value will appear at the bottom of the screen. Press ENTER to place the number in the list.

4. When all values in the list have been entered, press QUIT to exit from the list table.

Method II:
1. Press the key LIST
2. Select EDIT
3. To enter a new list, enter the new name after Name= . On the TI-86, the numbers may be stored in a prenamed list (xStat, yStat, fStat) or we can name a new list by going to an unnamed column.

To get a blank column, move the cursor up to the name line and move the cursor to the right to a new column.

Note: A value may be changed by moving the highlight in the table to that value, entering the correct value, and pressing ENTER.

4. Enter the values as elements of the list by moving the cursor to each position and entering a value. Press ENTER after each entry. When finished, press QUIT.

Note: A value may be changed by moving the cursor to that value and entering the new value.

Procedure C8. Selecting a list and viewing a set of numbers.

Select a List:
  On the TI-82, 83, or 84 Plus:
  Press the key for the name of the list,
  or
  on the TI-83 or 84 Plus:
  Press the key LIST and with NAMES highlighted at the top, select the name of the list, press ENTER
View a List:
  With name on the screen, press ENTER

Or

1. Press the key STAT
2. Select 1:EDIT
3. Move the highlight to the list.
4. Press QUIT to exit.

On the TI-85 or 86:
Enter the name of list and press ENTER.

or

1. Press the key LIST
2. Select NAMES, then select the list name and press ENTER

   or

   Select EDIT and move the cursor to the column desired.
   On the TI-85, a list name will need to be entered or selected.

Procedure C9. To perform operations on lists.

1. To add, subtract, multiply or divide lists, select the name of a list (see Procedure C8), press the appropriate key (+, −, ×, or ÷), select the second list name and press ENTER. Lists must be of the same length in order for these operations to be performed.
2. To perform an operation on all the elements of a list, enter a value, press the appropriate key (+, −, ×, or ÷), select the list name and press ENTER.
2. Functions such as finding the squares or square roots, may be performed on each element of a list by performing the function on the list or the list name.

Procedure C10. **To store a function.**

On the TI-82 or 83 or 84 Plus:

The TI-82 stores up to eight functions, using function names $Y_1, Y_2, Y_3, ... Y_8$. The TI-83 and 84 Plus store up to ten functions.

Use the key X, T, θ for the variable x.

Method I:
1. Enter inside of quotes the function to be stored.
2. Press the key STO▷
3. Press the key Y-VARS

   (On the TI-83 or 84 Plus, press the VARS key and highlight Y-VARS.)
4. Select 1:Function
5. Select a function name (This function will replace any previously stored function.)
6. Press ENTER

Example: $"x^2 + 2" \rightarrow Y_1$

Method II:
1. Display the function table by:
   Press the key $Y=$
2. Move the cursor to the line of the function.
3. Enter the function to the right of a function name.
4. Press QUIT

On the TI-85 or 86:

The TI-85 or 86 may store functions using names that may contain up to eight characters. The first character of a name must be a letter.

Use the key x-VAR for the variable x.

Method I:
1. Enter the name of the function.
   (Names may be up to eight characters long.)
2. Press the key =
3. Enter the function.
4. Press ENTER

To see a stored function, press the key RCL, enter the name of the function, and press ENTER.

Method II:
(Stores functions under the names $y1, y2, y3, ... y_{99}$)
1. Display the function table by:
   Pressing the key GRAPH, then selecting y(x)=
2. Enter the function after a function name.
3. Press QUIT

Note: To see stored functions, display the function table by pressing the key $Y=$. Pressing CLEAR while the cursor is on the same line as the function will erase the function. Be sure to QUIT the function list to return to the home screen.

Procedure C11. **To evaluate a function or create tables.**

A. To evaluate a function

Functions may be evaluated at a single value or at several values by entering the values in a list.

On the TI-82 or 83 or 84 Plus:

1. Obtain the function name
   (a) Press the VARS key and highlight Y-VARS at the top of the screen. (On the TI-82 press the Y-VARS key.)
   (b) Select 1:Function...
   (c) Select the name of the function.
2. After the function name enter, in parentheses, the x value, a list of x values, or a list name at which the function is to be evaluated.
   (Example: $Y_1(4)$ or $Y_1(\{1,2,3,4\})$ or $Y_1(L_1)$)
3. Press ENTER

The results will be given as a value or a corresponding list.

On the TI-85 or 86:

1. Obtain the function evalF
   (a) Press the key CALC
   (b) Select evalF
2. Enter either the function or the function name then a comma.
3. Enter the variable for which the function is being evaluated followed by a comma.

The function names $y1$, $y2$, etc. may be entered from the keyboard (y must be lower case)

or

may be selected by:
   (a) Pressing the key VARS
   (b) From menu, select EQU (after pressing MORE).
   (c) Select the variable name (may need page↓ or page↑).
4. Enter the value, a list of values, or a list name at which the function is to be evaluated.
   (Example: evalF($y1$, x, 4) or
               evalF($y1$,x,$\{1,2,4\}$))
5. Press ENTER

Note: In the evaluation of a function, instead of placing a list of values inside of braces {}, we may enter the list name.
See Procedure C7 for entering lists.

B. To create tables:
The TI-82, 83, 84 Plus, and 86 have the ability to create a table of values for a function by pressing the key TABLE. Parameters for the table are set by pressing the key TblSet on the TI-82, 83, and 84 Plus or by selecting TBLST on the TI-86.

In TblSet:
TblMin or TblStart is the value of the first x value in the table.
ΔTbl is the difference between x-values in the table.
Indpnt: Auto Ask allows the table to be created automatically if Auto is highlighted or by the calculator asking for values of the independent variable if Ask is highlighted.

Pressing (or selecting) TABLE will show a table of values of the independent variable and values for all functions that are turned on in the function table.

By using the cursor keys, we can scroll through the table.

Procedure C12. To store a constant for an alpha character.

1. Enter the value to be stored.
2. Press the key STO▷
   (STO▷ shows as → on screen.)
3. Enter the alpha character.
4. Press ENTER
(Example: 6.35 → *H*)
Note: To see the value stored for an alpha character, enter the alpha character
        and press ENTER.

Procedure C13. To obtain relation symbols (=, ≠, >, ≥, <,   and ≤).

1. Press the key: TEST
2. Select the desired symbol from the menu.

Procedure C14. To change from radian mode to degree mode and vice versa.

1. Press the key: MODE
2. Move the cursor to either Radian or Degree on the third line depending on
   which mode is desired. (The mode, which is highlighted, is the active mode.)
   Press ENTER.
3. Press CLEAR or QUIT to return to the home screen.

Procedure C15. Conversion of Coordinates

Make sure calculator is in correct mode -- radians or degrees.
A. To change from **polar to rectangular**:

On the TI-82 or 83 or 84 Plus:
1. Press the key ANGLE
2. To find *x*:
   Select 7:P▷R*x*(
   To find *y*:
   Select 8:P▷R*y*(
3. Enter the values of *r* and θ
   separated by commas and press
   ENTER.
Example: P▷R*x*(2.0, 45)
          gives 1.4 [in degree mode]

On the TI-85 or 86:
1. In square bracket [],
   Enter the magnitude and angle
   separated by ∠ .
2. Press the key VECTR
3. Select OPS
4. Select ▷Rec (after pressing
   MORE)
5. Press ENTER
Example: [2.0∠45]▷Rec
          gives [1.4, 1.4]
          in degree mode.

B. To change from **rectangular to polar**:

On the TI-82 or 83 or 84 Plus:

1. Press the key ANGLE
2. To find r:
    Select 5:R▷Pr(
    To find θ:
    Select 6:R▷Pθ(
3. Enter the values of x and y separated by commas and press ENTER.

Example: R▷Pθ(1.0, 1.0)
    gives 45 [in degree mode]
Note: On the TI-82, it may be just as easy to use the conversion formulas (Equation 3) directly -- especially when converting from polar to rectangular.

On the TI-85 or 86:

1. Enter the values of x and y inside of [ ] separated by a comma.
2. Press the key VECTR
3. Select OPS
4. Select ▷Pol
5. Press ENTER

Example: [1.0, 1.0] ▷Pol
    gives [1.4 ∠45]
    in degree mode.
Note: On the TI-85 or 86, vectors are entered or displayed as a pair of numbers inside of square brackets, [ ]. When the numbers are separated by a comma they are in rectangular form and when separated by ∠ are in polar form.

Procedure C16.   Finding the sum of vectors in polar form.

On the TI-82 or 83 or 84 Plus:
1. Find the x- and y-components of each vector.
    (see Procedure C15)
2. Find the sum of the x-components and the sum of the y-components.
3. Find the magnitude and direction of the resultant vector.
    (see Procedure C15)

or

Find the sum
    $A\cos\theta_A + B\cos\theta_B +...$
and store this for x
Find
    $A\sin\theta_A + B\sin\theta_B +...$
and store this for y,
then convert $[x, y]$ to polar form.

On the TI-85 or 86:
Vectors may be entered directly in polar form and then added together if the calculator is in correct mode. Press the key MODE and be sure that in the third line, Degree is highlighted and in the seventh line, CylV is highlighted.

Example:

    [12 ∠120] + [15 ∠75]
    gives [25 ∠95] when rounded.

Note: On the TI-85 or 86, vectors may be stored under variable names by entering the vector, pressing the key STO▷, entering the variable name and pressing ENTER.

or

On some calculators it may be more convenient to treat vectors as complex numbers.

Vectors may also be entered or modified by first pressing the key VECTR, then selecting EDIT.
The sum of vectors may be found by finding the sum of the variables.

The form of the output of a vector on the screen may be changed by pressing the key MODE and in the seventh line highlighting
    RectV for rectangular form
    CylV for polar form

Procedure C17. Conversion of complex numbers.

Make sure calculator is in correct mode -- radians or degrees.

A. To change from **polar to rectangular** form:

On the TI-82 or 83or 84 Plus:

Think of the complex number $r \angle \theta$ as the vector $[r \angle \theta]$ and use Procedure C15.
or
On the TI-83 or 84 Plus, the mode can be set to an output of the form $a + bi$ (under MODE key) and the number entered in the form $re^{\wedge}\theta i$ using the key for $e^{\wedge}$ and the key for i, (not the letter I) and then press ENTER. In this case it is first necessary to make sure that θ is in radians.

On the TI-85 or 86:

1. Enter the magnitude and angle separated by $\angle$ in ( ).
2. Press the key CPLX
3. Select $\triangleright$Rec (after pressing MORE) and press ENTER

Example: (2.0 $\angle$45)$\triangleright$Rec
    gives (1.4, 1.4) in
    degree mode.

B. To change from **rectangular to polar** form:

On the TI-82 or 83 or 84 Plus:

Think of the complex number $x + yj$ as the vector $[x, y]$ and use Procedure C15.

or

On the TI-83 or 84 Plus, the mode can be set to an output of the form $re^{\wedge}\theta i$ (under MODE key) and the number entered in the form $a + bi$ and then press ENTER.

On the TI-85 or 86:

1. Enter the values of x and y inside of ( ) separated by a comma.
2. Press the key CPLX
3. Select $\triangleright$Pol
4. Press ENTER

Example: (1.0, 1.0) $\triangleright$Pol
    gives (1.4 $\angle$45)
    in degree mode.

Whether θ is in radians or degrees will depend on whether the calculator is set to radians or degrees.

Note: On the TI-85, a complex number is expressed as a pair of numbers inside of parentheses ( ). When the numbers are separated by a comma they are in rectangular form and when separated by $\angle$ they are in polar form.

Procedure C18. **Operations on complex numbers.**

On the TI-82 or 83 or 84Plus:

1. If not already in rectangular form, change the complex number into rectangular form as when changing polar coordinates to rectangular coordinates (see Procedure C15).

    On the TI-83 and 84, complex numbers may be entered in the form $a + bi$ or $re^{\wedge}\theta i$.

2. Perform the operation on the rectangular forms of the complex numbers.

3. If the answer is to be in polar form, convert the result back to polar form as when changing rectangular coordinates to polar coordinates (see Procedure C15).

Note: On the TI-83 and 84 Plus the operations may be done leaving the complex numbers in the form $re^{\wedge}\theta i$ if the values for θ are in radians.

On the TI-85 or 86:

Operations are performed by entering the complex numbers in either rectangular or polar form using the operation symbols $+$, $-$, $\times$, or $\div$ between the numbers. (See note on Procedure C17.)

Note: Powers and roots of complex number may be found by raising the complex number to the desired power using $\wedge$.

The form of output of the complex number on the screen may be changed by pressing the key MODE and in the fourth line highlighting
    RectC for rectangular mode
    PolarC for polar mode

Procedure C19. **To determine the numerical derivative from a function.**

On the TI-82 or 83 or 84 Plus:
1. Press the key MATH
2. Select 8:nDeriv(
    The form is
        nDeriv(*y, variable, value*)

On the TI-85 or 86:
1. Press the key CALC
2. From the menu, select nDer
    The form is
        nDer(*y, variable, value*)

where *y* is the function itself or the name of the function, *variable* is the independent variable used in the function, and *value* is the number at which we wish to determine the derivative.

Example: nDeriv(Y_1,x,2)
The precision to which the derivative is calculated may be changed by indicating the precision as a fourth value inside the parentheses. If the fourth value is not given, the precision is 0.001. (This value determines the Δx.)

Example:
    nDeriv(Y_1,x,2,0.0001)

Example: nDer(y1,x,2)
The precision to which the derivative is calculated may be changed.
On the TI-85, press the key TOLER and on the TI-86, press the key MEM and select TOL, Then change the value for δ (this determines the Δx).

Note: The TI-85 and 86, also have the function der1 which determines the numerical first derivative as exactly as is possible and der2 which determines the numerical second derivative as exactly as possible.

Procedure C20. Function to calculate definite integral.

The form of the function if **fnInt**(y, *variable, a, b*)

On the TI-82, 83, or 84 Plus:
1. Press the key MATH
2. Select 9:fnInt(

On the TI-85 or 86:
1. Press the key CALC
2. Select fnint

where y is a function or name of a function, variable is the variable used in the function, and a and b are the lower and upper limits of integration.

(Example: fnInt(x^3, x, 1,3) is used to evaluate $\int_1^3 x^3\,dx$.)

A.3 Graphing Procedures

Procedure G1. To enter and graph functions.

1. Enter the function to be graphed according to Procedure C10.
2. To graph entered functions press the key: GRAPH
    3. To return to the home screen while a graph is on the screen, press QUIT or CLEAR.
Notes: 1. Functions entered by this procedure will remain in the calculator's memory until changed or erased or until the memory is cleared.
    2. While a graph is on the screen, pressing QUIT or CLEAR will clear the screen and return to the home screen. However, the graph is still in the calculator's memory and can be seen again by pressing GRAPH.

3. When a function has been entered, the equal sign for that function becomes highlighted indicating that the function is turned ON - that is, it is an active function and will be graphed if the GRAPH key is pressed. Only functions that are turned ON will be graphed.

To turn a function ON or OFF, with the function table on the screen:

On the TI-82 or 83 or 84 Plus:

(a) Move the cursor so that it falls on top of the equal sign of the function we wish to turn ON or OFF.

(b) Press the ENTER key. If the function was ON it will be turned OFF and if it was OFF it will be turned ON.

On the TI-85 or 86:

(a) Move the cursor so it is on the same line as the function to be turned ON or OFF.

(b) From the secondary menu, select SELCT. If the function was ON it will be turned OFF and if it was OFF it will be turned ON.

4. On the TI-83 and 84 Plus, different appearances may be obtained for graphs by moving the cursor to the symbol that appears to the left of a function name in the function table. To change this symbol, move the cursor to the symbol and press ENTER until the desired symbol is obtained.

Procedure G2. To change or erase a function.

1. Display the function table on the screen (see Method II of Procedure C10).
2. (a) To change a function: Use the arrow keys to move the cursor to the desired location and make the changes by inserting, deleting, or changing the desired characters.
   (b) To erase a function: With the cursor on the same line as the function, press the key: CLEAR
3. Select GRAPH to graph the function or QUIT to exit to the home screen.

Procedure G3. To obtain the standard viewing rectangle.

To obtain the standard viewing rectangle;

On the TI-82 or 83 or 84 Plus:

1. Press the key ZOOM
2. From the menu, select 6:Zstandard

On the TI-85 or 86:

1. Press the key GRAPH
2. From the menu, select ZOOM
3. From the secondary menu, select ZSTD.

For the standard viewing rectangle, the x-value at the left of the screen is -10 and at the right of the screen is 10 and the y-value at the bottom of the screen is -10 and at the top of the screen is 10. Each mark on the axes represents one unit.

Procedure G4. To use the trace function and find points on a graph.

1. Select the trace function:

   On the TI-82 or 83 or 84 Plus: On the TI-85 or 86:

   Press the key TRACE   With the graph menu on the screen, select TRACE.

2. As the right and left arrow keys are pressed, the cursor moves along the graph and, at the same time, the coordinates of the cursor are given on the bottom of the screen.

3. If more than one graph is on the screen, pressing the up or down cursor keys will cause the cursor to jump from one graph to another. The function will appear in the upper right of the screen. On the TI-82, the number of the function will appear at the upper right of the screen.

4. If the graph goes off the top or bottom of the screen, the cursor will continue to give coordinates of points on the graph.

5. If the cursor is moved off the right or left of the screen, the graph scrolls (moves) right or left to keep the cursor on the screen.

Note: If the cursor is moved by using the cursor keys without first pressing the trace key, then the cursor can be moved to any point on the screen. The coordinates of this screen point will be given at the bottom of the screen.

Procedure G5. To see or change viewing rectangle.

1. To see values for the viewing rectangle on the screen.

   On the TI-82 or 83 or 84 Plus: On the TI-85 or 86:

   Press the key WINDOW   With the graph menu on the screen, select RANGE

2. To change viewing rectangle values:

   Make sure the cursor is on the quantity to be changed and enter a new value for that quantity. To keep a value and not change it, either (a) just press the key ENTER **or** (b) use the cursor keys to move the cursor to a value that is to be changed.

3. To see the graph, after the new values have been entered, select GRAPH

   To return to the home screen, press the key: QUIT

In changing the scales for the axes, Xmin must be smaller than Xmax and Ymin must be smaller than Ymax. If either Xscl or Yscl is 0, no marks will appear on that axis.

Note: The standard viewing rectangle has the following values:

$$Xmin = -10 \qquad Ymin = -10$$
$$Xmax = 10 \qquad Ymax = 10$$
$$Xscl = 1 \qquad\; Yscl = 1$$

Procedure G6: To obtain preset viewing rectangles.

1. Bring the zoom menu to the screen.

    On the TI-82 or 83 or 84 Plus: On the TI-85 or 86:

    Press the key ZOOM   With the graph menu on the screen,
                                         select ZOOM

2. Select the desired preset viewing rectangle. (The left-hand column gives the selection from the ZOOM menu for the TI-82 or 83 or 84 Plus and the right hand column for the TI-85 or 86.)

    (a) For the **standard viewing rectangle**:
    $$Xmin = -10, Xmax = 10, Xscl = 1,$$
    $$Ymin = -10, Ymax = 10, Yscl = 1$$

         Select: 6:ZStandard   ZSTD

    (b) For the **square viewing rectangle**:
    The square viewing rectangle keeps the y-scale the same and adjusts the x-scale so that one unit on x-axis equals one unit on y-axis. For a square viewing rectangle the ratio of y-axis to x-axis is about 2:3.
         Select: 5:ZSquare   ZSQR (after pressing MORE)

    (c) For the **decimal viewing rectangle**:
    The decimal viewing rectangle makes each movement of the cursor (one pixel) equivalent to one-tenth of a unit
         Select: 4:ZDecimal   ZDECM (after pressing MORE)

    (d) For the **integer-viewing rectangle**:
    The integer-viewing rectangle makes each movement of the cursor (one pixel) equivalent to one unit. After selecting the integer-viewing rectangle, move the cursor to the point that is to be located at the center of the screen.
         Select: 8:ZInteger   ZINT (after pressing MORE)
         Then press ENTER.

    Note: To have one movement of the cursor in the x-direction differ by k units set:
    $$Xmin = -94k/2 \qquad Xmin = -126k/2$$
    $$Xmax = 94k/2 \qquad Xmax = 126k/2$$

    (e) For the **trigonometric viewing rectangle**:
    The trigonometric viewing rectangle sets the x-axis up in terms of π or degrees (depending on mode) and sets $Xscl = \dfrac{\pi}{2}$ (or 90°).
         Select: 7:ZTrig   ZTRIG

Procedure G7. To graph functions on an interval.

1. To graph a function on the interval $x < a$ or on the interval $x \le a$ for some constant a:
    In the function table after the equal sign, enter the function $f(x)$ followed by either $(x<a)$ or $(x \le a)$.

(Example: To graph $y = x^2$ on the interval $x<2$, enter for the function: $x^2(x<2)$))

2. To graph a function on the interval $a < x < b$ or on the interval $a \leq x \leq b$ for some constants a and b:

   In the function table after the equal sign, enter the function $f(x)$ followed by either $(x>a)(x<b)$ or $(x\geq a)(x\leq b)$.

   (Example: To graph $y = x^2$ on the interval $-3\leq x\leq 2$, enter for the function: $x^2(x\geq -3)(x\leq 2)$)

3. To graph a function on the interval $x > a$ or on the interval $x \geq a$ for some constant a:

   In the function table after the equal sign, enter the function $f(x)$ followed by either $(x>a)$ or $(x\geq a)$.

   (Example: To graph $y = x - 5$ on the interval $x>2$, enter for the function: $(x-5)(x>2)$)

Note: The above forms may be combined by writing their sum.

   Example: $(x+5)(x<2) + (7-x^2)(x\geq 2)$

   Where the function contains more than one term, enclose the function in parentheses.

Procedure G8. Changing graphing modes.

1. To change modes:

On the TI-82 or 83 or 84 Plus:	On the TI-85 or 86:
Press the key MODE	With the graph menu on the screen, select FORMT (after pressing MORE)

   Then, use the cursor keys to highlight the desired option.

2. To draw graphs connected or unconnected (with dots)

Highlight Connected to connect plotted points with lines.	Highlight DrawLine to connect plotted points with lines.
Highlight Dot to just plot individual points.	Highlight DrawDot to just plot individual points.

3. To draw graphs simultaneously or sequentially

Highlight Sequential to graph each function in sequence.	Highlight SeqG to graph each function in sequence.
Highlight Simul to graph all selected functions at the same time.	Highlight SimulG to graph all selected functions at the same time..

4. Highlight the desired option in each line then press ENTER.

5. To return to the home screen, press CLEAR or QUIT.

Procedure G9. To zoom in or out on a section of a graph.

1. With a graph on the screen,

On the TI-82 or 83 or 84 Plus:	On the TI-85or 86:
Press the key ZOOM	With the graph and graph menu on the screen, select ZOOM

2. From the menu:

To zoom in, select 2:Zoom In	To zoom in, select ZIN
or	**or**
To zoom out, select 3:Zoom Out	To zoom out, select ZOUT

3. The cursor keys may be used to move the cursor near the point of interest. After zooming, the point at the location of the cursor will be near the center of the screen.
4. Press ENTER
5. Repeated zooms may be made (if no other key is pressed in the meantime) by moving the cursor near the point of interest and pressing ENTER.

Note: It is a good idea to use the TRACE function together with ZOOM.

Procedure G10.   **To change zoom factors**

1. Obtain the zoom menu:

On the TI-82 or 83 or 84 Plus:	On the TI-85 or 86:
Press the key ZOOM	With the graph menu on the screen, select ZOOM

2. Then from the menu:

Highlight MEMORY in the top row and select 4:SetFactors.	Select ZFACT (after pressing MORE twice).

3. Change the factors as desired.

   There are factors for magnification in both the x- and y-directions. The preset factors are 4 in both directions. Make changes to XFact for the magnification in the x-direction and to YFact for magnification in the y-direction.
   Caution: Do not set factors too large.
4. To exit this screen press the key: QUIT

Procedure G11.   **Solving an equation in one variable.**

1. Write the equation so that it is in the form $f(x) = 0$, let $y = f(x)$ and graph the function. Use a viewing rectangle so that the x-intercept of interest appears on the screen.
2. Determine a solution by:
   (a) Using Procedure G9 to zoom in on the x-intercept.
   or
   (b) Using the built in procedure:

On the TI-82 or 83 or 84 Plus:	On the TI-85 or 86:
(1) Press the key CALC	(1) From the graph menu, select MATH (after pressing MORE).
(2) From the menu, select 2:zero (2:root on the TI-82) The words Left Bound? (or Lower Bound?) appear.	(2) From the secondary menu, select ROOT.

(3) Move the cursor near, but to the left of the intercept and press ENTER. The words Right Bound? (or Upper Bound?) appear.

(4) Move the cursor near, but to the right of the intercept and press ENTER.
The word Guess? appears.

Move the cursor near the desired point. (It is usually sufficient to leave the cursor at its last position.)

(5) Press ENTER again. The root (*x*-intercept) will appear at the bottom of the screen.

(3) On the TI-86 the words LeftBound? Appear. Follow steps (3), (4), and (5) for The TI-82 and 83.

On the TI-85, move the cursor near the intercept and press ENTER. The root (*x*-intercept) will appear at the bottom of the screen.

Procedure G12. Finding maximum and minimum points.

1. Graph the function and adjust the viewing rectangle so that the desired local maximum or local minimum point is on the screen.

2. Determine the local maximum or local minimum point by:

   (a) Using Procedure G9 to zoom in on the point

   or

   (b) Use the built-in procedure:

   On the TI-82 or 83 or 84 Plus:

   (1) Press the key CALC

   (2) From the menu, select 3:minimum or 4:maximum. The words Left Bound? (or Lower Bound?) appear.

   (3) Move the cursor near, but to the left of the desired point and press ENTER. The words Right Bound? (or Upper Bound?) appear.

   (4) Move the cursor near, but to the right of the desired point and press ENTER.
   The word Guess? appears.

   Move the cursor near the desired point. (It is usually sufficient to leave the cursor at the last position.)

   On the TI-85 or 86:

   (1) Adjust the viewing rectangle so that the point of interest is the highest point or lowest point on the screen.

   (2) From the graph menu, select MATH (after pressing MORE).

   (3) From the secondary menu, select FMIN for a local minimum or FMAX for a local maximum point (after pressing MORE).

   (4) On the TI-86 the words LeftBound? appear. Follow steps (3), (4), and (5) for the TI-82 and 83.

   On the TI-85 move the cursor near the desired point and press ENTER.

(5) Press ENTER again. The coordinates of the point will appear at the bottom of the screen.

The coordinates of the point will appear at the bottom of the screen.

Note: On the TI-85 LOWER and UPPER may be used to select lower and upper bounds on which the maximum and minimum values are to be found.

Procedure G13. Find the value of a function at a given value of *x*.

1. Graph the function and adjust the viewing rectangle so that the given *x*-value is within the restricted domain of the screen.

2. Determine the value of the function by:

   (a) Using Procedure G9 to zoom in on the point

   or

   (b) Use the built-in procedure:

   On the TI-82 or 83 or 84 Plus:
   (1) Press the key CALC
   (2) From the menu, select 1:value. *X=* appears on the screen.
   (3) Enter the given *x*-value and press ENTER. The value of the function will appear at the bottom of the screen.

   On the TI-85 or 86:
   (1) From the graph menu, select EVAL (after pressing MORE twice).
   (2) The words Eval *x* = appear on the screen.
   (3) Enter the *x*-value and press ENTER. The value of the function will appear at the bottom of the screen.

Procedure G14. To zoom in using a box.

1. Obtain the zoom menu:
   On the TI-82 or 83 or 84 Plus:
   Press the key ZOOM

   On the TI-85 or 86:
   With the graph menu on the screen, select ZOOM

2. Then
   Select 1:Zbox

   From the secondary menu select BOX

3. Use the cursor keys to move the cursor to a location where one corner of the box is to be placed and press ENTER.

4. Use the cursor keys to move the cursor to the location for the opposite corner of the box and press ENTER. As the cursor keys are moved a box will be drawn and when ENTER is pressed, the area in the box is enlarged to fill the screen.

Procedure G15. Finding an intersection point of two graphs.

and adjust the viewing rectangle so the region of the graph that contains the intersection point is on the screen. Then,

1. Enter the functions into the calculator as two different functions (Y_1 and Y_2).

2. Graph the functions on the same screen and adjust the viewing rectangle so that the region containing the intersection point is on the screen.

3. Determine the point of intersection by:

   (a) Using Procedure G9 to zoom in on the point. To obtain the desired degree of accuracy, using the trace function move the cursor just to the left and then just to the right of the intersection point after each zoom. A solution to the desired accuracy is obtained when the x-values, on either side of the intersection point, rounded off to the desired significant digits are equal.

   or

   (b) Use the built-in procedure:

       (1) Select the correct mode to find the point of intersection.

On the TI-82, 83 or 84 Plus:	On the TI-85 or 86:
Press the key CALC	With the graph menu on the screen, select MATH (after pressing MORE). Then, select ISECT (after pressing MORE).
Select 5:intersect	

       (2) Move the cursor near the point of intersection.

       (3) If the cursor is on one of the graphs, continue. Otherwise, use the up and down cursor keys to move the cursor to the correct graph. (On the TI-82, 83 or 84 Plus, and 86 the words "First curve?" appear on the screen.)

       (4) Press ENTER. The cursor will jump to another graph.

       (5) If the cursor is on the correct second graph, continue. Otherwise, use the up and down cursor keys to change the cursor to the correct graph. (On the TI-82, 83, 84 Plus and 86 the words "Second curve?" appear.)

       (6) Press ENTER.

       (7) Obtain the coordinates of the intersection point.

On the TI-82, 83, 84 Plus or 86:	On the TI-85:
The word "Guess?" appears on the screen. Make sure the cursor is near the point of intersection and press ENTER. The word Intersection appears. The coordinates of the intersection point appear on the screen.	The word ISECT appears..

Procedure G16. **To graph each of two functions and then their sum.**

1. Set the correct values for the viewing rectangle.
2. Enter the first function into the function table for one function (Y_1).
3. Enter the second function into the function table for a second function (Y_2).
4. Move the cursor after the equal sign for a third function (Y_3). Press the Y-VARS key (or VARS, then select Y-VARS) and select the first function name (Y_1) from the menu.
5. Press the key: +
6. Press the Y-VARS key (or VARS, then select Y-VARS) and select the second function name (Y_2) from the menu.
   (The third line could now appear as: $Y_3 = Y_1 + Y_2$)
7. Press GRAPH

Procedure G17. **To change graphing modes for regular functions, parametric, or polar equations.**

1. Set the graphing mode:
   Press the key MODE.
   Make active:

On the TI-82 or 83 or 84 Plus:	On the TI-85 or 86:
(a) For graphing in **rectangular** coordinates:	
In fourth line: Func	In fourth line: RectC
	In fifth line: Func
(b) For graphing **parametric** equations:	
In fourth line: Par	In fourth line: RectC
	In fifth line: Param
(c) For graphing in **polar** coordinates:	
In fourth line: Pol	In fourth line: PolarC
	In fifth line: Pol

   Then, press QUIT or CLEAR.

2. Set the graphing format:

On the TI-82:	On the TI-85 or 86:
Press the key WINDOW	Press the key GRAPH
Highlight FORMAT in the top row.	From the menu, select FORMT (after pressing MORE)
On the TI-83 or 84 Plus:	
Press the key FORMAT	
Then:	Then:

   Make active:
   (a) For graphing in **rectangular** coordinates or for **parametric** equations: RectGC
   (b) For graphing in **polar** coordinates: PolarGC
   Then, press QUIT or CLEAR.
   (To make an item active, move the cursor to that item and press ENTER. The active item is highlighted.)

Procedure G18.   **To graph parametric equations.**

1. Make sure the calculator is in correct mode for parametric equations. (Procedure G17)

2. To enter the equations:

    On the TI-82 or 83 or 84 Plus: On the TI-85 or 86:
    Press the key $Y=$ Press the key GRAPH.
    On the screen will appear: From the menu, select $E(t) =$

    $X_{1t} =$ On the screen will appear:
    $Y_{1t} =$ $xt1=$
    $X_{2t} =$ $yt1=$
    $Y_{2t} =$ (enter up to 99 pairs of equations)

    (etc.)
    (enter up to 6 pairs of equations)

    Functions are not turned on (the equal signs are not highlighted) until functions have been entered for both x and y. Functions may be turned on and off as with functions in rectangular coordinates (see Note 3, Procedure G1). Only functions turned on will be graphed.

3. Enter the first equation for x in the first row and the corresponding equation of y in the second row. If there are other sets of functions you wish to graph, enter them for the other x's and y's.

    To obtain the variable t

    On the TI-82 or 83 or 84 Plus: On the TI-85 or 86:
    Press the key X,T,θ   Select t from the menu.

4. Press or select GRAPH to see the graph or QUIT to exit.

Procedure G19.   **To change values for the parameter t.**

With the calculator in the mode for parametric equations:
1. Follow Procedure G5 to see the values for the viewing rectangle.
2. Enter new values for Tmin, Tmax, and Tstep and press ENTER or move to the next item by using the cursor keys.
    The standard viewing rectangle values for t are:

    $$\text{Tmin} = 0, \text{Tmax} = 6.28...(2\pi), \text{ and Tstep} = 0.13...(\pi/24)$$
    $$(\text{or Tmax} = 360 \text{ and Tstep} = 7.5 \text{ if in degree mode}).$$

    Tstep determines how often values for x and y are calculated. The size of Tstep will affect the appearance of the graph. Tmax should be sufficiently large to give a complete graph.
3. Press QUIT to exit.

Procedure G20. To graph equations involving y^2.

    1. Solve the equation for y. There will be two functions: $y = f(x)$ and $y = g(x)$. In many cases it will be true that $g(x) = -f(x)$.

    2. Store $f(x)$ under one function name and $g(x)$ for the second function name (see Procedure C10).   (Example: $Y_1 = f(x)$, $Y_2 = g(x)$)
       or
       If $g(x) = -f(x)$, we may do the following:
       (a) Store $f(x)$ after a function name (Example: Y_1).
       (b) Move the cursor after the equal sign for a second function name (Example: Y_2).
       (c) Enter the negative sign: $(-)$
       (d) Obtain and enter the name of the first function. (Example: $Y_2 = -Y_1$) (See Procedure C11.)

    3. Graph the functions.

Procedure G21. To obtain special Y-variables.

On the TI-82 or 83 or 84 Plus:

    1. To obtain window quantities (Ymin, Ymax, Xmin, or Xmax): press the key: VARS
       From the menu:
        Select: Window..
        Then select desired quantity.

    2. To obtain variable names Y_1, Y_2, etc.:
       On the TI-82,
        press the key: *Y*-VARS
       On the TI-83 or 84 Plus,
        press the key: VARS,
        Highlight *Y*-VARS at the top
       From the menu:
        select: Function
        then, select the function name.

On the TI-85 or 86:
The best way perhaps is to just enter the name from the keyboard. When entering the name from the keyboard, be sure the correct upper or lower case is entered.
or

    1. Press the key VARS
    2. Select ALL
    3. Press F1 to page down till desired variable is on screen.
    4. Use cursor keys to select desired variable.
    5. Press ENTER

Procedure G22. To shade a region of a graph.

    1. Enter the desired function for a function name. Make sure all functions not wanted are deleted or turned off.

    2. Select appropriate window values and graph the function.

3. Select the Shade command:

On the TI-82, 83, or 84 Plus:
Press the key: DRAW
From the menu, select 7:Shade(

On the TI-82 the form is
Shade(*L, U, D, Lt, Rt*)
On the TI-83 the form is
Shade(*L, U, Lt, Rt, Pat, D*)

On the TI-85 or 86:
From the GRAPH menu,
select DRAW (after pressing
MORE) and Select Shade

On the TI-85, the form is
Shade(*L, U, Lt, Rt*)
On the TI-86, the form is
Shade(*L, U, Lt, Rt, Pat, D*)

Where:

L is lower boundary (value or function) of the shaded area.

U is upper boundary (value or function) of the shaded area.

D is the density value (a digit from 1 to 8) that determines the spacing of the vertical lines the calculator draws to shade the region. The higher the value for the density, the more widely spaced the shading lines. If the density value is omitted, the shading is solid. (optional)

Lt is the left most x-value. (optional)

Rt is the right most x-value. (optional)

On the TI-83, *Pat* is the pattern. (optional)

(A digit from 1 to 4 – each digit gives a different fill pattern.).

[*D, Lt, Rt, Pat* are optional - but on the TI-82, if *Lt* and *Rt* are included, *D* must be also included.]

(Example: Shade(–10, Y_1) shades in the area above $y = -10$ and

below the graph of Y_1)

4. Enter the appropriate values for L, U, and/or D and left and right x-values. Then, press ENTER.

5. To clear the shading:

On the TI-82, 83 or 84 Plus:
Press the key: DRAW
From the menu,
Select 1:ClrDraw

On the TI-85 or 86:
From the GRAPH menu,
select DRAW (after pressing
MORE) and Select CLDRW
(After press MORE twice).

Note: To shade above the function represented by Y_1, use Shade(Y_1 , Ymax).

To shade below the function represented by Y_1, use Shade(Ymin, Y_1).

The TI-83 Plus and TI-84 Plus may have an application program called Inequalz that allows the insertion of inequality symbols for the equal signs in the function table. To run this or to turn it off, press the key APPS and select :Inequalz and press ENTER. The inequality symbols may be selected by using the function keys ($F_1, F_2, etc.$)

Procedure G23. **To graph in polar coordinates**

      1. Make sure the calculator is in correct mode for polar coordinates. (Procedure G17)

      2. To enter polar equations:

On the TI-82 or 83 or 84 Plus: On the TI-85 or 86:

Press the key $Y=$ Press the key GRAPH.
On the screen will appear: From the menu, select $r(\theta) =$
$r1=$ On the screen will appear: $r1=$
$r2=$ (up to 99 polar equations may be
$r3=$ entered as $r1, r2, r3, ...$)
(etc.)
(up to 6 equations may be entered)
Functions may be turned on and off as with functions in rectangular
coordinates (see Procedure G1, Note 3).

      3. To enter the equation for r.
For the variable θ
On the TI-82 or 83 or 84 Plus   On the TI-85 or 86:
Press the key X,T,θ Select θ from the menu.

      4. Select GRAPH to graph the function or QUIT to exit.

Procedure G24. **To change viewing rectangles values for θ.**

With the calculator in the mode for polar coordinates:

      1. Follow Procedure G5 to see the values for the viewing rectangle.

      2. Enter new values for θmin, θmax, and θstep and press ENTER or move to the next item by using the cursor keys.

The standard viewing rectangle values are

        θmin = 0, θmax = 6.28... (2π), and θstep = 0.13...($\pi/24$)
        (or θmax = 360 and θstep = 7.5 if in degree mode).

θstep determines how often points are calculated and may affect the appearance of the graph. Leaving θstep at approximately 0.1 is generally sufficient, but θstep may have to be changed if there is a major change in the viewing rectangle. θmax should be sufficiently large to give a complete graph.

      3. Press QUIT to exit.

Procedure G25. **Drawing a tangent line at a point.**

      1. Store the function for a function name.

      2. Draw the tangent line using the following:

On the TI-82 or 83 or 84 Plus: On the TI-85 or 86:

(a) While on the home screen, press   (a) With the graph on the screen
     the key DRAW.   select DRAW from the GRAPH
                                     menu (after pressing MORE)

(b) Select: 5:Tangent(
The form is
    Tangent(y, *value*)
Where y is the function or
function name and *value* is the
x-value.
Example: Tangent(Y_1,2)

(c) Press ENTER

or

The tangent line may be drawn by:

(a) With the graph on the screen, use
trace function to place the cursor
at the correct point, then press
DRAW.

(b) Select 5:Tangent(and press
ENTER to return to graph.

(c) Press ENTER a second time.
The tangent line will appear on
the graph and the slope of the
line will be given at the bottom
of the screen.

(b) From the secondary menu,
select TanLn (after pressing
MORE twice). The form is
    TanLn(y, *value*).
where y is the function or
function name and *value* is the
x-value.
Example: TanLn(y1,3)

or

The tangent line may be drawn by:

(a) With the graph and the GRAPH
menu on the screen, select MATH
(after pressing MORE).

(b) From the secondary menu, select
TANLN (after pressing MORE
twice).

(c) Move the cursor to the desired
point and press ENTER. The
tangent line will be drawn on the
graph and the slope of the tangent
line given at the bottom of the
screen.

All tangent lines will appear on the screen until cleared. To clear tangent lines
from the screen use the same process as in Procedure G22, Step 5.

Procedure G26. Determining the value of the derivative at points on a graph.

On the TI-82 or 83 or 84 Plus:

1. With a graph on the screen, press
   the key CALC.
2. Select 6:*dy/dx*.
3. Move the cursor to the desired
   point and press ENTER. The
   value of the numerical derivative
   appears on the bottom of the
   screen.

On the TI-85 or 86:

1. With the graph and the graph
   menu on the screen, press the key
   MATH (after pressing MORE).
2. From the secondary menu select
   dy/dx.
3. Move the cursor to the desired
   point and press ENTER. The
   value of the numerical derivative
   appears at the bottom of the
   screen.

Procedure G27. To find the constant of integration for an indefinite integral.

This procedure assumes that the general antiderivative is known.
1. Store a value of 0 for C. ($0 \rightarrow C$)
2. Store the antiderivative function with C attached for a function name in the
    function table and graph.

3. Given the information that $y = y_0$ when $x = x_0$ (for the given point on the graph (x_0, y_0)) and with the graph of the antiderivative (for $C = 0$) on the screen, evaluate the antiderivative function at $x = x_0$ (Procedure G13).   With the cursor on the graph of the antiderivative, read the corresponding y-value, y_a. The constant of integration is the difference between the given y-value y_0 and the value from the graph y_a. That is $C = y_0 - y_a$.

4. To check your answer, store this value for C, graph the antiderivative function, and evaluate at $x = x_0$.

Procedure G28.   Function to determine the area under a graph.

With the graph of a function on the screen:

On the TI-82, 83, or 84 Plus:
1. Press the key CALC
2. Select 7: $\int f(x) dx$
3. Move the cursor to the first x-value (Lower Limit) and press ENTER.
4. Move the cursor to the second x-value (Upper Limit) and press ENTER.

On the TI-85 or 86:
1. Select MATH (after pressing MORE)
2. Select $\int f(x)$
3. Move the cursor to first x-value (Lower Limit) and press ENTER.
4. Move the cursor to the second x-value (Upper Limit) and press ENTER.

The area between the curve, the x-axis, and the two x-values will be shaded and will be given at the bottom of the screen.

A.4 Programming Procedures

Procedure P1.   To enter a new program.

1. Press the key: PRGM
2. Enter programming mode:

On the TI-82 or 83 or 84 Plus:
Highlight NEW in the top row and select 1:Create New

On the TI-85 or 86:
Select EDIT.

3. After Name=, enter a name for the program. The name may be up to eight characters long. (Notice that the calculator is ready for alpha characters.) Then, press ENTER.

4. The word PROGRAM appears on the top row followed by the name of the program and the cursor is on the second row preceded by a colon (:).   The calculator is now in programming mode and is ready for the first statement in the program.

5. Enter the steps of the program one line at a time, pressing ENTER after each step. (On some calculators, more than one instruction may be placed on the same line if the instructions are separated by a colon (:). Care needs to be exercised in doing this.)

6. When finished entering the program, press QUIT to return to the home screen.

Procedure P2. To execute, edit, or erase a program.

1. Press the key: PRGM

2. To **execute** a program:

On the TI-82 or 83 or 84 Plus:	On the TI-85 or 86:
With EXEC highlighted in the top row, select the name of the program to be executed. Then, press ENTER.	Select NAMES From the secondary menu, select the name of the program. Then, press ENTER.

A program may be executed additional times by pressing ENTER if no other keys have been pressed in the meantime.

3. To **edit** a program:

On the TI-82 or 83 or 84 Plus:	On the TI-85 or 86:
Highlight EDIT in the top row and select the name of the program to be edited.	Select EDIT. From the secondary menu, select the name of the program to be edited.

Make changes in the program as desired. The cursor keys may be used to move the cursor to any place in the program. The calculator is now in programming mode.

4. To **erase** a program:

On the TI-82 or 83 or 84 Plus:	On the TI-85 or 86:
Press the key MEM.	Press the key MEM
On the TI-82 or 83:	Select DELET
Select 2:Delete...	Select PRGM
Select 7:Prgm (or 6:prgm on the 82)	(after pressing MORE)
On the TI-83 Plus or 84 Plus	
Select 2:Mem Mgmt/Del...	
Then select 7:Prgm	

Move the marker on the left until it is next to the name of the program to be erased and press DEL. (On the TI-82 or TI-83, press ENTER.) The program will be erased from memory. (Caution: Before erasing a program, make sure that you really want to erase it.)

Procedure P3. **To input a value for a variable.**

Entering a value for a variable uses the keywords: Input or Prompt.
The form of the command is: **Input** *v* or **Prompt** *v*
(*v* represents the variable name)

With the calculator in programming mode:
1. To access the word Input:

On the TI-82 or 83 or 84 Plus:	On the TI-85 or 86:
Press the key PRGM	From the menu, select I/O
Highlight I/O in the top row.	From the secondary menu,
Select 1:Input **or** 2:Prompt	Select Input **or** Promp

2. Enter the variable name and press ENTER.
   (Example: Input *C* **or** Prompt *C*)

When the program is executed the word Input followed by a variable will cause a question mark to appear on the screen while the word Prompt will cause the name of the variable followed by the question mark to appear. Any number entered is stored for that variable and when the variable is later used in the program that value is used in the calculation.

Procedure P4. **To evaluate an expression and store that value to a variable.**

With the calculator in programming mode:

On the TI-82 or 83 or 84 Plus:	On the TI-85 or 86:
Enter the expression, press the key STO▷, and then, enter a variable name.	Enter the expression, press the key STO▷ and then, enter a variable name.
	or
	Enter a variable name, press the key =, and then enter the expression.

The expression on the left will be evaluated at values that have been stored for the variables used on the left and then, stored for the variable name on the right.   (Example: $(9/5)C + 32 \rightarrow F$)

Procedure P5. **To display quantities from a program.**

Either the value for a variable or an alpha expression may be displayed by using the keyword Disp.

The form of the command is: **Disp** *v*   (*v* represents the variable name)

The quantity *v* may be a variable name, a number, or a set of characters enclosed inside of quotes. If *v* is a quantity enclosed in quotes, the quantity will be printed exactly as it appears.

With the calculator in programming mode:
1. To access the word Disp:

On the TI-82 or 83 or 84 Plus:	On the TI-85 or 86:
Press the key PRGM	From the menu, select I/O
Highlight I/O in the top row.	From the secondary menu,
Select 3:Disp.	select Disp

2. Enter the variable name and press ENTER.
    (Examples: Disp *F* or Disp "ANSWER IS")

When the program is executed, the value stored for the variable or the expression will be displayed on the screen.

Note: Quotes (") are located on the keyboard of the TI-82, 83 and 84 plus. On the TI-85 and 86, they are located under the I/O menu when in programming mode.

Procedure P6. To add or delete line from a program.

With the calculator in programming mode, select EDIT, then:
1. To add a line:
    (a) To insert a line before a given line move the cursor to the first character of the line and to insert a line after a given line move the character to the last character in the given line.
    (b) Press the key INS, then press ENTER This will create a blank line either before or after the given line
    (c) Place the cursor in the desired line. An instruction may now be entered.
2. To delete a line
    (a) Move the cursor to the desired line
    (b) Press the key CLEAR, then press DEL.

Procedure P7. Making a decision (If... command).

The form of the If command is **If** *condition*

To access If...
While in programming edit mode (at a step in program):

On the TI-82 or 83 or 84 Plus:	On the TI-85 or 86:
To select the If command:	To select the If command
(a) Press PRGM	(a) Select CTL
(b) With CTL highlighted,	(b) Select If
select 1:If	

Following the word *If* enter the *condition*. (Examples: x < 0 or T + 5 ≥ 20)

If *condition* is true, the statement following the If command is executed next. If *condition* is not true, the second statement following the If command is the next statement executed.

The *condition* may involve one of the mathematical symbols <, ≤, >, ≥, =, or ≠ to compare two variables and/or constants. To obtain these symbols, press the key TEST and select from the menu.

Example: Input x
            If $x < 0$
            Disp "NEGATIVE"
            If $x \geq 0$
            Disp "NON NEGATIVE"

Procedure P8. Making a decision with options (If...Then...Else... commands).

The form of this combination is

> (program statements before)
> **If** condition
> **Then**
> (program statements)
> **Else**
> (program statements)
> **End**
> (program statements after)

To access If... Then... Else...
While in programming edit mode (at a step in program):

On the TI-82 or 83 or 84 Plus:	On the TI-85 or 86:
1. Press PRGM and with CTL highlighted	1. Select CTL
2. (a) For the If command Select 1:If	2. (a) For the If command Select If
(b) For the Then command Select 2:Then	(b) For the Then command Select Then
(c) For the Else command Select 3:Else	(c) For the Else command Select Else
(d) For the End command Select 7:End	(d) For the End command Select End

If the condition is true, the statements following the Then command are executed. If the condition is not true, the statements following the Else command are executed. After either set of statements is executed, the statements after the END statement are executed.

The condition involves one of the mathematical symbols $<, \leq, >, \geq, =,$ or $\neq$ to compare two variables and/or constants. To obtain these symbols, press the key TEST and select from the menu.
Else is optional (the Else section may be omitted).

Example: Input x
            If $x < 0$
            Then
            Disp "NEGATIVE"
            Else
            Disp "NON NEGATIVE"
            End

Procedure P9.   Finding the area under a graph.

1. The desired function should be stored under a function name in the function table. (Procedure C6)
2. Enter the following program into your calculator to find the area between two values B and C:

   (It is assumed here that the function is stored under the function name Y_5)

```
PROGRAM:AREA
Disp "ENTER FIRST Z"
Input B
Disp "ENTER SECOND Z"
Input C
fnInt( Y₅ ,X,B,C)→I
Disp "AREA IS:"
Disp I
```

The form of the definite integral function that is used in this program is
fnInt(*function name, variable, smaller value, larger value*)
(This function may also be used without being in a program.)

To obtain the definite integral function fnInt:

On the TI-82 or 83 or 84 Plus:	On the TI-85 or 86:
Press the key MATH	Press the key CALC
From the menu, select 9:fnInt(	From the menu, select fnInt

(See also Procedure C20.)

Procedure P10.   Creating a While... loop (While, End commands).

The construction of the loop is of the form:

```
... (program statements before loop)
While condition
... (program statements within loop if condition true)
End
... (programming statements after loop)
```

To access the While and End statements,
While in programming edit mode (at a step in program):

On the TI-82 or 83 or 84 Plus:

1. To select the While command:
   (a) Press PRGM
   (b) With CTL highlighted, select 5:While
2. To select the End command:
   (a) Press PRGM
   (b) With CTL highlighted, select 7:End

On the TI-85 or 86:

1. To select the While command:
   (a) Select CTL
   (b) Select While (after pressing MORE)
2. To select the End command
   (a) Select CTL
   (b) Select End

In this case, if *condition* is a true statement the statements following the word While are performed and if *condition* is not true, the program will then transfer to the statements following the word End. For information on *condition* see Procedure P7.

Example: $1 \to K$
              While $(K{<}5)$
              $K + 1 \to K$
              End
              Disp K

When this program is run, it will cycle through the loop until K is no longer less than 5, then it will display the result. The result of this program is that $K = 5$.

Procedure P11. Creating a Repeat... loop (Repeat, End commands).

The construction of the loop is of the form:

      ... (program statements before loop)
      Repeat *condition*
      ... (statements within loop repeated until condition is true)
      End
      ... (programming statements after loop)

To access Repeat and End statements,
While in programming edit mode (at a step in program):

On the TI-82 or 83 or 84 Plus:	On the TI-85 or 86:

1. To select the Repeat command:
(a) Press PRGM
(b) With CTL highlighted,
    select 6:Repeat
	1. To select the Repeat command:
	(a) Select CTL
	(b) Select Repeat (after
	    pressing MORE)

2. To select the End command: (a) Press PRGM (b) With CTL highlighted, select 7:End	2. To select the End command (a) Select CTL (b) Select End

In this case, if *condition* is a false statement the statements following the word Repeat are performed until the condition becomes true. The program will then transfer to the statements following the word End.

Example: $10 \to K$
              Repeat $(K{<}5)$
              $K - 1 \to K$
              End
              Disp K

K is initially set to the value 10, but each time through the loop one is subtracted from K until K is less than 5. At that point the loop ends and the value of K is displayed. The result of this program is that $K = 4$.

Procedure P12. Creating a For... loop (For, End commands).

The construction of the loop is of the form:

> ... (program statements before loop)
> **For**(*var*, *beg*, *end*, *inc*)
> ... (statements within loop repeated if *end* not exceeded)
> **End**
> ... (programming statements after loop)

To access For and End statements,
While in programming edit mode (at a step in program):

On the TI-82 or 83 or 84 Plus:	On the TI-85 or 86:

1. To select the For command:
   (a) Press PRGM
   (b) With CTL highlighted,
       select 4:For
	1. To select the For command:
	   (a) Select CTL
	   (b) Select For

2. To select the End command:
   (a) Press PRGM
   (b) With CTL highlighted,
       select 7:End
	2. To select the End command
	   (a) Select CTL
	   (b) Select End

Where   *var* is a variable name
          beg is a beginning value
          end is an ending value
          inc is an increment value and is optional
             (Increment is 1 if this is omitted.)

In this case, the first time through the loop, *var* has the beginning value, and the next time through the loop *var* has the value of *beg* + *inc*. Each time through the loop the value for *var* is increased by *inc*. The loop stops when the value for *var* exceeds the value for *end* and the program will then transfer to the statements following the word End.

Example: $0 \rightarrow K$
             For(I,1,5)
             $K + I \rightarrow K$
             End
             Disp K

The first time through the loop, I has a value of 1 that is added to K giving 1. The second time I has a value of 2, which is added to the previous K giving 3. Etc. The program loops 5 times (as *I* goes from 1 to 5). The result of this program is that $K = 15$.

Procedure P13. Creating an If...Goto... loop (Lbl, If, Goto commands).

The construction of the loop is of the form:

> ... (program statements before loop)
> **Lbl** *label*
> ... (program statements within loop)
> **If** *condition*
> **Goto** *label*
> ... (programming statements after loop)

To access Lbl, If, and Goto statements,
W hile in programming edit mode (at a step in program):

On the TI-82 or 83 or 84 Plus:	On the TI-85 or 86:

1. Select the label (Lbl) command:
  (a) Press PRGM
  (b) With CTL highlighted, select 9:Lbl
	1. Select the label (Lbl) command:
	  (a) Select CTL
	  (b) Select Lbl (after pressing MORE)

2. Select the If command:
  (a) Press PRGM
  (b) With CTL highlighted, select 1:If
	2. Select If command
	  (a) Select CTL
	  (b) Select If

3. Select the Goto command:
  (a) Press PRGM
  (b) With CTL highlighted, select 0:Goto
	3. Select the Goto command
	  (a) Select CTL
	  (b) Select Goto (after pressing MORE)

Note: A label may be a letter or a numerical digit.	Note: A label may be a name of up to 8 characters, beginning with a letter.

The form of the label command is Lbl *label* (where *label* is a character)
The form of the If command is If *condition*
The form of the Goto command is Goto *label*
The statements within the loop are executed as long as condition is true. When condition ceases to be true, the program will transfer to statements following the Goto statement.

Example: $0 \rightarrow K$
Lbl G
$K + 1 \rightarrow K$
If $(K<5)$
Goto G
Disp K

The program loop through the Goto G statement back up to the Lbl G statement until K is no longer less than 5. It then displays the value for K. The result of this program is that $K = 5$.

Procedure P14. **To display a graph from within a program.**

While in programming edit mode (at a step in program):

On the TI-82 or 83 or 84 Plus:	On the TI-85 or 86:
1. Press the key PRGM	1. Select I/O
2. Select I/O	2. Select DispG
3. Select 4:DispGraph	

Procedure P15. **Drawing a line on the screen from within a program (Line command).**

The form of the line command is **LINE**(*a, b, c, d*)

While in programming edit mode (at a step in program):

On the TI-82 or 83 or 84 Plus:	On the TI-85 or 86:
1. Press the key DRAW	1. Press the key GRAPH
2. With DRAW highlighted, select 2:LINE	2. Select DRAW (after pressing MORE)
	3. Select LINE

where a and b are the coordinates of one point and *c* and *d* are the coordinates of a second point. This command draws a line from (*a, b*) to (*c, d*).

Example: LINE(0,2,X,Y_1) draws a line from the point (0,2) to the point (x, $f(x)$).

Procedure P16. **To temporarily halt execution of a program. (Pause command)**

While in programming edit mode (at a step in program):

On the TI-82 or 83 or 84 Plus:	On the TI-85 or 86:
1. Press the key PRGM	1. Select CTL
2. With CTL highlighted, select 8:PAUSE	2. Select Pause (after pressing MORE twice)

The Pause command when used in a program halts execution of the steps of a program until the ENTER key is pressed.

A.5 Matrix Procedures

Procedure M1. **To enter or modify a matrix.**

1. Press the key MATRX. This gives the matrix menu.
2. Enter EDIT mode.

On the TI-82 or 83 or 84 Plus:	On the TI-85or 86:
Highlight EDIT in the top row.	Select EDIT from the menu.

3. Select a name for the matrix.

On the TI-82 or 83 or 84 Plus:

Select one of the matrices [A], [B], [C], ... This will be the new matrix or the matrix to be edited.

On the TI-85 or 86:

Enter the name for a new matrix or select the name of a matrix to be edited, then press ENTER. The name may be up to eight characters long (only 5 letters will show in name box).

4. The first line on the screen will contain the word MATRIX followed by the name of the matrix and two numbers separated by ×. These two numbers represent the size of the matrix. First enter the number of rows and press ENTER, then the numbers of columns and press ENTER. To keep the values the same, use the cursor keys to move the cursor or just press ENTER.

5. Enter the elements of the matrix.

On the TI-82 or 83 or 84 Plus:

The notation of the element of the matrix to be entered or changed is highlighted and is shown at the bottom of the screen by the row and column number of the element followed by its value.

As a value is entered, the number at the bottom of the screen will change. To enter this value into the matrix, press ENTER. Values are entered by rows. Continue until all elements are correctly entered.

The cursor keys may be used to move the highlight to particular element(s) to be changed.

On the TI-85 or 86:

On the left of the screen are two numbers separated by a comma. These are row and column numbers of elements of the matrix. Enter one value at a time. After each value is entered, press ENTER to go to the next value. Values are entered by row. (As each value is entered the calculator will jump to the next column.)

The cursor keys may be used to go from one element to the next if only certain elements are to be changed.

On the TI-85, the menu selections ◁Col and Col▷ may be used to change columns.

6. Exit from matrix edit mode by pressing the key: QUIT. (Pressing CLEAR will clear the value for a given element.)

Procedure M2. To select the name of a matrix and to view the matrix.

1. Press the key MATRIX. This gives the matrix menu.
2. Select the name of the matrix.

On the TI-82 or 83 or 84 Plus:
With NAMES highlighted in the top row, select one of the matrices [A], [B], [C], ...

On the TI-85 or 86:
Select NAMES from the matrix menu. From the secondary menu, select the name of the matrix.

3. View the matrix by pressing ENTER. In some cases the matrix may extend off the screen to the right or to the left. To see that part of the matrix not on the screen use the right or left cursor keys to scroll (move) the unseen part onto the screen.

<u>Procedure M3</u>. Addition, subtraction, scalar multiplication or multiplication of matrices.
        See Procedure M2 for how to select the names of matrices.

1. Addition of two matrices:
   Select the name of the first matrix, press the addition sign, select the name of the second matrix, then press ENTER.
   (Example: $[A] + [B]$)

2. Subtraction of two matrices:
   Select the name of the first matrix, press the subtraction sign, select the name of the second matrix, then press ENTER.
   (Example: $[A] - [B]$)

3. Multiplication of matrix by a scalar:
   Enter the scalar, select the name of the matrix and press ENTER.
   (Example: $2.5 [A]$ or $2.5 \times [A]$)

4. Multiplication of two matrices:
   Select the name of the first matrix, select the name of the second matrix, and press ENTER.
   (Example: $[A][B]$ or $[A] \times [B]$)

Many of these operations may be combined into one statement. The result of a matrix calculation is stored under ANS. An error will occur if the operation is not defined for the matrices selected.

<u>Procedure M4</u>. To find the determinant, inverse, and transpose of a matrix.

1. To find the determinant:

On the TI-82 or 83 or 84 Plus:	On the TI-85 or 86:
Press the key MATRX	Press the key MATRX
Highlight MATH in the top row.	From the matrix menu, select MATH.
From the menu, select det.	

   Then select the name of the matrix (Procedure M2) and press ENTER.
   (Example: det($[A]$) or det A}

2. To find the inverse:
   Select the name of the matrix.
   Press the key x^{-1} and press ENTER.

   (Example: $[A]^{-1}$ or A^{-1})

3. To find the transpose:
   Select the name of the matrix, then

On the TI-82 or 83 or 84 Plus:	On the TI-85 or 86:
Press the key MATRX	Press the key MATRX
Highlight MATH in the top row.	From the matrix menu, select MATH.

   From the menu, select ^{T} and press ENTER.   (Example: $[A]^T$ or A^T)

<u>Procedure M5</u>. To store a matrix.

1. Select the name of a matrix or enter a matrix expression.
2. Press the key STO▷
3. Select the name of the second matrix and press ENTER.
   (Example: $[A] + [B] \rightarrow [C]$ or $A + B \rightarrow C$)

On the TI-85 or 86, we may also store a matrix by using the equal sign. Example: $C = A + B$.

Note: Storing a matrix for a second matrix, erases the previous contents of the second matrix.

Procedure M6: Row operations on a matrix.

1. Obtain the matrix operations menu by first pressing the key MATRX, then:

On the TI-82, 83, or 84 Plus:	On the TI-85 or 86:
Highlight MATH in the top row.	From the matrix menu, select OPS (then press MORE).

2. Select the desired operation from the menu. Names of matrices are selected as in Procedure M1.

   In the following, the left hand column gives the selection, form, and example for the TI-82, 83, or 84 Plus and the right hand column given the selection, form, and example for the TI-85 or TI-86. To obtain some of these operations, it may be necessary to scroll down the given list by using the bottom cursor key.

   (a) To interchange two rows on a matrix, select

rowSwap(	RSwap
Form:	Form:
rowSwap(*name, row1, row2*)	**RSwap**(*name, row1, row2*)
Example: rowSwap([*A*], 1, 3)	Example: Rswap(*A*, 1, 3)

   (Interchanges rows 1 and 3 of matrix *A*.)

   (b) To multiply a row by a constant, select:

*row(	MultR
Form:	Form:
***row**(*multiplier, name, row*)	**MultR**(*multiplier, name, row*)
Example: rowSwap(1/2, [*A*], 3)	Example: multR(1/2, *A*, 3)

   (Multiplies row 3 of matrix *A* by ½.)
   To divide a row by a constant, multiply by the reciprocal.

   (c) To multiply a row by a constant and add to another row, select

*row+(	MRAdd
Form:	Form:
***row+**(*mult,name,row1,row2*)	**mRAdd**(*mult,name,row1,row2*)
Example: *row+(−5,[*A*], 1, 3)	Example: mRAdd(−5, *A*, 1, 3)

   (Multiplies row 1 of matrix *A* by −5 and adds the result to row 3 and replaces row 3.)

3. Press ENTER.

Note: The matrix operation *ref* gives the row echelon form directly for a matrix and the operation *rref* gives the reduced row echelon form directly for a matrix. The forms are: ref (*A*) and rref(*A*) where *A* is the name of the matrix to be reduced.

A.6 Statistical Procedures

Procedure S1. To enter or change single variable statistical data.

On the Texas Instruments graphing calculators statistical data is stored as lists of numbers.
(See also Procedure C7)
1. Press the key STAT
2. Enter edit mode.

On the TI-82 or 83or 84 Plus:
With EDIT highlighted in the top row, select 1:Edit...

On the TI-85 or 86:
Select EDIT from the menu.

3. Enter the data values and frequencies. All frequencies must be entered as integers.

On the TI-82 or 83 or 84 Plus:
Enter the data values under one list name. The list names are
$L_1, L_2, L_3, L_4, L_5,$ and L_6.
key is pressed. After entering each data value, press ENTER.

The data values are entered at the bottom of the screen and don't appear in the list until the ENTER

The related frequencies are entered in a similar manner under a second list name. After entering each frequency, press ENTER.

Make sure the frequencies are entered in the same order as the data items so that they correspond. Corresponding items should be in the same row of the table.

On the TI-85 or 86:
Select EDIT from the menu.
On the TI-85:
The calculator will ask for a xlist name and a ylist name. The xlist is for the data values and the ylist is for the frequencies. Enter a name (up to eight characters) for xlist and press ENTER and enter a name for ylist and press ENTER.

Enter each data value as a x-value and the corresponding frequency as a y-value. (y_1 is the frequency for data item x_1, y_2 for x_2, etc.)
On the TI-86:
There are three built in names for lists: xSstat, yStat, or fStat. Data may also be entered in other lists where the user selects the name. For the user to select the name, move the cursor to the name row and move to the right to an unnamed column.

Data is entered in a column by moving the cursor to that column (the column name and element number appears at the bottom of the screen), entering a value, and pressing ENTER.

Make sure the frequencies are entered in the same order as the data items so that they correspond.

Note: On the TI-83 and 84 Plus other lists may be placed in the STAT edit table by using the SetUpEditor available on the menu.

Note: When there are intermediate data values with a frequency of zero, it is best, for graphing purposes, to enter these data values and the zero frequencies.

4. To edit data.

<table>
<tr><td>On the TI-82, 83, or 84 Plus:
Move the cursor to the values to be changed, make the change, and press ENTER. As the data item in a list is highlighted, the value of that item appears at the bottom of the screen.</td><td>On the TI-85 or 86:
Move the cursor to the values to be changed and make the desired change.</td></tr>
</table>

5. To exit, after all data have been entered and the cursor appears on the next value, press QUIT.

6. To view a list, see Procedure C8.

Procedure S2. To graph single variable statistical data.

1. Make sure the viewing rectangle is appropriate for the data to be graphed, that the graph display has been cleared, and that all functions in the function table have been turned off.

2. Be sure the data to be graphed is stored in lists in the calculator. (Procedure S1)

3. Define the plot and draw the graph:

On the TI-82 or 83 or 84 Plus:

(a) Press the key STAT PLOT

(b) Select one of the three plots.
The screen will show the plot number selected, followed by several options. The highlighted options are active. Select the plot number desired and press ENTER. There will be several options for the plot number chosen.

(c) Highlight ON to turn that plot on and press ENTER.

(d) Highlight the type of graph. The first symbol following the word Type is for a scatter plot, the second for a xyLine.

On the TI-82, the third is for a box plot, and the fourth for a histogram.

On the TI-83 or 84 Plus, the third is for a histogram, the fourth and fifth for boxplots, and the sixth for a normal probability plot.

On the TI-85 or 86:

(a) Press the key STAT

(b) From the menu, select DRAW

(c) From the secondary menu, to draw a graph, select:
HIST for a histogram.
SCAT for a scatter graph
xyLine for a frequency polygon.

On the TI-86 there are also the options:
BOX for a box plot
MBOX for a modified box plot.

On the TI-86, before plotting, the plot must be set up as follows:
After pressing STAT, from the secondary menu, select PLOT. Then, select PLOT1, PLOT2, or PLOT3. The highlighted options are active.

(a) Highlight ON
(b) From the menu, select the type of plot.

(e)   After Xlist enter the list name of the data values. After Ylist (or Freq), enter the name of the data frequencies (On the TI-82, highlight the name of the list of frequencies.). After each entry press ENTER.

(f)   For Mark, highlight the symbol to be used in plotting data.

(g)   To exit, press QUIT

(h)   To see the graph, press GRAPH.

(c)   Select the names for the Xlist and Ylist. In this case, the Ylist will be the frequencies.

(d)   Select the symbol to be used.

(e)   To exit, press QUIT.

(f)   To see the graph, press GRAPH key and select GRAPH.

Be sure to turn the plot off when finished.

Be sure to turn the STAT PLOT off when finished.

Notes:
1.   For a histogram the value of Xscl will determine width of bars.
2.   One type of graph should be cleared before displaying another.
3.   When graphing statistical data, all functions listed in the function table should be turned off.

Procedure S3.   To clear the graphics screen.

1.   All functions and plots should be deleted or turned off.
2.   Clear the screen:

On the TI-82 or 83 or 84 Plus:
Press the key DRAW.
With DRAW highlighted,
Select 1:ClrDraw.
Press ENTER.

On the TI-85 or 86:
Press the key STAT
From the menu, select DRAW.
From the secondary menu, select CLDRW.

On the TI-83, 84, or 86, a drawing cleared while a plot is turned on will be redrawn the next time the GRAPH key is pressed.

Procedure S4.   To obtain mean, standard deviation and other statistical information.

Be sure one-variable data has been stored as in Procedure S1.
1.   Press the key STAT
2.   Obtain the statistical information:

On the TI-82 or 83 or 84 Plus:
(a)   Highlight CALC in the top row and from the menu select 1:1 Var Stats

On the TI-85:
(a)   From the menu, select CALC.

(b) 1-Var Stats will appear on the home screen. After these words first place the list name for the data and if frequencies are given in a second list name, the list name for the frequencies. Example:

$$\text{1-Var Stats } L_1, L_2$$

(c) Press ENTER

or

(a) On the TI-82, with CALC highlighted at the top of the screen, select: 3:Setup Then, under 1-Var Stats, select the list name to be used for the data values and the list name to be used for the frequencies. (If each has a frequency of one, select 1.) and press QUIT

(b) Again press STAT and highlight CALC in the top row.

(c) From the menu, select 1:1-Var Stats, then press ENTER

(b) Make sure the xlist name is correct and press ENTER and the ylist name is correct and press ENTER.

(c) From the secondary menu, select 1-VAR.

On the TI-86:

(a) From the menu, select CALC

(b) From the secondary menu, select OneVa. On the home screen OneVar appears. After these words enter the name of the data list, comma, the name of the frequency list. If the name of the frequency list is omitted the frequencies are taken as 1. If both list names are omitted the data list is taken as the list named xStat and the frequency is taken as the list named fStat.

Note: On the TI-86, to get list names, press the key VARS, select STAT (after pressing MORE twice), move the indicator to the variable name desired and press ENTER.

(c) Press ENTER

The following statistical quantities appear:

$\bar{x}$ is the mean of all the x-values (the first list name)

$\sum x$ is the sum of all the x-values

$\sum x^2$ is the sum of all the squares of the x-values

Sx is the s-standard deviation of the x-values

σx is the σ-standard deviation of the x-values

n is the total number of x-values

3. To clear the screen.
   Press CLEAR to clear the screen.

On the TI-85 or 86:
Press QUIT to exit. Then CLEAR to clear screen.

Procedure S5. To clear statistical data from memory.

1. Press the key: MEM.
2. From the menu, select Delete or MemMgmt/Del....
3. Then select   List...
4. Use the cursor keys to move the indicator to the name of the list to be deleted and press ENTER. On the TI-83 or 84 Plus, press the DELETE key will delete the entire list and name.

    On the TI-82 or 83 or 84 Plus, data is best deleted by pressing the key STAT, selecting 4:ClrList, entering the names, separated by commas, of the lists to be cleared (by pressing the keys for those names), and then pressing ENTER. This will not clear the list names.
5. Press QUIT to exit.

Note: **Caution:** After pressing the MEM key, do not select RESET – this will erase everything in the calculator.

Procedure S6. To enter or change two variable statistical data.

On the TI-82, 83, 84 Plus, 85, and 86 calculators, data is stored as lists of numbers. (See Procedure C7) Data stored in one list are the x-values and data stored in a second list are the y-values. The two lists must correspond in values.

1. First press the STAT.
2. Enter edit mode to edit data stored in lists.

    On the TI-82 or 83 or 84 Plus: On the TI-85 or 86:
    With EDIT highlighted in the top   Select EDIT from the menu.
    row, select 1:Edit...

3. Enter the x-values as one list and the y-values as a second list:

    On the TI-82 or 83 or 84 Plus: On the TI-85:
    Enter the x-values under a list name.   The calculator will ask for an xlist
    After each value is entered, press   name and a ylist name. The xlist is
    ENTER.   for the x-values and the ylist is for
    Enter the related y-values under a   the y-values. Enter a name (up to
    second list name. After each value is   eight characters) for xlist and press
    entered, press ENTER.   ENTER and enter a name for ylist
    Make sure the y-values are entered in   and press ENTER .
    the same order as the x-values so that   Enter the first x-value and the
    they correspond by being in the same   corresponding y-value, the second x-
    row.   value and the second y-value, etc.

    Note: On the TI-83 and 84 Plus   On the TI-86:
           other lists may be placed in   There are three built in names for
           the STAT edit table by using   lists: xStat, yStat, or fStat. Data may
           the SetUpEditor available on   be also be entered in these lists or in
           the menu.   other lists where the user selects the
                                                 name. For the user to name a list,
                                                 move the cursor up to the name row
                                                 and move to a column to the right.

Data is entered in a column by moving the cursor to that column (the column name and element number appears at the bottom of the screen), entering a value, and pressing ENTER.

Make sure the *y*-values are entered in the same order as the *x*-values items so that they correspond.

Normally the *x*-values would be entered under xStat, the *y*-values under yStat, and the frequency of each point under fStat.

4. To edit data.

On the TI-82 or 83 or 84 Plus:	On the TI-85 or 86:
Move the cursor to the values to be changed, make the change, and press ENTER. As the data item in a list is highlighted, the value of that item appears at the bottom of the screen.	Move the cursor to the values to be changed and make the desired change.

5. To exit, after all data have been entered and the cursor appears on the next value, press QUIT.

6. To view a list, see Procedure C8.

Procedure S7. To determine a regression equation.

1. Be sure two-variable data has been stored as in Procedure S6.
2. Obtain the proper menu:

On the TI-82 or 83 or 84 Plus:

(a) Press the key STAT

(a) Highlight CALC in the top row and from the menu select the type of regression equation required.

(b) The words indicating the type of regression equation to be selected will appear on the home screen. After these words enter the list name for the *x*-values and secondly, after a comma, place the list name for the *y*-values.

On the TI-85 or 86:

(a) Press the key STAT

(b) From the menu, select CALC.

Then, on the TI-85:

(c) Make sure the xlist name is correct and press ENTER and the ylist name is correct and press ENTER.

(d) Select the type regression equation from the menu.

or

A third list name may be added to indicate the frequencies of each pair of values.
Examples:
$$\text{LinReg } L_1, L_2$$
$$\text{QuadReg } L_1, L_2, L_3$$

(c) Press Enter to see the constants in the equation.

or

(a) On the TI-82, with CALC highlighted at the top of the screen, select: 3:Setup
Then, under 2-Var Stats, select the list name to be used for the x-values and the list name to be used for the y-values. Then select the list name to be used as a frequency of each point.
To exit, press QUIT

(b) Again press STAT and highlight CALC in the top row.

(c) From the menu, select the regression equation required, then press ENTER

On the TI-86:

(c) From the secondary menu, select the type of regression equation required. On the home screen the words indicating the type of equation selected will appear. After these words enter the name of the x-value list, comma, and the name of the y-value list. A third list name may be added to indicate the list that contains the frequencies of each point. If the name of the frequency list is omitted the frequencies are taken as 1.

Example:
$$\text{LinReg LISTA, LISTB}$$
If the list names are omitted the x-value list is taken as the list named xStat, the y-value list is taken as the list named yStat and the frequency is taken as the list named fStat.

Note: On the TI-86, to get list names, press the key VARS, select STAT (after pressing MORE twice), move the indicator to the variable name desired and press ENTER.

(d) Press ENTER

3. The types of regression equations are listed below. (The form of the menu item for the TI-82, TI-83, and TI-84 Plus is on the left and for the TI-85 and TI-86 is on the right.)

(a) For linear regression ($y = a + bx$), select:
LinReg($ax + b$) or   LinR
LinReg($a + bx$)

(b) For logarithmic regression ($y = a + b \ln x$), select
LnReg   LnR

(c) For Exponential Regression ($y = ab^x$), select
ExpReg   ExpR

(d) For Power Regression ($y = ax^b$), select
PwrReg   PwrR

(e) For Quadratic Regression ($y = ax^2 + bx + c$), select
QuadReg   P2Reg

(f) For Cubic Regression ($y = ax^3 + bx^2 + cx + d$), select
    CubicReg   P3Reg

(g) For Quartic Regression ($y = ax^4 + bx^3 + dx^2 + ex + f$), select
    QuartReg   P4Reg

Note: The TI-83, TI-84 Plus, and TI-86 also have a sinusoidal regression using the formula $y = a\sin(bx + c) + d$ and a logistic regression of the form

$$y = \frac{c}{1 + ae^{-bx}} \text{ for the TI-83 and 84 Plus and } y = \frac{a}{1 + be^{cx}} + d \text{ on the TI-86.}$$

The calculator gives values for constants in the equations and in some cases, the correlation coefficient as *corr* or *r*. To see the *r*-value on the TI-83, it is necessary to turn diagnostics on. Do this by pressing the key CATALOG, moving down the list until the words DiagnosticOn appears and press ENTER, then press ENTER again.

Procedure S8. To graph data and the related regression equation.

1. Make sure the viewing rectangle (Window values) is set properly to graph the set of data, that the graph display has been cleared (see Procedure S3), and that all functions in the function table have been removed or turned off and that all statistical plot graphs have been turned off.
2. If the data has not already been entered into the calculator, it will need to be entered (Procedure S6) and the regression equation calculated (Procedure S7).
3. Store the regression equation for a function name and graph:

On the TI-82, 83, or 84 Plus:
(a) Press the key $y=$
(b) Move the cursor after an unused function name.
(c) Press the key VARS
(d) From the menu, select Statistics...

(e) Move the highlight in the top row to EQ
(f) Select RegEQ
(g) Press the key GRAPH

On the TI-85 or 86:
(a) Press the key GRAPH
(b) From the menu, select $y(x)=$
(c) Move the cursor so it comes after an unused function name.
(d) Press the key VARS

On the TI-85 or 86:
(a) Press the key GRAPH
(b) From the menu, select $y(x)=$
(c) Move the cursor so it comes after an unused function name.
(d) Press the key VARS

This graphs the regression equation.

Note: On the TI-83, TI-84 Plus and TI-86, the regression equation may be stored directly for a function name by adding the function name to the lists when the equation is calculated. (Example: LinReg L1, L2, $Y1$)

4. Define the plot and draw a scatter graph:

On the TI-82, 83, or 84 Plus:

(a) Press the key STAT PLOT

(b) Select one of the three plots. (Press ENTER) The screen will highlight the plot number selected, followed by several options. The highlighted options are active. (To make an option active, move the cursor to that option and press ENTER.)

(c) Highlight the option ON.

(d) Highlight the type of graph. The first symbol following the word Type is for a scatter plot.

(e) On the TI-82, for Xlist, highlight the name of the x-values. For Ylist, highlight the name of the y-values.

On the TI-83 or 84 Plus enter names for the Xlist and for the Ylist.

(f) For Mark, make active the symbol to be used in plotting data.

(g) To exit, press QUIT

(h) To see the graph with the data points, press GRAPH

Be sure to turn the plot off when finished.

On the TI-85:

(a) Press the key STAT

(b) From the menu, select DRAW

(c) From the secondary menu, to draw a graph, select: SCAT for a scatter graph.

On the TI-86, before plotting, the plot must be set up.

Do this as follows:

After pressing STAT, from the secondary menu, select PLOT

Then, select PLOT1, PLOT2, or PLOT3. The highlighted options are active.

(a) Highlight ON

(b) From the menu, select the type of plot.

(c) Select the names for the Xlist and Ylist.

(d) Select the symbol to be used.

(e) To exit, press QUIT.

(f) To see the graph, press GRAPH key and select GRAPH.

Be sure to turn the plot off when finished.

Appendix B

Procedures for the TI-89 Graphing Calculator

The Texas Instruments TI-89 Calculator is somewhat different from the other graphing calculators. It is a very powerful calculator and has the ability to do symbolic algebra and calculus. However, in this manual, we shall only consider procedures for the TI-89 that are comparable to the procedures that we considered for the other calculators.

An obvious difference of the TI-89 Calculator is the home screen. The home screen is divided into sections. At the top of the screens the **Toolbar** which gives actions that are associated with the function keys. The middle part of the screen is the **History Area** and gives recent entries and answers. As the screen becomes filled, these scroll off the top of the screen. The bottom line on the home screen is the **Status Line** and shows the current status of the calculator. The second line from the bottom is the **Entry Line.** All expressions or instructions are made on the Entry Line. Once an entry has been made and the ENTER key pressed, the entry and results appear in the History Area. Any previous entry may be moved back to the Entry Line by highlighting the entry and pressing ENTER.

Note: All the calculator functions and instructions are listed in alphabetical order in an area called the **Catalog**. There are some functions and instructions that may only be found in the catalog. The best way to access an item in the Catalog is to press the key CATALOG. Then, press the key for the first letter of the item you desire (do not press the ALPHA key first). Use the top and bottom cursor keys to scroll through the list of items until the mark (▶) on the left is in front of the item desired. The format of the selected item will be given at the bottom of the screen. Press ENTER to copy the item to the HOME screen.

Example: For the greatest integer function (int): With the cursor on the Entry Line of the HOME screen, press CATALOG, then press I and use the bottom cursor key to move the mark down to int(and press ENTER. Enter a number, the close the set of parentheses, and press ENTER. (Example: int(–2.15) = –2)

B.1 Basic Calculator Procedures

<u>Procedure B1</u>. **To turn calculator on and off.**

> 1. To turn the calculator on, press the key: ON (in the lower left corner).
> 2. To turn the calculator off, first press the key: 2nd then the key: ON

<u>Procedure B2</u>. **To clear all memory in the calculator.**

> **WARNING: The following steps will erase ALL calculator memory.**
> 1. Turn the calculator on.
> 2. Press the key: MEM   (Do this by pressing the key: 2nd then the key: 6)

A list of the amount of memory that is used in various areas of the calculator appears. To reset the calculator and erase all memory, press the F1 key. To exit, press the ESC key.

Procedure B3. Adjusting the contrast of the screen.

To adjust the contrast of the screen you will need to use the key containing the green diamond ($\boxed{\blacklozenge}$) and the addition (+) or subtraction (−) keys.
1. To make lighter: Press the diamond key ($\boxed{\blacklozenge}$), then press the + key.
2. To make darker: Press the diamond key ($\boxed{\blacklozenge}$), then press the − key.

The contrast change takes place in a series of increments. Thus, the above steps may have to be repeated to obtain the contrast you desire. The diamond key ($\boxed{\blacklozenge}$) must be pressed each time before pressing the arrow key.

Procedure B4. To Clear the Calculator Screen

1. With the cursor on the Entry Line, press CLEAR to clear the Entry Line.
2. With the cursor in the History Area, press CLEAR to erase a highlighted item.

B.2 General Calculator Procedures

Procedure C1.   To perform secondary operations or obtain alpha characters.

1. To perform secondary operations:
    (a) Press the key: 2nd
    (b) Then, press the key for the desired action
        (Example: To find the square root of a number, first press the key 2nd followed by the multiplication key ×. (The square root symbol $\sqrt{}$ appears above the multiplication × key.) Enter the number, close the set of parentheses, and press ENTER.)
2. To obtain alpha characters:
    (a) Press the key: ALPHA
    (b) Then, press the key for the desired character.
    (For example: To display the letter *a* on the screen, press the key: ALPHA then press the key that has the letter *A* above and to its right.)
    On the TI-89, pressing the key: 2nd before pressing the ALPHA key will lock in lower case letters. Press the Shift key $\boxed{\blacktriangle}$ before the ALPHA key to lock in upper case letters. To return to numeric mode, press ALPHA again.
3. For other function and symbols that appear in green on the keyboard:
    (a) Press the green diamond key ($\boxed{\blacklozenge}$).
    (b) Then, press the key for the desired item.

Procedure C2. Selection of menu items

Some menu items are selected by first pressing a function key (F1, F2, F3, F4, F5, F6, F7, or F8). The action of each function key is given at the top of the screen. Menu items may also appear in lists on the screen. To select from such a list, either highlight the item and press ENTER **or** press the number key that corresponds to the item in the list. In many cases, pressing the right cursor key will accomplish the same task.

(Example: To find the absolute value of the number –2.15, press the key: MATH, then select 1:Number. From the menu, select 2:abs(, enter the value –2.15, close the set of parentheses and press ENTER. The result is 2.15.)

Procedure C3. To perform arithmetic operations.

1. To add, subtract, multiply or divide press the appropriate key: $+, -, \times, $ or $\div$. On the graphing calculator screen, the multiplication symbol will appear as * and the division symbol will appear as /. If parentheses are used to denote multiplication it is not necessary to use the multiplication symbol. (The calculation $2(3 + 5)$ is the same as $2 \times (3 + 5)$.)

2. To find exponents use the key $\wedge$. (Example: 2^4 is entered as 2^4)

3. After entering the expression press the key ENTER to perform the calculation.

Notes: (a) On the graphing calculator, there is a difference between a   negative number and the operation of subtraction. The key (-) is used for negative numbers and the key – is used for subtraction.

(b) Enter numbers in scientific notation by using the key EE.

(Example: To enter 2.15×10^5, enter 2.15, press EE, enter 5.)

Procedure C4. To correct an error or change a character within an expression.

1. (a) To make changes in the current expression:
Use the cursor keys to move the cursor to the position of the character to be changed.

(b) To make changes in the last expression entered, after the ENTER key has been pressed, obtain a copy of the expression by pressing the key ENTRY. (Continue pressing of the key ENTRY will give earlier expressions entered.) Then, use the cursor keys to move the cursor to the character to be changed.

2. Characters may now be replaced, inserted, or deleted at the cursor position.

(a) To replace a character at the cursor position, just press the new character.

(b) To insert a character or characters in the position the cursor occupies, the calculator should be in Insert Mode as indicated by a narrow cursor. (When not in Insert Mode, the cursor is wide.) To change in and out of Insert Mode, press the key: INS. When in Insert Mode, the desired character(s) may be entered at the cursor position.

(c) To delete a character in the position the cursor occupies, press the key: DEL

Procedure C5. Selection of special mathematical functions.

1. To raise a number to a power (including squares):
(a) Enter the number
(b) Press the key ^
(c) Enter the power
(d) Press ENTER.

2. To find the root of a number (other than square root):

Use fractional exponents. (Example: To find $\sqrt[3]{27}$, find 27^(1/3).)

3. To find the factorial of a positive integer:
   (a) Enter the number
   (b) Press the key MATH and select 7:Probability.
   (c) Select 1:!
   (d) Press ENTER

5. For the number π, press the key for π on the keyboard.

6. To find the absolute value of a number:
   (a) Press the key MATH and select 1:Number
   (b) Select 2:abs(
   (c) Enter the number and add the closing parenthesis.
   (d) Press ENTER

Procedure C6. Changing the mode of your calculator.

Press the key: MODE. You will see several lines. Each line pertains to different areas of the calculator. To change the mode for a particular area, move the highlight to that area and press the right or left cursor keys. Select from the menu that appears. To go back to a previous screen, press ESC.

1. To change the way numbers are displayed on the screen, highlight the quantity after Display Digits and press the right or left cursor keys. This will give a choice of the number of digits to be displayed and whether to fix the number of decimal places or to use floating point numbers. Select the type desired and press ENTER. (Normally, select FLOAT and press ENTER.)

2. To change the type of notation used to display numbers, highlight the quantity following Exponential Format and press the right or left cursor keys. This gives a choice of normal, scientific, or engineering notation. Select the type desired and press ENTER.

3. To change the type of angular measurement used, highlight the quantity following the word Angle and press the right or left cursor keys. Select the desired unit and press ENTER.

4. After changes have been made, press ENTER again to save these changes.

(Some of the other lines on this screen will be considered later.)

Procedure C7. Storing a set of numbers as a list.

On the Entry Line on the Home Screen,
Method I.
1. Press the key {
2. Enter the numbers to be in the list separated by commas.
3. Press the key }
4. Press the key STO▷
5. Enter a name of up to eight characters for the list and press ENTER.
Or

Method II
1. Press the key APPS.
2. Select Data/Matrix Editor.
3. Select 3:New... to enter a **new list**,
   (a) With the word after Type highlighted, press the right cursor key.
   (b) Select 3:List
   (c) Move the highlight down to the box following the word Variable.
   (d) Enter a variable name in the box. This becomes the name of the list. The name can be up to eight characters and the first character must be a letter. It cannot be a name used for anything else. Press ENTER, then press ENTER again.
   (e) Place the data elements in only one column.

   or

   Select 2:Open... to open an **old list**,
   (a) With the word after Type highlighted, press the right cursor key.
   (b) Select 3:List
   (c) Move the cursor down so the word after Variable: is highlighted and press the right arrow key. Select a list from the names and press ENTER.

   or

   Select 1:Current to open **last list** used.
4. The data elements will be in one column. Changes may be made to any element by moving the cursor to the element and making the changes.
5. To exit, after all data have been entered or changed, press HOME or QUIT.

Procedure C8. Viewing a set of numbers.

On the Entry Line of the Home Screen, type the name of a list and press ENTER.

or

Follow the procedure of Method II in Procedure C7.

Procedure C9. To perform operations on lists.

On the Entry Line on the Home Screen,
1. To add, subtract, multiply or divide lists, enter the name of a list, press the appropriate key ($+$, $-$, $\times$, or $\div$), enter the name of the second list and press ENTER. Lists must be of the same length in order for these operations to be performed.
2. To perform an operation on all the elements of a list, enter a value, press the appropriate key ($+$, $-$, $\times$, or $\div$), enter the name of a list and press ENTER.
3. Functions such as finding the squares or square roots, may be performed on each element of a list by performing the function of the list or list name.

Procedure C10. To store a function.

Functions may be stored in three ways: In the $Y=$ Editor, using the STO▷ key, or using the word Define to define a function.

Method I:
Use the $Y=$ Editor which can store up to 99 functions. This is the best way to enter functions for graphing.
1. Press the $Y=$ key.
2. Move the cursor to a function name.
3. Enter the function, the cursor will move to the Entry Line
4. When finished press ENTER
3. Press QUIT or HOME to return to the home screen.
To see a stored function, press the key $Y=$.

or

Method II:
On the Entry Line on the Home Screen,
1. Enter the function to be stored.
2. Press the key STO▷
3. Enter a function name followed by a variable in parentheses (such as $y5(x)$).
4. Press ENTER
To see a function stored in this way, press RCL, enter the function name without parentheses (such as (f5)), and press ENTER twice.

or

Method III:
On the Entry Line on the Home Screen,
1. Use the command Define. (Press F4 and select Define from the menu.)
2. Enter a function name followed by a variable in parentheses (such as $y5(x)$).
3. Press the key =.
4. Enter the function and press ENTER.
To see a function stored in this way, press RCL, enter the function name without parentheses (such as (y5)), and press ENTER twice.

Procedure C11. To evaluate a function or create tables.

Evaluating a function:
Functions may be evaluated at a single value or at a list containing several values. Enter the function name followed by parentheses containing a number, a list of values for evaluation, or a list name. And press ENTER.
Example: y1(2) or y1({1,2,3,4}) or y1(*a list name*)

Creating tables:
The TI-89 has the ability to create a table of values for a function by pressing the key TABLE. Parameters for the table are set by pressing the key TblSet.

In TblSet:
  tblStart is the value of the first x value in the table.
  Δtbl is the difference between table values.
  Graph<–>Table: May be Off or On
    With Off the values are calculated based on tblStart and Δtbl values
    With On the values are calculated based on Graph pixel values for x.
  Independent: May be Auto or Ask
    With auto allows the table to be created automatically or with Ask by the
    calculator providing values for y when x values are entered in the table.

Pressing (or selecting) TABLE will show a table of values of the independent variable and values for all functions that are turned on in the function table. By using the cursor keys, we can scroll either direction through the table.

Procedure C12. **To store a constant for an alpha character.**

On the Entry Line on the Home screen,
1. Enter the value to be stored.
2. Press the key STO▷
    (STO▷ shows as → on screen.)
3. Enter the alpha character.
4. Press ENTER
(Example: $6.35 \to H$)
Note: To see the value stored for an alpha character, enter the alpha character
    and press ENTER.

Procedure C13. **To obtain relation symbols (=, ≠, >, ≥, <, and ≤).**

1. Press the key: MATH
2. Select 8:Test
3. Select the desired symbol from the menu.

Procedure C14. **To change from radian mode to degree mode and vice versa.**

1. Press the key: MODE
2. Highlight the quantity following Angle.
3. Press the right cursor key and select either Radian or Degree.
4. Press ENTER, then press ENTER again..

Procedure C15. **Conversion of Coordinates**

Make sure calculator is in correct mode -- radians or degrees.
Press the key MATH and select Angle.
1. To change from **polar to rectangular**:
    (a) To find x: Select 3:P▷Rx(
        To find y: Select 4:P▷Ry(
    (c) Enter the values of r and θ separated by commas, add the closing
        parenthesis and press ENTER.
    Example: P▷Rx(2.0, 60) gives the x-value 1 [in degree mode]

2. To change from **rectangular to polar**:
    (a) To find *r*:   Select 5:R▷Pr(
         To find *θ*:   Select 6:R▷Pθ(
    (b) Enter the values of *x* and *y* separated by commas, add the closing parentheses and press ENTER.
    Example: R▷Pθ(1.0, 1.0) gives 45 [in degree mode]

Procedure C16. Finding the sum of vectors in polar form.

Make sure calculator is in correct mode -- radians or degrees and that the Vector Format is Cylindrical (To get this, press MODE, highlight the word following the words Vector Format, press right cursor key, select Cylindrical and press ENTER. If Rectangular were selected, output would be in terms of x and y.)

Vectors may be entered directly in polar form ($[r, \angle \theta]$) and then added together. Example: [12, ∠120] + [15, ∠75] gives [25, ∠95] when rounded.

Note: On the TI-89, vectors may be stored under variable names by entering the vector, pressing the key STO▷, entering the variable name and pressing ENTER. The sum of vectors may be found by finding the sum of the variables.

Procedure C17. Conversion of complex numbers.

Make sure calculator is in correct mode -- radians or degrees.
To change the format in which complex numbers appear on the calculator, press MODE, highlight the word following the words Complex Format, press the right cursor key, and select the form. Here i = $\sqrt{-1}$ and is entered on the calculator as [*i*] (with the square brackets).

1. To change from **polar to rectangular** form:
    Change the Complex Format to rectangular, enter the number in the form ($r \angle \theta$), and press ENTER.

    or

    Think of the complex number $r \angle \theta$ as the vector $[r, \angle \theta]$ and use Procedure C15.

2. To change from **rectangular to polar** form:
    Change the Complex Format to polar, enter the number in the form *a* +*b*[*i*], and press ENTER.

    or

    Think of the complex number *x* + *yj* as the vector [*x*, *y*] and use Procedure C15.

Notes: On the TI-89, a complex number is expressed or entered in the
    form *a* + *bi*, in the form $re^{i\theta}$, or in the form ($r \angle \theta$).

Procedure C18. Operations on complex numbers.

Operations are performed by entering the complex numbers in either rectangular or polar form using the operation symbols +, −, ×, or ÷ between the numbers.

Note: Powers and roots of complex number may be found by raising the complex number to the desired power using ^.
The form of output of the complex number is determined by the mode of the calculator. (See Procedure C17.)

Procedure C19. To determine the numerical derivative from a function.

1. Store a value for the *variable* that is used in Step 4 below.
2. Press the key MATH
3. Select A:Calculus
4. Select A:nDeriv(
   Enter in the form: **nDeriv**(*function, variable,h*)
   where *function* is a function or name of a function, *variable* is the variable used in the function and *h* is the increment used in the calculation and should be a small number.
5. Press ENTER.
   (Example: nDeriv($x^{2,}x$,0.1)   (Gives a value of 6 if 3 is first stored for x.))

or

1. Press the key *d*   (Press the key 2nd, then the key 8, not the letter D.)
2. Enter in the form: **d**(*function, variable*) | *variable = value*
   where *function* is a function or name of a function, *variable* is the variable used in the function and *value* is the number at which the derivative is to be evaluated.
3. Press ENTER.   (Example: d(x^2, x) | x=5 (Gives a value of 10.))

Procedure C20. Function to calculate definite integral.

1. Press the key MATH
2. Select A: Calculus
3. Select B:nInt(or   Press the key ∫   (Press the key 2nd, then the key 7)
4. Enter in the form **nInt** (*function, variable, lower, upper*)
   or ∫ (*function, variable, lower, upper*)
   where *function* is a function or name of a function, *variable* is the variable used in the function, *lower* and *upper* are the lower limit and upper limits of integration.
5. Press ENTER.   (Example: ∫ (3$x^{2,}x$,1,2)   (Gives a value of 7.))

B.3 Graphing Procedures

Procedure G1. To enter and graph functions.

Functions are entered in accordance with the procedures given in Procedure C10.
For graphing it is best to use Method I (The Y= Editor).
To graph selected function or functions, press the key GRAPH.